1/3/84

TRIGONOMETRY:
Triangles and Functions

THIRD EDITION

M. L. Keedy
PURDUE UNIVERSITY

Marvin L. Bittinger
**INDIANA UNIVERSITY–
PURDUE UNIVERSITY AT INDIANAPOLIS**

ADDISON-WESLEY PUBLISHING COMPANY
Reading, Massachusetts ■ Menlo Park, California
London ■ Amsterdam ■ Don Mills, Ontario ■ Sydney

Sponsoring Editor: *Patricia Mallion*
Production Editor: *Martha Morong*
Designer: *Vanessa Piñeiro*
Illustrator: *VAP International Communications Ltd.*
Art Coordinator: *Joseph Vetere*
Cover Design: *Vanessa Piñeiro*
Cover Illustrator: *Bob Trevor*

Library of Congress Cataloging in Publication Data

Keedy, Mervin Laverne
  Trigonometry: triangles and functions.

  Includes index.
  1. Trigonometry. I. Bittinger, Marvin L. II. Title.
QA531.K33 1982   516.2'4   81-14974
ISBN 0-201-13408-X   AACR2

Reprinted with corrections, May 1983

PREFACE

There are substantial differences between the second edition of this text and the present third edition. Following is a list of features that characterize this edition.

1. *Flexibility of topic organization.* There are several paths that one can take through the material. The emphasis can be on triangles, or the emphasis can be analytic. Some of the possibilities are detailed on pp. v–vi.

2. *Exercises.* This edition contains many new exercises. In response to comments from users, we have added exercises that require something of the student other than an understanding of the immediate objectives of the lesson at hand, yet are not necessarily highly challenging. The challenge exercises of the second edition have been augmented here. Thus the first exercises in an exercise set are very much like the examples in the text for that section. These exercises are graded in difficulty and are paired. That is, each even-numbered exercise is very much like the one that precedes it. The next exercises (marked ☆) require the student to go beyond the immediate objectives. They may, for example, ask the student to synthesize objectives in the section, or those from preceding chapters with those of this section. For the first two types of exercises, answers to the odd-numbered ones are given at the back of the book. The instructor can therefore easily make an assignment that is varied in terms of availability of answers. The challenge exercises are marked ★, and some of them are quite difficult. Answers to the challenge exercises are not in the text, but are given in the answer booklet.

3. *Keying of objectives.* For each section, the objectives (listed in the margins) are identified as [i], [ii], [iii], and so on. The text material

in which the objectives are developed is marked similarly: [i], [ii], [iii], and so on. In the exercise sets, those exercises that pertain to a specific objective are also marked in like fashion. The questions in the tests and reviews are keyed to the section and subsection to which they apply. For example, a question pertaining to objective [iii] in Section 4.2 will be marked [4.2, iii].

A student having difficulty with an exercise set or a test/review can use the keying system to find text material, worked examples, and margin exercises that pertain to the particular area of difficulty.

4. *Calculators.* Many exercises, as well as some parts of the development, are designed with the calculator in mind. Although it is perfectly feasible to use this text without a calculator, we have indicated those exercises, examples, or sections in which the use of a calculator is recommended by the symbol ▦. A calculator will be found useful in many other places as well.

Most of the calculator exercises in the second edition were much like the other exercises, except that the numbers were more complicated. In this edition the use of the calculator has been made much more comprehensive.

5. *Readability and understandability.* Although readability and understandability are related, they are actually separate features of a text. It is easy to write prose that is easy to read but impossible to understand. Therefore, we discuss these items separately.

With respect to *readability,* we have striven to say what we feel needs to be said, but without excess verbiage. The goal was to make the reading level of this text quite low, without sounding condescending. This book is written *to the student.* Theorems, principles, and procedures are stated for maximum student understanding, yet the tone of the book is still mature.

With respect to *understandability,* the goal was to produce a sequence in which each topic is developed, step by carefully-described step. At each appropriate point, examples are given, sufficient in coverage that the routine part of the homework exercises is thoroughly covered. *Cautions* are frequently given to the student: for example, "Don't make the mistake of thinking that $\sqrt{a^2 + b^2} = a + b$."

6. *Functions and transformations.* This text applies the concepts of function, relation, and transformation quite thoroughly. The idea of transformations makes Chapter 1 unique and sets up the study of later material.

Chapter 1 may be a bit long. Users have given us valuable feedback with respect to this chapter. Some have recommended that it be broken up, with some of the topics imbedded in other chapters, whereas others feel quite positive about teaching the entire chapter. After careful consideration, we have decided to leave the chapter intact, but with some rearrangement of topics. Although the topics can be taught in various orders, there is no *best* order. Moreover, for reference purposes, it is desirable to leave all the material together in its own chapter. Some possible reorderings of the topics are given below.

7. *Supplementary materials.* The following supplementary materials are available.

- An Answer Booklet containing the answers to the even-numbered exercises and to the more challenging exercises.

- A Student Solutions Booklet containing worked-out solutions to selected odd-numbered exercises.

- A Test Booklet containing five classroom-ready tests of various types for each chapter and five final examinations.

TOPIC SELECTION AND ORDERING

There are numerous topics that can be omitted, depending on the nature of a particular class and/or course. Chapter 7 can easily be omitted entirely, for example.

Trigonometry

There are several different possible tracks through the trigonometry. We detail three such tracks, one with analytic emphasis, the others with triangle emphasis.

Track I (Analytic emphasis)

Proceed through Chapters 2–5 as written, omitting Section 2.6. This track gives a thorough grounding in trigonometry with an initial emphasis on the analytic aspects.

Section 2.1, which is an introduction to triangle trigonometry, could be postponed until the beginning of Chapter 3. Certain topics, such as vectors and polar coordinates, can be omitted.

Track II (Triangle emphasis)

Section 2.1
Chapter 3 (omitting 3.6)   } This is triangle trigonometry
Chapter 5

Remainder of Chapter 2   } This is analytic trigonometry
Chapter 4

Track III (Minimal course, triangle emphasis)

Section 2.1
Chapter 3 (omitting 3.6)
Chapter 5 (omitting certain   } This is triangle trigonometry
 optional sections if desired)

Section 3.6
Perhaps Section 2.7   } This is analytic trigonometry
Chapter 4

Note that in the minimal course, most of Chapter 2 is omitted. There are other variations possible. One could teach Chapters 2, 3, 5 and then 4, for example.

Functions and Transformations

Some users consider Chapter 1 a bit long. The following will help such users see how to break it up.

Track I (*Analytic emphasis*)

Sections 1.1–1.6 are needed before teaching Chapter 2, and should be taught initially, except in a class that has a good background in functions and graphing. In such a class, Sections 1.2 and 1.3 might be omitted. Section 1.7 on *inverses* could be postponed, and taught just prior to introducing the inverses of the trigonometric functions.

Track II (Triangle emphasis)

Sections 1.1–1.3 should be taught initially, except for those classes that are already familiar with functions and graphing. For those classes it is possible to begin the study of trigonometry without teaching any of Chapter 1.

After teaching the triangle trigonometry, one should proceed with Sections 1.1–1.6, and then the analytic trigonometry. Section 1.7 can be postponed and taught just before introducing the inverses of the trigonometric functions.

Among the salient features of the text that have been retained are the following.

1. *Use of margins.* The margins contain objectives for each lesson for student reference. They also contain developmental exercises, which have proved to be extremely effective. The text refers the student to these exercises at appropriate places. When students come to these exercises, they are to stop reading and do them. Then the answers can be checked at the back of the text. Thus students receive reinforcement, guidance, and practice before continuing with the text development. The exercises in the first part of the exercise sets are very similar to these developmental exercises in the margins. The margin materials constitute a built-in study guide.

2. *Flexibility of teaching method.* There are many ways in which to use this book. The instructor who wishes to use it in a traditional way should simply ignore the margins and have students ignore them. The instructor who wishes to use the lecture method primarily but also wants to introduce some student-centered activity into the class, can easily do so by interrupting the lecture and having students work the exercises in the margins at appropriate times. On the other hand, the book is well suited for use in math labs or other systems of individualized instruction, or in any approach that is essentially self-study. The book can be used with minimal instructor guidance, so it is particularly effective for use in large classes.

The authors wish to thank the following reviewers, who helped with the development of this new edition: Al H. Chew, Central Arizona College; and B. Y. Fein, Oregon State University.

January 1982   M. L. K.
                                                               M. L. B.

CONTENTS

1

$f(f(f(x)))$

2

3

viii CONTENTS

1

RELATIONS, FUNCTIONS, AND TRANSFORMATIONS

$f\bigl(f(f(x))\bigr)$

$$f\Bigl(f\bigl(f(x)\bigr)\Bigr)$$

OBJECTIVES

You should be able to:

[i] Find Cartesian products of small sets.

[ii] Indicate or find certain relations.

[iii] Find the domain and range of a relation.

1. Let $A = \{d, e, f\}$ and $B = \{1, 2\}$. List all the ordered pairs in $A \times B$.

2. List all of the ordered pairs in the Cartesian square of $\{1, 2, 3, 4\}$. Save the list for later use.

1.1 RELATIONS AND ORDERED PAIRS

[i] Cartesian Products

Consider the sets $A = \{1, 2, 3\}$ and $B = \{a, b\}$. From these sets we can pick numbers and form ordered pairs: for example,

$$(1, a), \qquad (2, a), \qquad \text{and} \qquad (3, b).$$

DEFINITION

> The Cartesian product of two sets A and B, symbolized $A \times B$, is defined as the set of *all* ordered pairs having the first member from set A and the second member from set B.

Example 1 Find the Cartesian product $A \times B$, where $A = \{1, 2, 3\}$ and $B = \{a, b\}$.

The Cartesian product, $A \times B$, is as follows:

$$(1, a), \quad (2, a), \quad (3, a),$$
$$(1, b), \quad (2, b), \quad (3, b).$$

> CAUTION! **Be sure in forming Cartesian products that in each pair of $A \times B$ the first member is taken from A and the second member is taken from B.**

DO EXERCISE 1.

The sets A and B may be the same.

Example 2 Consider the set $\{2, 3, 4, 5\}$. Find the Cartesian product of this set by itself.

The Cartesian product of this set by itself is called a *Cartesian square* and is as follows:

```
5 | (2, 5)  (3, 5)  (4, 5)  (5, 5)
4 | (2, 4)  (3, 4)  (4, 4)  (5, 4)
3 | (2, 3)  (3, 3)  (4, 3)  (5, 3)
2 | (2, 2)  (3, 2)  (4, 2)  (5, 2)
  -------------------------------
     2       3       4       5
```

The headings at the bottom and at the left are for reference only. The Cartesian square consists only of the ordered pairs.

DO EXERCISE 2.

[ii] Relations

In a Cartesian product we can pick out ordered pairs that make up common relations, such as $=$ or $<$ as in the following examples.

Example 3 In the Cartesian product of Example 2 indicate all ordered pairs for which the first member is the same as the second. This set of ordered pairs is the relation =.

$$(2, 5) \quad (3, 5) \quad (4, 5) \quad (5, 5)$$
$$(2, 4) \quad (3, 4) \quad (4, 4) \quad (5, 4)$$
$$(2, 3) \quad (3, 3) \quad (4, 3) \quad (5, 3)$$
$$(2, 2) \quad (3, 2) \quad (4, 2) \quad (5, 2)$$

Example 4 In the Cartesian product of Example 2 indicate all ordered pairs for which the first member is less than the second. This set of ordered pairs is the relation <.

$$(2, 5) \quad (3, 5) \quad (4, 5) \quad (5, 5)$$
$$(2, 4) \quad (3, 4) \quad (4, 4) \quad (5, 4)$$
$$(2, 3) \quad (3, 3) \quad (4, 3) \quad (5, 3)$$
$$(2, 2) \quad (3, 2) \quad (4, 2) \quad (5, 2)$$

DO EXERCISES 3 AND 4.

There are also many relations that do not have common names and relations with which we are not already familiar. Any time we select a set of ordered pairs from a Cartesian product, we have selected some relation. This is true even if we make the selection at random.

Example 5 The following set of ordered pairs is a relation, but is not a familiar one. It has no common name.

$$(2, 5) \quad (3, 5) \quad (4, 5) \quad (5, 5)$$
$$(2, 4) \quad (3, 4) \quad (4, 4) \quad (5, 4)$$
$$(2, 3) \quad (3, 3) \quad (4, 3) \quad (5, 3)$$
$$(2, 2) \quad (3, 2) \quad (4, 2) \quad (5, 2)$$

We shall now make our definition of relation.

DEFINITION

A *relation* from a set A to a set B is defined to be any set of ordered pairs in $A \times B$.

[iii] Domain and Range

DEFINITION

The set of all first members in a relation is called its *domain*. The set of all second members in a relation is called its *range*.

Example 6 Find the domain and range of the relation = in Example 3.

Domain: $\{2, 3, 4, 5\}$; Range: $\{2, 3, 4, 5\}$

3. In the Cartesian square of Exercise 2, indicate all ordered pairs for which the first member is the same as the second. This is the relation =.

4. In the Cartesian square of Exercise 2, indicate all ordered pairs for which the first member is greater than the second. This is the relation >.

5. Find the domain and range of the relation in Exercise 3.

6. Find the domain and range of the relation in Exercise 4.

7. Consider the relation whose ordered pairs are (2, 2), (1, 1), (1, 2), and (1, 3). Find the domain and range.

Example 7 Find the domain and range of the relation $<$ in Example 4.

Domain: $\{2, 3, 4\}$;   Range: $\{3, 4, 5\}$

Example 8 Find the domain and range of the relation in Example 5.

Domain: $\{2, 4, 5\}$;   Range: $\{2, 3, 5\}$

DO EXERCISES 5–7.

EXERCISE SET 1.1

[i]

1. List all ordered pairs in the Cartesian product $A \times B$, where $A = \{0, 2\}$ and $B = \{a, b, c\}$. Remember that first coordinates come from A and second coordinates from B.

2. List all ordered pairs in the Cartesian product $A \times B$, where $A = \{1, 3, 5, 9\}$ and $B = \{d, e, f\}$. Remember that first coordinates come from A and second coordinates from B.

[ii] For Exercises 3–8, consider the set $\{-1, 0, 1, 2\}$.

3. Find the set of ordered pairs in the relation $<$ (is less than).

4. Find the set of ordered pairs in the relation $>$ (is greater than).

5. Find the set of ordered pairs in the relation $\leq$ (is less than or equal to).

6. Find the set of ordered pairs in the relation $\geq$ (is greater than or equal to).

7. Find the set of ordered pairs in the relation $=$.

8. Find the set of ordered pairs in the relation $\neq$.

[iii]

9. a) List all the ordered pairs in the Cartesian square $D \times D$, where $D = \{-1, 0, 1, 2\}$.

b) Graph the relation whose ordered pairs are (0, 0), (1, 1), (0, 1), and (1, 2).

c) List the domain and the range of this relation.

10. a) List all the ordered pairs in the Cartesian square $E \times E$, where $E = \{-1, 1, 3, 5\}$.

b) Graph the relation whose ordered pairs are (−1, 1), (1, 1), (−1, 3), and (1, 3).

c) List the domain and the range of this relation.

OBJECTIVES

You should be able to:

[i] Determine whether an ordered pair of numbers is a solution of an equation with two variables.

[ii] Graph certain equations.

[iii] Given the graph of a relation, describe the domain and the range.

1.2 GRAPHS OF EQUATIONS

Relations in Real Numbers

We are most interested in relations involving $R \times R$, where R is the set of real numbers. The set R is infinite. Thus relations involving R may be infinite, and therefore cannot be indicated by listing the ordered pairs one at a time. We usually indicate such relations with some sort of picture in $R \times R$. This kind of picture is called a *graph*.

Points in the Plane and Ordered Pairs

On a number line each point corresponds to a number. On a plane each point corresponds to a number pair from $R \times R$. To represent $R \times R$ we draw an x-axis and a y-axis perpendicular to each other. Their intersection is called the *origin* and is labeled 0. The arrows show the positive directions.

This is called a Cartesian coordinate system. The first member of an ordered pair is called the *first coordinate*. The second member is

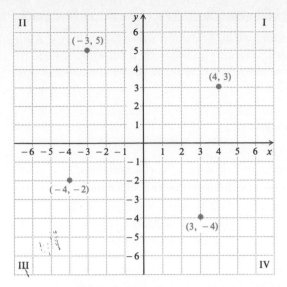

1. Use graph paper.

   a) Draw and label an x-axis and a y-axis.
   b) Label the quadrants.
   c) Plot the points in the relation $(3, 2)$, $(-5, -2)$, $(-4, 3)$.
   d) Find the domain and range of the relation.

called the *second coordinate.* Together these are called the *coordinates of a point.* The axes divide the plane into four *quadrants,* indicated by the Roman numerals.

DO EXERCISE 1.

2. Determine whether $(1, 7)$ is a solution of $y = 2x + 5$.

[i] Solutions of Equations

If an equation has two variables, its solutions are ordered pairs of numbers. A *solution* is an ordered pair which when substituted alphabetically for the variables produces a true equation.

3. Determine whether $(-1, 4)$ is a solution of $y = 2x + 5$.

Example 1 Determine whether the following ordered pairs are solutions of the equation $y = 3x - 1$: $(-1, -4)$, $(7, 5)$.

$$
\begin{array}{c|l}
\multicolumn{2}{c}{y \;=\; 3x - 1} \\
\hline
-4 & 3(-1) - 1 \\
-4 & -3 - 1 \\
 & -4
\end{array}
$$
We substitute -1 for x and -4 for y (alphabetical order of variables).

The equation becomes true: $(-1, -4)$ is a solution.

4. Determine whether $(-2, 5)$ is a solution of $y = x^2$.

$$
\begin{array}{c|l}
\multicolumn{2}{c}{y \;=\; 3x - 1} \\
\hline
5 & 3 \cdot 7 - 1 \\
5 & 21 - 1 \\
 & 20
\end{array}
$$
We substitute.

The equation becomes false: $(7, 5)$ is not a solution.

5. Determine whether $(4, -5)$ is a solution of $x^3 - y^2 = 39$.

DO EXERCISES 2–5.

---

*The first coordinate is sometimes called the *abscissa* and the second coordinate the *ordinate.*

[ii] Graphs of Equations

The solutions of an equation are ordered pairs and thus constitute a relation. To *graph* an equation, or a relation, means to make a drawing of its solutions. Some general suggestions for graphing are as follows.

Graphing suggestions

a) **Use graph paper.**

b) **Label axes with symbols for the variables.**

c) **Use arrows to indicate positive directions.**

d) **Scale the axes; that is, mark numbers on the axes.**

e) **Plot solutions and complete the graph. When finished write down the equation or relation being graphed.**

Example 2 Graph $y = 3x - 1$.

We find some ordered pairs that are solutions. To find an ordered pair, we choose *any* number that is a sensible replacement for x and then determine y. For example, if we choose 2 for x, then $y = 3(2) - 1$, or 5. We have found the solution (2, 5). We continue making choices for x, and finding the corresponding values for y. We make some negative choices for x, as well as positive ones. We keep track of the solutions in a table.

x	y
0	-1
1	2
2	5
-1	-4
-2	-7

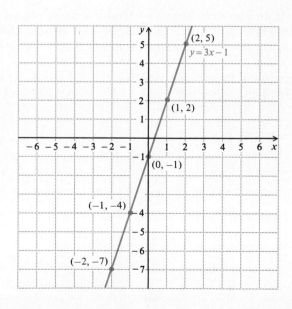

The table gives us the ordered pairs $(0, -1)$, $(1, 2)$, $(2, 5)$, and so on. Next, we plot these points. If we had enough of them, they would make a solid line. We can draw the line with a ruler, and label it $y = 3x - 1$.

Note that the equation $y = 3x - 1$ has an infinite (unending) set of solutions. The *graph of the equation* is a drawing of the relation that is made up of its solutions. Thus the relation consists of all pairs (x, y) such that $y = 3x - 1$ is true. That is, $\{(x, y)\,|\,y = 3x - 1\}$.

DO EXERCISE 6.

Example 3 Graph $y = x^2 - 5$.

We select numbers for x and find the corresponding values for y. The table gives us the ordered pairs $(0, -5)$, $(-1, -4)$, and so on.

x	y
0	-5
-1	-4
1	-4
-2	-1
2	-1
-3	4
3	4

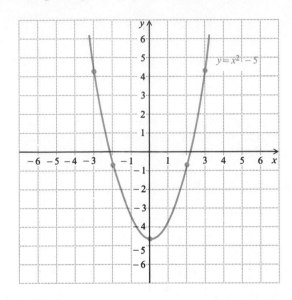

Next we plot these points. We note that as the absolute value of x increases, $x^2 - 5$ also increases. Thus the graph is a curve that rises gradually on either side of the y-axis, as shown above.

This graph shows the relation $\{(x, y)\,|\,y = x^2 - 5\}$.

DO EXERCISE 7.

Example 4 Graph the equation $x = y^2 + 1$.

Here we shall select numbers for y and then find the corresponding values for x. This time we arrange them in a horizontal table.

x	2	2	1	5	5
y	-1	1	0	-2	2

We must remember that x is the first coordinate and y is the second coordinate. Thus the table gives us the ordered pairs $(2, -1)$, $(2, 1)$, $(1, 0)$, and so on. We note that as the absolute value of y gradually increases, the value of x also gradually increases. Thus the graph is a curve that stretches farther and farther to the right as it gets farther from the x-axis.

6. Graph $y = -3x + 1$.

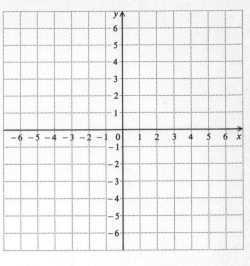

7. Graph $y = 3 - x^2$.

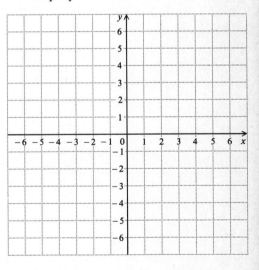

8. Graph $x = y^2 - 5$. [*Hint:* Select values for y and then find the corresponding values of x. When you plot, be sure to find x (horizontally) first.] Compare it with the graph of Example 3.

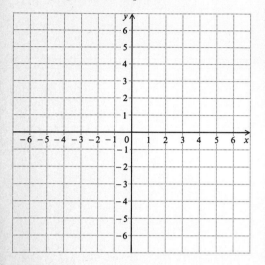

This graph shows the relation $\{(x, y) \mid x = y^2 + 1\}$.

DO EXERCISE 8.

You can always use a calculator to find as many values as desired. This can be especially helpful when you are uncertain about the shape of a graph.

Example 5 Graph the equation $xy = 12$.

We find numbers that satisfy the equation.

x	1	−1	2	−2	3	−3	4	−4	6	−6	12	−12
y	12	−12	6	−6	4	−4	3	−3	2	−2	1	−1

We plot these points and connect them. To see how to do this, note that neither x nor y can be 0. Thus the graph does not cross either axis. As the absolute value of x gets small, the absolute value of y must get large, and vice versa. Thus the graph consists of two curves, as follows.

9. Graph $xy = 1$.

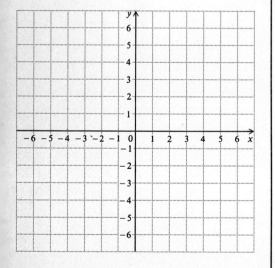

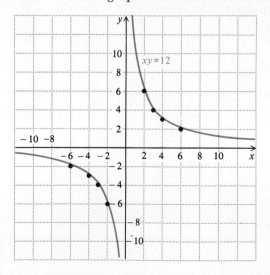

DO EXERCISE 9.

Example 6 Graph the equation $y = |x|$.

We find numbers that satisfy the equation.

x	0	1	−1	2	−2	3	−3	4	−4
y	0	1	1	2	2	3	3	4	4

We plot these points and connect them. To see how to do this, note that as we get farther from the origin, to the left or right, the absolute value of x increases. Thus the graph is a curve that rises to the left and right of the y-axis. It actually consists of parts of two straight lines, as follows.

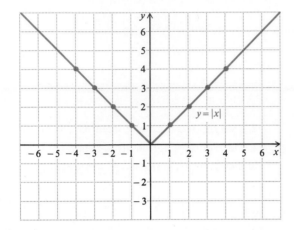

DO EXERCISE 10.

[iii] Domains and Ranges

In each example we see a relation in real numbers, together with its domain and range.

Example 7

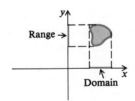

Example 8

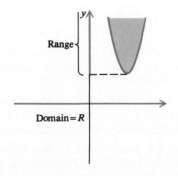

10. Graph the equation $x = |y|$. Compare it with the graph of Example 6.

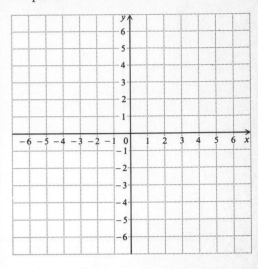

On each diagram in Exercises 11–13:

a) shade (on the x-axis) the domain;
b) shade (on the y-axis) the range.

11.

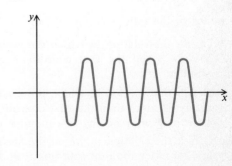

12.

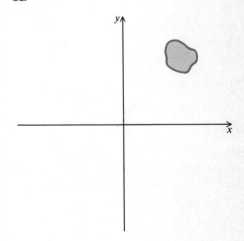

13.

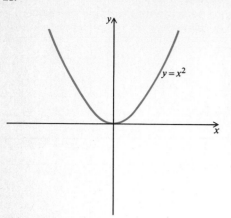

Example 9

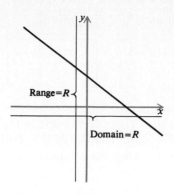

DO EXERCISES 11–13 ON PP. 9 AND 10.

EXERCISE SET 1.2

[i]

1. Determine whether $(-1, -3)$ is a solution of $y = 5x + 2$.

2. Determine whether $(-2, 8)$ is a solution of $y = 3x - 7$.

3. Determine whether $(-2, 7)$ is a solution of $4x + 3y = 12$.

4. Determine whether $(0, -2)$ is a solution of $2x - 3y = 6$.

5. Determine whether $(3, 0)$ is a solution of $x^2 - 2y = 6$.

6. Determine whether $(-2, -1)$ is a solution of $x^2 - 2y = 6$.

[ii] Graph.

7. $y = x + 3$　　　**8.** $y = x - 2$

9. $y = 3x - 2$　　　**10.** $y = -4x + 1$

11. $y = x^2$　　　**12.** $y = -x^2$

13. $y = x^2 + 2$　　　**14.** $y = x^2 - 2$

15. $x = y^2 + 2$　　　**16.** $x = y^2 - 2$

17. $y = |x + 1|$　　　**18.** $y = |x - 1|$

19. $x = |y + 1|$　　　**20.** $x = |y - 1|$

21. $xy = 10$　　　**22.** $xy = -18$

Graph and compare.

23. $y = x^2 + 1$ and $y = (-x)^2 + 1$

24. $y = x^2 - 2$ and $y = 2 - x^2$

[iii]

25. Graph a relation as follows.

a) Draw a circle with radius of length 2, centered at (4, 3). Shade the circle and its interior.

b) Shade (on the x-axis) the domain. Describe the domain.

c) Shade (on the y-axis) the range. Describe the range.

26. Graph a relation as follows.

a) Draw a triangle with vertices at (1, 1), (4, 2), and (3, 6). Shade the triangle and its interior.

b) Shade (on the x-axis) the domain. Describe the domain.

c) Shade (on the y-axis) the range. Describe the range.

☆

In each of the following, all relations to be considered are in $R \times R$, where R is the set of real numbers.

27. Graph the relation in which the second coordinate is always 2 and the first coordinate may be any real number.

28. Graph the relation in which the first coordinate is always -3, and the second coordinate may be any real number.

29. Graph the relation in which the second coordinate is always 1 more than the first coordinate, and the first coordinate may be any real number.

30. Graph the relation in which the second coordinate is always 1 less than the first coordinate, and the first coordinate may be any real number.

31. Graph the relation in which the second coordinate is always twice the first coordinate, and the first coordinate may be any real number.

32. Graph the relation in which the second coordinate is always half the first coordinate, and the first coordinate may be any real number.

33. Graph the relation in which the second coordinate is always the square of the first coordinate, and the first coordinate may be any real number.

34. Graph the relation in which the first coordinate is always the square of the second coordinate, and the second coordinate may be any real number.

Graph the following equations.

35. $y = |x| + x$

36. $y = x|x|$

37. $y = |x^2 - 4|$

38. $y = x^3$

39. $y = \sqrt{x}$

40. $y = |x^3|$

41. $|y| = |x|$

42. $|x| + |y| = 0$

43. $|xy| = 1$

44. Graph the relation $\{(x, y) | 1 < x < 4 \text{ and } -3 < y < -1\}$.

★ ─────────────────────────────────────

45. Graph $|x| + |y| = 1$.

46. Graph the relation $\{(x, y) | |x| \le 1 \text{ and } |y| \le 2\}$.

▦ Graph. In each case use your calculator and substitute at least twenty values of x between -5 and 5.

47. $y = \frac{1}{3}x^3 - x + \frac{2}{3}$

48. $y = \frac{1}{3}x^3 - \frac{1}{2}x^2 - 2x + 1$

1.3 FUNCTIONS

[i] Recognizing Graphs of Functions

A *function* is a special kind of relation. It is defined as follows.

DEFINITION

A *function* **is a relation in which no two ordered pairs have the same first coordinate and different second coordinates.**

In a function, given a member of the domain (a first coordinate) there is one and only one member of the range that goes with it (the second coordinate). Thus each member of the domain *determines* exactly one member of the range. It is easy to recognize the graph of a relation that is a function. If there are two or more points of the graph on the same vertical line, then the relation is not a function. Otherwise it is. Here are some graphs of functions. In graph (c), the solid dot indicates that $(-1, 1)$ belongs to the graph. The open dot indicates that $(-1, -2)$ does not belong to the graph. Thus no vertical line crosses the graph more than once.

OBJECTIVES

You should be able to:

[i] Recognize a graph of a function.

[ii] Use a formula to find function values.

[iii] Find the domain of a function, given by a formula.

[iv] Compose pairs of functions f and g, finding formulas for $f \circ g$ and $g \circ f$.

a)
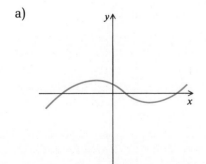

b)
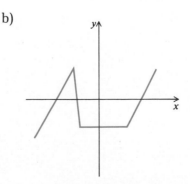

c)
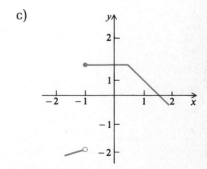

1. Which of the following are graphs of functions?

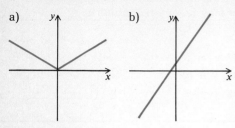

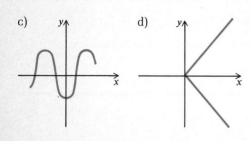

The following are not graphs of functions, because they fail the so-called *vertical line test*. That is, we can find a vertical line that meets the graph in more than one point.

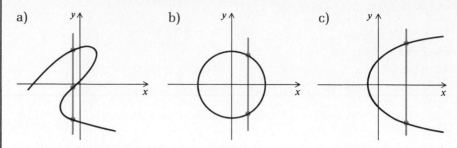

The vertical line test. **If it is possible for a vertical line to meet a graph more than once, the graph is not the graph of a function.**

DO EXERCISE 1.

[ii] Notation for Functions

Functions are often named by letters, such as f or g. A function f is thus a set of ordered pairs. If we represent the first coordinate of a pair by x, then we may represent the second coordinate by $f(x)$. The symbol $f(x)$ is read "f of x" or "f at x." The number represented by $f(x)$ is called the "value" of the function at x. [*Note:* "$f(x)$" does *not* mean "f times x."]

Example 1 Below let us call the function in color g.

(1, 4)	(2, 4)	(3, 4)	(4, 4)
(1, 3)	(2, 3)	(3, 3)	(4, 3)
(1, 2)	(2, 2)	(3, 2)	(4, 2)
(1, 1)	(2, 1)	(3, 1)	(4, 1)

Here $g(1) = 4$, $g(2) = 3$, $g(3) = 2$, and $g(4) = 4$.

Example 2 Let us call the function graphed below f. To find function values, we locate x on the x-axis, and then find $f(x)$ on the y-axis.

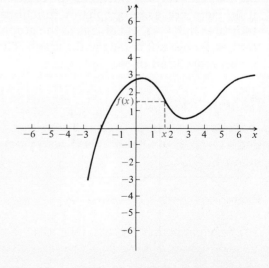

From the graph we can see that $f(-2) = 0$, that $f(0)$ is about 2.7, and that $f(3)$ is about 0.6.

DO EXERCISES 2 AND 3.

Some functions in real numbers can be defined by formulas, or equations. Here are some examples:

$$g(s) = 3, \qquad p(t) = \frac{1}{t}, \qquad f(x) = 3x^2 + 4x - 5, \qquad u(y) = |y| + 3.$$

Function values can be obtained by making substitutions for the variables.

A function such as g above is called a *constant function* because all its function values are the same. The range contains only one number, 3.

Example 3 Let $f(x) = 2x^2 - 3$. We can find function values as follows:

a) $\qquad f(0) = 2 \cdot 0^2 - 3 = -3;$

b) $\qquad f(-1) = 2(-1)^2 - 3 = 2 \cdot 1 - 3 = -1;$

c) $\qquad f(5a) = 2(5a)^2 - 3 = 2 \cdot 25a^2 - 3 = 50a^2 - 3;$

d) $\qquad f(a - 4) = 2(a - 4)^2 - 3 = 2(a^2 - 8a + 16) - 3$
$$= 2a^2 - 16a + 29;$$

e) $\dfrac{f(a + h) - f(a)}{h} = \dfrac{[2(a + h)^2 - 3] - [2a^2 - 3]}{h}$

$$= \frac{2a^2 + 4ah + 2h^2 - 3 - 2a^2 + 3}{h}$$

$$= \frac{4ah + 2h^2}{h} = 4a + 2h.$$

If you have trouble finding function values when a formula is given, think of the formula, in the case of Example 3, as

$$f(\quad) = 2(\quad)^2 - 3.$$

Then whatever goes in the blank on the left between parentheses goes in the blank on the right between parentheses.

DO EXERCISE 4.

[iii] Finding the Domain of a Function

When a function in $R \times R$ is given by a formula, the domain is understood to be the set of all real numbers that are sensible replacements.

Example 4 Find the domain of $g(x) = \dfrac{x}{x^2 + 2x - 3}$.

The formula makes sense as long as a replacement for x does not make the denominator 0. To find those replacements that do make the de-

2. In this function, what is $f(1)$? $f(2)$? $f(3)$? $f(4)$?

(1, 4)	(2, 4)	(3, 4)	(4, 4)
(1, 3)	(2, 3)	(3, 3)	(4, 3)
(1, 2)	(2, 2)	(3, 2)	(4, 2)
(1, 1)	(2, 1)	(3, 1)	(4, 1)

What is the domain of this function? What is the range?

3. From the graph, find the following function values approximately: $f(2)$, $f(0)$, $f(-3)$.

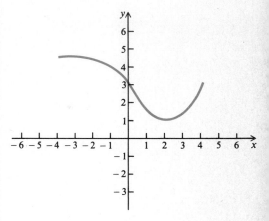

4. $f(x) = 3x^2 + 1$. Find:

a) $f(0)$;

b) $f(1)$;

c) $f(-1)$;

d) $f(2a)$;

e) $f(a + 1)$;

f) $\dfrac{f(a + h) - f(a)}{h}$.

Find the domain of each function.

5. $f(x) = \dfrac{x + 1}{3x^2 + 10x + 8}$

6. $g(x) = \sqrt{10x + 25}$

7. $p(x) = x^3 - 4x^2 + 2x + 8$

nominator 0 we solve $x^2 + 2x - 3 = 0$:

$$x^2 + 2x - 3 = 0$$
$$(x - 1)(x + 3) = 0$$
$$x - 1 = 0 \quad \text{or} \quad x + 3 = 0$$
$$x = 1 \quad \text{or} \quad x = -3.$$

Thus the domain consists of the set of all real numbers except 1 and -3. We can name this set $\{x \mid x \neq 1 \text{ and } x \neq -3\}$.

Example 5 Find the domain of $f(x) = \sqrt{5x - 3}$.

The formula makes sense as long as a replacement makes the radicand nonnegative (no negative number has a square root). Thus to find the domain we solve the inequality $5x - 3 \geq 0$:

$$5x - 3 \geq 0$$
$$5x \geq 3$$
$$x \geq \frac{3}{5}.$$

The domain is $\{x \mid x \geq \frac{3}{5}\}$.

Example 6 Find the domain of $t(x) = x^3 + |x|$.

There are no restrictions on the numbers we can substitute into this formula. We can cube any real number, we can take the absolute value of any real number, and we can add the results. Thus the domain is the entire set of real numbers.

DO EXERCISES 5–7.

Functions, Mappings, and Machines

Functions can be thought of as mappings. A function f *maps* the set of first coordinates (the domain) to the set of second coordinates (the range).

As in the diagram each x in the domain corresponds to (or is mapped onto) just one y in the range. That y is the second coordinate of the ordered pair (x, y).

Example 7 Consider the function f for which $f(x) = 2x + 3$.

Since $f(0)$ is 3, this function maps 0 to 3.

Since $f(3) = 9$, this function maps 3 to 9.

The concept of function is illuminated somewhat by considering a so-called "function machine." In this drawing we see a function machine designed, or programmed, to do the mapping (function) f. The inputs acceptable to the machine are the members of the domain of f. The outputs are, of course, members of the range of f.

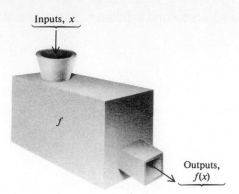

Inputs, x

Outputs, $f(x)$

We may sometimes refer to an ordered pair $(x, f(x))$ as an input-output pair. The graph of a function consists of all the input–output pairs.

[iv] Composition of Functions

Functions can be combined in a way called *composition* of functions. Consider, for example,

$$f(x) = x^2 \quad \text{This function squares each input.}$$

and

$$g(x) = x + 1. \quad \text{This function adds 1 to each input.}$$

We define a new function that first does what g does (adds 1) and then does what f does (squares). The new function is called the *composition* of f and g, and is symbolized $f \circ g$. Let us think of hooking two function machines f and g together to get the resultant function machine $f \circ g$.

DEFINITION

The composed function $f \circ g$ is defined as follows: $f \circ g(x) = f(g(x))$.

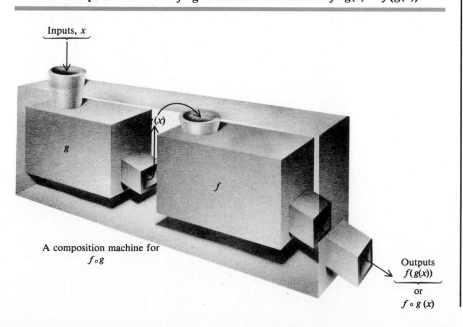

Inputs, x

A composition machine for $f \circ g$

Outputs $f(g(x))$

or $f \circ g(x)$

8. Let $f(x) = x - 2$ and $g(x) = 2x^2$. Find $f \circ g(x)$ and $g \circ f(x)$.

Now let us see how to find a formula for $f \circ g$, and also the reverse composition $g \circ f$.

Example 8 Given that $f(x) = x^2$ and $g(x) = x + 1$, find formulas for $f \circ g(x)$ and $g \circ f(x)$.

By definition of $\circ$,

$$f \circ g(x) = f(g(x)).$$

Now

$$
\begin{aligned}
f(g(x)) &= f(x + 1) && \text{Substituting } x + 1 \text{ for } g(x) \\
&= (x + 1)^2 && \text{Substituting } x + 1 \text{ for } x \text{ in} \\
& && \text{the formula for } f(x) \\
&= x^2 + 2x + 1.
\end{aligned}
$$

Similarly,

$$
\begin{aligned}
g \circ f(x) &= g(x^2) && \text{Substituting } x^2 \text{ for } f(x) \\
&= x^2 + 1. && \text{Substituting } x^2 \text{ for } x \text{ in the} \\
& && \text{formula for } g(x)
\end{aligned}
$$

Note that the function $f \circ g$ is not the same as the function $g \circ f$.

In order for functions to be composable, such as f and g above, the outputs of g must be acceptable as inputs for f. In other words, if any number $g(x)$ is not in the domain of f, then x is not in the domain of $f \circ g$. Note also the order of happenings in $f \circ g$. The function $f \circ g$ does *first* what g does, *then* what f does.

9. Let $u(x) = 2x^2$ and $v(x) = 3x + 2$. Find $u \circ v(x)$ and $v \circ u(x)$.

DO EXERCISE 8.

Example 9 Let $f(x) = 2x$ and $g(x) = x^2 + 1$. Find $f \circ g(x)$ and $g \circ f(x)$.

By definition of $\circ$,

$$
\begin{aligned}
f \circ g(x) &= f(g(x)) \\
&= f(x^2 + 1) && \text{Substituting } x^2 + 1 \text{ for } g(x) \\
&= 2(x^2 + 1) && \text{Substituting } x^2 + 1 \text{ for } x \\
& && \text{in the formula for } f(x) \\
&= 2x^2 + 2.
\end{aligned}
$$

Similarly,

$$
\begin{aligned}
g \circ f(x) &= g(f(x)) \\
&= g(2x) && \text{Substituting } 2x \text{ for } f(x) \\
&= (2x)^2 + 1 && \text{Substituting } 2x \text{ for } x \text{ in} \\
& && \text{the formula for } g(x) \\
&= 4x^2 + 1.
\end{aligned}
$$

DO EXERCISE 9.

EXERCISE SET 1.3

[i]

1. Which of the following are graphs of functions?

a)

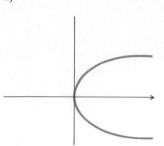

b)

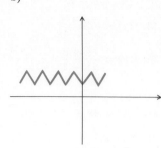

c)

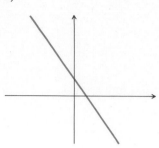

d)

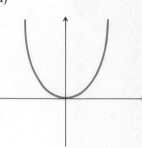

2. Which of the following are graphs of functions? An open circle indicates that the point does not belong to the graph.

a)

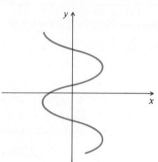

b)

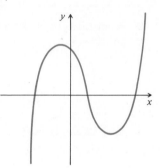

c)

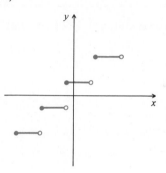

d)

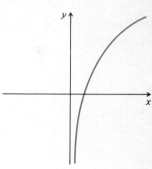

[ii]

3. $f(x) = 5x^2 + 4x$. Find:
   a) $f(0)$;   b) $f(-1)$;
   c) $f(3)$;   d) $f(t)$;
   e) $f(t-1)$;   f) $\dfrac{f(a+h) - f(a)}{h}$.

4. $g(x) = 3x^2 - 2x + 1$. Find:
   a) $g(0)$;   b) $g(-1)$;
   c) $g(3)$;   d) $g(t)$;
   e) $g(a+h)$;   f) $\dfrac{g(a+h) - g(a)}{h}$.

5. $f(x) = 2|x| + 3x$. Find:
   a) $f(1)$;   b) $f(-2)$;
   c) $f(-4)$;   d) $f(2y)$;
   e) $f(a+h)$;   f) $\dfrac{f(a+h) - f(a)}{h}$.

6. $g(x) = x^3 - 2x$. Find:
   a) $g(1)$;   b) $g(-2)$;
   c) $g(-4)$;   d) $g(3y)$;
   e) $g(2+h)$;   f) $\dfrac{g(2+h) - g(2)}{h}$.

7. ▦ $f(x) = 4.3x^2 - 1.4x$. Find:
   a) $f(1.034)$;   b) $f(-3.441)$;
   c) $f(27.35)$;   d) $f(-16.31)$.

8. ▦ $g(x) = \sqrt{2.2|x| + 3.5}$. Find:
   a) $g(17.3)$;   b) $g(-64.2)$;
   c) $g(0.095)$;   d) $g(-6.33)$.

9. $f(x) = \dfrac{x^2 - x - 2}{2x^2 - 5x - 3}$. Find:
   a) $f(0)$;   b) $f(4)$;
   c) $f(-1)$;   d) $f(3)$.

10. $s(x) = \sqrt{\dfrac{3x - 4}{2x + 5}}$. Find:
   a) $s(10)$;   b) $s(2)$;
   c) $s(1)$;   d) $s(-1)$.

[iii] In Exercises 11–20, find the domain of each function.

11. $f(x) = 7x + 4$

12. $f(x) = |3x - 2|$

13. $f(x) = 4 - \dfrac{2}{x}$

14. $f(x) = \sqrt{x - 3}$

15. $f(x) = \sqrt{7x + 4}$

16. $f(x) = \dfrac{1}{9 - x^2}$

17. $f(x) = \dfrac{1}{x^2 - 4}$

18. $f(x) = \dfrac{2x + 6}{x^3 - 4x}$

19. $f(x) = \dfrac{4x^3 + 4}{4x^2 - 5x - 6}$

20. $f(x) = \dfrac{x^3 + 8}{x^2 - 4}$

[iv] In Exercises 21–26, find $f \circ g(x)$ and $g \circ f(x)$.

21. $f(x) = 3x^2 + 2, \quad g(x) = 2x - 1$

22. $f(x) = 4x + 3, \quad g(x) = 2x^2 - 5$

23. $f(x) = 4x^2 - 1, \quad g(x) = \dfrac{2}{x}$

24. $f(x) = \dfrac{3}{x}, \quad g(x) = 2x^2 + 3$

25. $f(x) = x^2 + 1, \quad g(x) = x^2 - 1$

26. $f(x) = \dfrac{1}{x^2}, \quad g(x) = x + 2$

☆ _____

27. From this graph, find approximately $g(-2)$, $g(-3)$, $g(0)$, and $g(2)$.

28. From this graph, find approximately $h(-2)$, $h(0)$, $h(3)$, and $h(-3)$.

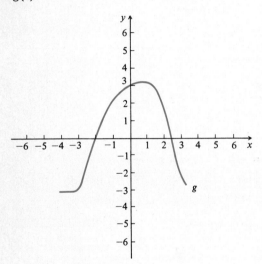

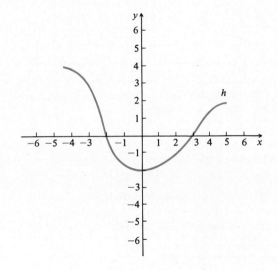

For each function find $\dfrac{f(x + h) - f(x)}{h}$.

29. $f(x) = \dfrac{1}{x}$

30. $f(x) = \dfrac{1}{x^2}$

31. $f(x) = \sqrt{x}$ (Rationalize the numerator.)

Find the domain of each function.

32. $f(x) = \dfrac{\sqrt{x}}{2x^2 - 3x - 5}$

33. $f(x) = \dfrac{\sqrt{x + 3}}{x^2 - x - 2}$

34. $f(x) = \dfrac{\sqrt{x + 1}}{x + |x|}$

35. $f(x) = \sqrt{x^2 + 1}$

36. For $f(x) = \dfrac{1}{1 - x}$, find $f \circ f(x)$ and $f \circ [f \circ f(x)]$.

37. In Exercise 23, find the domain of $f \circ g$ and $g \circ f$.

38. Determine whether the relation $\{(x, y) \mid xy = 0\}$ is a function.

★ _____

39. The *greatest integer function* $f(x) = [x]$ is defined as follows: $[x]$ is the greatest integer that is less than or equal to x. For example, if $x = 3.74$, then $[x] = 3$; and if $x = -0.98$, then $[x] = -1$. Graph the greatest integer function for values of x such that $-5 \leq x \leq 5$.

40. Graph the equation $[y] = [x]$. See Exercise 39. Is this the graph of a function?

1.4 SYMMETRY

Symmetry with Respect to a Line

In the figure points P and P_1 are said to be *symmetric* with respect to line ℓ. They are the same distance from ℓ.

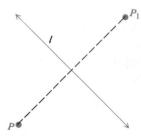

DEFINITION

> Two points P and P_1 are *symmetric with respect to a line ℓ* if and only if ℓ is the perpendicular bisector of the segment $\overline{PP_1}$. The line ℓ is known as the *line of symmetry*.

We also say that the two points P and P_1 are *reflections* of each other across the line. The line is therefore also known as a *line of reflection*.

Now consider a set of points (geometric figure) as in the colored curve below. This figure is said to be symmetric with respect to the line ℓ, because if you pick any point Q in the set, you can find another point Q_1 in the set such that Q and Q_1 are symmetric with respect to ℓ.

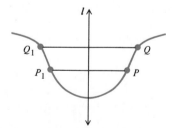

DEFINITION

> A figure, or set of points, is *symmetric with respect to a line ℓ* if and only if for each point Q in the set there exists another point Q_1 in the set for which Q and Q_1 are symmetric with respect to line ℓ.

1. a) Plot the point (3, 2). Let the
   y-axis be a line of symmetry.
   Plot the point symmetric to
   (3, 2). What are its coordinates?

   b) Plot the point (−4, −5). Let the
   y-axis be a line of reflection.
   Plot the image of (−4, −5).
   What are its coordinates?

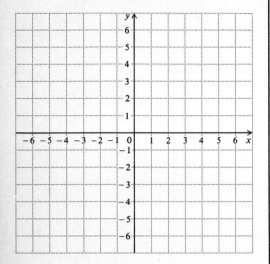

Imagine picking the preceding figure up and flipping it over. Points P and P_1 would be interchanged. Points Q and Q_1 would be interchanged. These are, then, pairs of symmetric points. The entire figure would look exactly like it did before flipping. This means that *each* point of the figure is symmetric with *some* point of the figure. Thus the figure is symmetric with respect to the line. A point and its reflection are known as *images* of each other. Thus P_1 is the image of P, for example. The line ℓ is known as an *axis of symmetry*.

[i] Symmetry with Respect to the Axes

There are special and interesting kinds of symmetry in which a coordinate axis is a line of symmetry. The following example shows figures that are symmetric with respect to an axis and a figure that is not.

Example 1 In graph (a), flipping the figure about the y-axis would not change the figure. In graph (b), flipping the graph about the x-axis would not change the figure. In graph (c), flipping about either axis would change the figure.

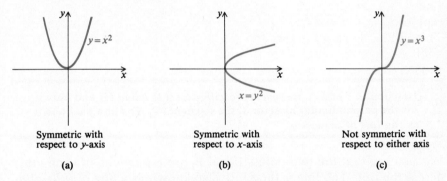

Let us consider a figure like graph (a), symmetric with respect to the y-axis. For every point of the figure there is another point the same distance across the y-axis. The first coordinates of such a pair of points are additive inverses of each other.

Example 2 In the relation $y = x^2$ there are points (2, 4) and (−2, 4). The first coordinates, 2 and −2, are additive inverses of each other, while the second coordinates are the same. For every point of the figure (x, y), there is another point (−x, y).

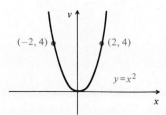

DO EXERCISE 1.

Let us consider a figure symmetric with respect to the x-axis. For every point of such a figure there is another point the same distance across the x-axis. The second coordinates of such a pair of points are additive inverses of each other.

Example 3 In the relation $x = y^2$ there are points $(4, 2)$ and $(4, -2)$. The second coordinates, 2 and -2, are additive inverses of each other, while the first coordinates are the same. For every point of the figure (x, y), there is another point $(x, -y)$.

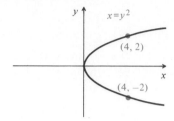

DO EXERCISE 2.

Let us consider a figure symmetric with respect to the y-axis, as in Example 2. Suppose it is defined by an equation. If in this equation we replace x by $-x$, we obtain a new equation, but we get the same figure. This is true because any number x gives us the same y-value as its additive inverse, $-x$.

Let us consider a figure symmetric with respect to the x-axis, as in Example 3. Suppose it is defined by an equation. If we replace y by $-y$ in the equation, we obtain a new equation, but we get the same figure. This is true because any number y gives us the same x-value as $-y$. We thus have a means of testing a relation for symmetry, when it is defined by an equation.

When a relation is defined by an equation:

1. **If replacing x by $-x$ produces an equivalent equation, then the graph is symmetric with respect to the y-axis.**

2. **If replacing y by $-y$ produces an equivalent equation, then the graph is symmetric with respect to the x-axis.**

Example 4 Test $y = x^2 + 2$ for symmetry with respect to the y-axis.

a) Replace x by $-x$.

$$y = x^2 + 2 \qquad (1)$$
$$y = (-x)^2 + 2$$

b) Simplify, if possible.

$$y = (-x)^2 + 2 = x^2 + 2 \qquad (2)$$

c) Is the resulting equation (2) equivalent to the original (1)?

Since the answer is yes, the graph is symmetric with respect to the y-axis.

2. Let the x-axis be a line of symmetry.

a) Plot the point $(4, 3)$. Plot the point symmetric to it. What are its coordinates?

b) Plot the point $(3, -5)$. Plot its image after reflection across the x-axis. What are its coordinates?

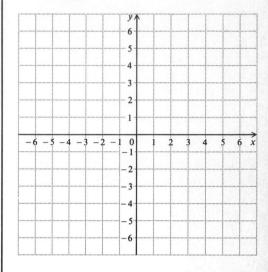

Test the following for symmetry with respect to the coordinate axes.

3. $y = x^2 - 3$

4. $y^2 = x^3$

5. $x^4 = y^2 + 2$

6. $3y^2 + 4x^2 = 12$

7. $a + 3b = 5$

8. $2p^3 + 4q^3 = 1$

Example 5 Test $y = x^2 + 2$ for symmetry with respect to the x-axis.

a) Replace y by $-y$.

$$y = x^2 + 2 \tag{1}$$
$$-y = x^2 + 2$$

b) Simplify, if possible.

The equation is simplified for the most part, although we could multiply on both sides by -1, obtaining

$$y = -(x^2 + 2). \tag{2}$$

c) Is the resulting equation (2) equivalent to the original (1)?

This answer may be obvious to you and is no. To be sure one might use some trial-and-error reasoning as follows. Suppose we substitute 0 for x in equation (1). Then

$$y = 0^2 + 2 = 2,$$

so $(0, 2)$ is a solution of equation (1). For equations (1) and (2) to be equivalent, $(0, 2)$ must also be a solution of equation (2). We substitute to find out:

$$
\begin{array}{c|c}
y & = -(x^2 + 2) \\
\hline
2 & -(0^2 + 2) \\
 & -2
\end{array}
$$

Thus $(0, 2)$ is not a solution of equation (2); hence the equations are not equivalent and the graph is not symmetric with respect to the x-axis.

Example 6 Test $x^2 + y^4 + 5 = 0$ for symmetry with respect to the x-axis.

a) Replace y by $-y$.

$$x^2 + y^4 + 5 = 0 \tag{1}$$
$$x^2 + (-y)^4 + 5 = 0$$

b) Simplify, if possible.

$$x^2 + (-y)^4 + 5 = x^2 + y^4 + 5 = 0 \tag{2}$$

c) Is the resulting equation equivalent to the first?

Since the answer is yes, the graph is symmetric with respect to the x-axis.

DO EXERCISES 3–8.

Symmetry with Respect to a Point

Two points are symmetric with respect to a point when they are situated as shown on the following page. That is, the points are the same distance from that point, and all three points are on a line.

DEFINITION

Two points P and P_1 are *symmetric with respect to a point Q* if and only if Q (the point of symmetry) is the midpoint of segment $\overline{PP_1}$.

A *set* of points is symmetric with respect to a point when each point in the set is symmetric with some point in the set. This is illustrated here. Imagine sticking a pin in this figure at O and then rotating the figure 180°. Points P and P_1 would be interchanged. Points Q and Q_1 would be interchanged. These are pairs of symmetric points. The entire figure would look exactly as it did before rotating. This means that *each* point of the figure is symmetric with *some* point of the figure. Thus the figure is symmetric with respect to the point O.

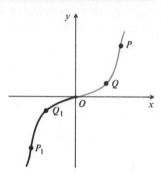

DEFINITION

A set of points is *symmetric with respect to a point B* if and only if for every point P in the set there exists another point P_1 in the set for which P and P_1 are symmetric with respect to B.

[ii] Symmetry with Respect to the Origin

A special kind of symmetry with respect to a point is symmetry with respect to the origin.

Example 7 In graphs (a) and (b), rotating the figure about the origin 180° would not change the figure. In graph (c) such a rotation would change the figure.

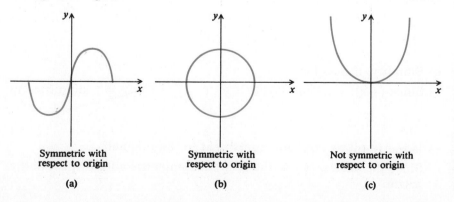

Symmetric with respect to origin	Symmetric with respect to origin	Not symmetric with respect to origin
(a)	(b)	(c)

9. Draw coordinate axes. Let the origin be a point of symmetry.

a) Plot the point (3, 2). Plot the point symmetric to it. What are its coordinates?

b) Plot the point (−4, 3). Plot the point symmetric to it. What are its coordinates?

c) Plot the point (−5, −7). Plot the point symmetric to it. What are its coordinates?

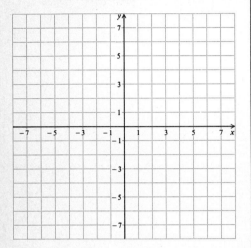

Let us consider figures like (a) and (b) above, symmetric with respect to the origin. For every point of the figure there is another point the same distance across the origin. The first coordinates of such a pair are additive inverses of each other, and the second coordinates are additive inverses of each other.

DO EXERCISE 9.

Example 8 The relation $y = x^3$ is symmetric with respect to the origin. In this relation are the points (2, 8) and (−2, −8). The first coordinates are additive inverses of each other. The second coordinates are additive inverses of each other. For every point of the figure (x, y), there is another point (−x, −y).

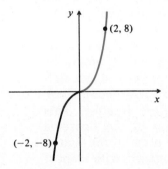

In a relation symmetric with respect to the origin, as in Example 8, if we replace x by −x and y by −y, we obtain a new equation, but we will get the same figure. This is true because whenever a point (x, y) is in the relation, the point (−x, −y) is also in the relation. This gives us a means for testing a relation for symmetry when it is defined by an equation.

When a relation is defined by an equation, if replacing x by $-x$ and replacing y by $-y$ produces an equivalent equation, then the graph is symmetric with respect to the origin.

Example 9 Test $x^2 = y^2 + 2$ for symmetry with respect to the origin.

a) Replace x by −x and y by −y.

$$x^2 = y^2 + 2$$
$$(-x)^2 = (-y)^2 + 2$$

b) Simplify, if possible.

Since $(-x)^2 = x^2$ and $(-y)^2 = y^2$, $(-x)^2 = (-y)^2 + 2$ simplifies to

$$x^2 = y^2 + 2.$$

c) Is the resulting equation equivalent to the original?

Since the answer is yes, the graph is symmetric with respect to the origin.

Example 10 Test $2x + 3y = 8$ for symmetry with respect to the origin.

a) Replace x by $-x$ and y by $-y$.

$$2x + 3y = 8$$
$$2(-x) + 3(-y) = 8$$

b) Simplify, if possible.

$$2(-x) + 3(-y) = -2x - 3y,$$

so

$$-(2x + 3y) = 8 \quad \text{and} \quad 2x + 3y = -8$$

c) Is the resulting equation equivalent to the original?

The equation $2x + 3y = -8$ is *not* equivalent to $2x + 3y = 8$, so the graph is not symmetric with respect to the origin.

DO EXERCISES 10–15.

Test the following for symmetry with respect to the origin.

10. $x^2 + 3y^2 = 4$

11. $x = y$

12. $x = -y$

13. $xy = 5$

14. $ab = -5$

15. $u = |v|$

EXERCISE SET 1.4

[i], [ii] In Exercises 1–12, test for symmetry with respect to the coordinate axes and the origin.

1. $3y = x^2 + 4$ **2.** $5y = 2x^2 - 3$ **3.** $y^3 = 2x^2$ **4.** $3y^3 = 4x^2$

5. $2x^4 + 3 = y^2$ **6.** $3y^2 = 2x^4 - 5$ **7.** $2y^2 = 5x^2 + 12$ **8.** $3x^2 - 2y^2 = 7$

9. $2x - 5 = 3y$ **10.** $5y = 4x + 5$ **11.** $3b^3 = 4a^3 + 2$ **12.** $p^3 - 4q^3 = 12$

[ii] In Exercises 13–24, test for symmetry with respect to the origin.

13. $3x^2 - 2y^2 = 3$ **14.** $5y^2 = -7x^2 + 4$ **15.** $5x - 5y = 0$ **16.** $3x = 3y$

17. $3x + 3y = 0$ **18.** $7x = -7y$ **19.** $3x = \dfrac{5}{y}$ **20.** $3y = \dfrac{7}{x}$

21. $y = |2x|$ **22.** $3x = |y|$ **23.** $3a^2 + 4a = 2b$ **24.** $5v = 7u^2 - 2u$

☆

Consider the following figure for Exercises 25–28.

25. Graph the reflection across the x-axis.

26. Graph the reflection across the y-axis.

27. Graph the reflection across the line $y = x$.

28. Graph the figure formed by reflecting each point through the origin.

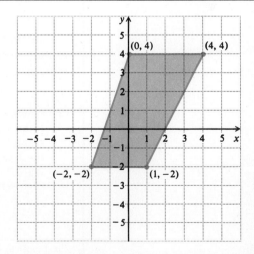

★

29. Consider symmetries with respect to the x-axis, the y-axis, and the origin. Prove that symmetry with respect to any two of these implies symmetry with respect to the other.

OBJECTIVE

You should be able to:

[i] Given the graph of a relation, graph its transformation under translations, reflections, stretchings, and shrinkings.

1.5 TRANSFORMATIONS

[i] Given a relation, we can find various ways of altering it to obtain another relation. Such an alteration is called a *transformation*. If such an alteration consists merely of moving the graph without changing its shape or orientation, the transformation is called a *translation*.

Vertical Translations

Consider the relations $y = x^2$ and $y = 1 + x^2$ whose graphs are shown below. The graph of $y = 1 + x^2$ has the same shape as that of $y = x^2$, but is moved upward a distance of 1 unit. Consider any equation $y = f(x)$. Adding a constant a to produce $y = a + f(x)$ changes each function value by the same amount, a, hence produces no change in the shape of the graph, but merely translates it upward if the constant is positive. If a is negative, the graph will be moved downward.

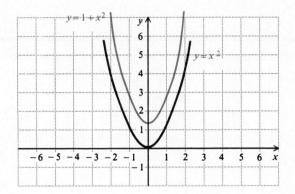

1. a) Graph $y = x^2$. Then graph $y = 2 + x^2$ and compare.
   b) Graph $y = -2 + x^2$ and compare.

DO EXERCISE 1.

Note that $y = 1 + x^2$ is equivalent to $y - 1 = x^2$. Thus the transformation described above amounts to replacing y by $y - 1$ in the original equation.

THEOREM 1

In an equation of a relation, replacing y by $y - a$, where a is a constant, translates the graph vertically a distance of $|a|$. If a is positive, the translation is upward. If a is negative, the translation is downward. *

---

*In practice, when working with functions, we are more inclined to write $y = a + f(x)$ instead of $y - a = f(x)$, but for relations in general we are more likely to replace y by $y - a$.

If in an equation we replace y by $y + 3$, this is the same as replacing it by $y - (-3)$. In this case the constant a is -3 and the translation is downward. If we replace y by $y - 5$, the constant a is 5 and the translation is upward.

Example 1 Sketch a graph of $y = |x|$ and then one of $y = -2 + |x|$.

The graph of $y = |x|$ is shown below. Now consider $y = -2 + |x|$. Note that $y = -2 + |x|$ is equivalent to $y + 2 = |x|$ or $y - (-2) = |x|$. This shows that the new equation can be obtained by replacing y in $y = |x|$ by $y - (-2)$, so the translation is downward, 2 units, as shown below.

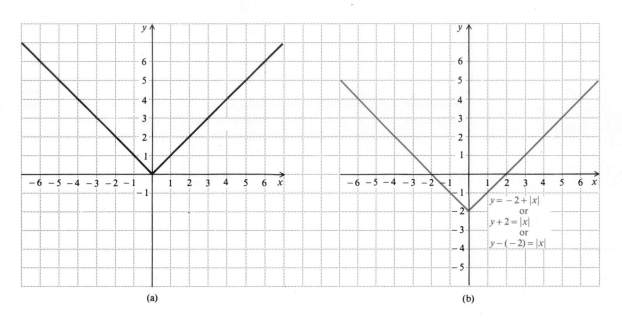

(a)   (b)

DO EXERCISES 2 AND 3.

Horizontal Translations

Replacing y by $y - a$ in an equation translates vertically a distance of $|a|$. The translation is in the positive direction (upward) if a is positive. A similar thing happens in the horizontal direction. If we replace x by $x - b$ everywhere x occurs in the equation, we translate a distance of $|b|$ horizontally. If b is positive, we translate in the positive direction (to the right). If b is negative, we translate in the negative direction (to the left).

THEOREM 2

In an equation of a relation, replacing x by $x - b$, where b is a constant, translates the graph horizontally a distance of $|b|$. If b is positive, the translation is to the right. If b is negative, the translation is to the left.

2. Sketch a graph of $y = -3 + x^2$.

3. Here is a graph of $y = f(x)$. There is no formula for it. Sketch a graph of $y = 3 + f(x)$.

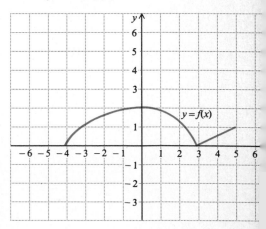

4. Sketch a graph of $y = (x + 3)^2$.

5. Here is a graph of $y = g(x)$. There is no formula for it. Sketch a graph of $y = g(x - 4)$.

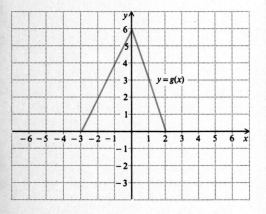

Example 2 Given a graph of $y = |x|$, sketch a graph of $y = |x + 2|$.

Here we note that x is replaced by $x + 2$, or $x - (-2)$. Thus $b = -2$, and the graph will be moved two units in the negative direction (to the left).

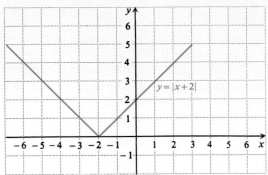

Example 3 A circle centered at the origin with radius of length 1 has an equation $x^2 + y^2 = 1$. If we replace x by $x - 1$ and y by $y + 2$, we translate the circle so that the center is at the point $(1, -2)$.

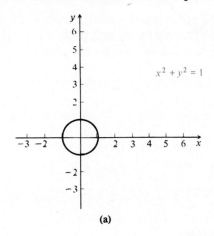

(a)

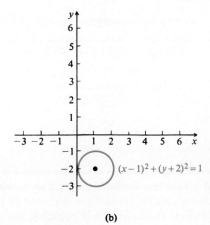

(b)

DO EXERCISES 4 AND 5.

Vertical Stretchings and Shrinkings

Consider the function $y = |x|$. We will compare its graph with that of $y = 2|x|$ and $y = \frac{1}{2}|x|$. The graph of $y = 2|x|$ looks like that of $y = |x|$ but every output is doubled, so the graph is stretched in a vertical direction. The graph of $y = \frac{1}{2}|x|$ is flattened or shrunk in a vertical direction since every output is cut in half.

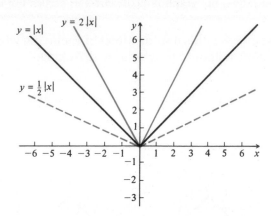

Consider any equation such as $y = f(x)$. Multiplying on the right by the constant 2 will double every function value, thus stretching the graph both ways from the horizontal axis. A similar thing is true for any constant greater than 1. If the constant is between 0 and 1, then the graph will be flattened or shrunk vertically.

DO EXERCISES 6 AND 7.

When we multiply by a negative constant, the graph is reflected across the x-axis as well as being stretched or shrunk.

Example 4 Compare the graphs of $y = |x|$, $y = -2|x|$, and $y = -\frac{1}{2}|x|$.

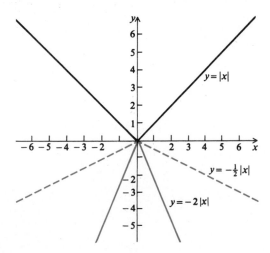

Multiplying by c on the right is, of course, equivalent to dividing by c on the left.

6. Graph $y = x^2$. Graph $y = 2x^2$ and compare.

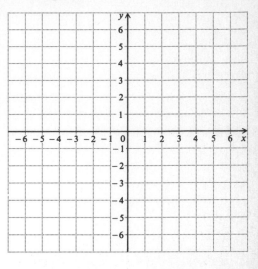

7. Graph $y = \dfrac{1}{2}x^2$. Compare with $y = x^2$.

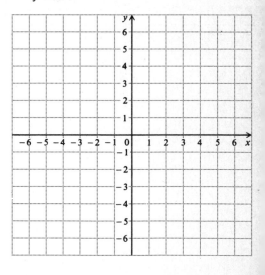

THEOREM 3

In an equation of a relation, dividing y by a constant c does the following to the graph.

1. If $|c| > 1$, the graph is stretched vertically.
2. If $|c| < 1$, the graph is shrunk vertically.
3. If c is negative, the graph is also reflected across the x-axis.*

Note that if $c = -1$, this has the effect of replacing y by $-y$ and we obtain a reflection without stretching or shrinking.

Example 5 Here is a graph of $y = f(x)$. Sketch a graph of $\dfrac{y}{2} = f(x)$ or $y = 2f(x)$.

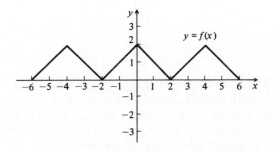

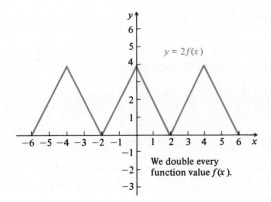

We double every function value $f(x)$.

Example 6 On the following page is a graph of $y = g(x)$. Sketch a graph of $\dfrac{y}{-\dfrac{1}{2}} = g(x)$ or $y = -\dfrac{1}{2}g(x)$.

---

*Again, with functions, we are more inclined to write $y = cf(x)$ instead of $y/c = f(x)$, but for relations in general we are more likely to replace y by y/c.

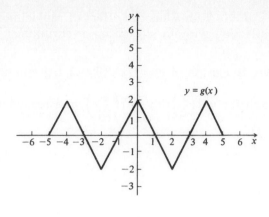

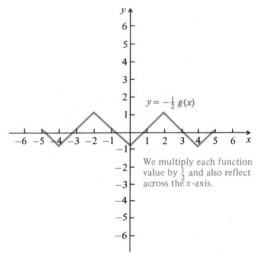

We multiply each function value by $\frac{1}{2}$ and also reflect across the x-axis.

DO EXERCISES 8 AND 9.

Horizontal Stretchings and Shrinkings

For vertical stretchings and shrinkings we divided y by a constant c. Similarly, if we divide x by a constant wherever it occurs, we will get a horizontal stretching or shrinking.

THEOREM 4

In an equation of a relation, dividing x wherever it occurs by d does the following to the graph.

1. If $|d| < 1$, the graph is shrunk horizontally.
2. If $|d| > 1$, the graph is stretched horizontally.
3. If d is negative, the graph is also reflected across the y-axis.*

---

*Again, with functions, we are more inclined to write $y = f(kx)$ instead of $y = f(x/d)$, but for relations in general we usually consider replacing x by x/d.

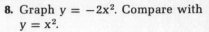

8. Graph $y = -2x^2$. Compare with $y = x^2$.

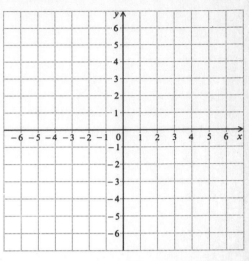

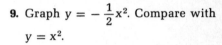

9. Graph $y = -\dfrac{1}{2}x^2$. Compare with $y = x^2$.

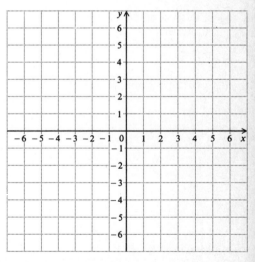

Here is a graph of $y = t(x)$. Use graph paper to sketch the following.

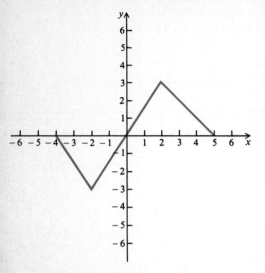

10. Sketch a graph of $y = \frac{1}{2}t(x)$.

11. Sketch a graph of $y = -2t(x)$.

12. Sketch a graph of $y = t(2x)$.

13. Sketch a graph of $y = t\left(-\frac{1}{2}x\right)$.

Note that if $d = -1$, this has the effect of replacing x by $-x$ and we obtain a reflection without stretching or shrinking.

Example 7 Here is a graph of $y = f(x)$. Sketch a graph of $y = f\left(\dfrac{x}{\frac{1}{2}}\right)$ or $y = f(2x)$, a graph of $y = f\left(\dfrac{x}{2}\right)$ or $y = f\left(\dfrac{1}{2}x\right)$, and a graph of $y = f\left(\dfrac{x}{-2}\right)$ or $y = f\left(-\dfrac{1}{2}x\right)$.

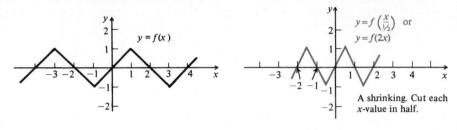

A shrinking. Cut each x-value in half.

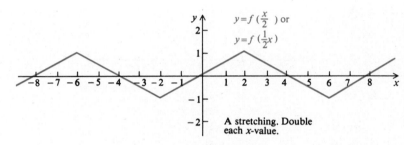

A stretching. Double each x-value.

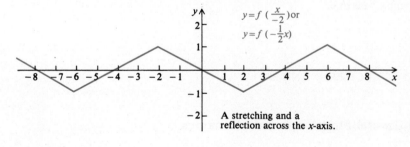

A stretching and a reflection across the x-axis.

DO EXERCISES 10–13.

EXERCISE SET 1.5

[i] In Exercises 1–14, sketch graphs by transforming the graph of $y = |x|$.

1. $y + 3 = |x|$ **2.** $y = 2 + |x|$ **3.** $y = |x - 1|$ **4.** $y = |x + 2|$

5. $y = -4|x|$ **6.** $\dfrac{y}{3} = |x|$ **7.** $y = \dfrac{1}{3}|x|$ **8.** $y = -\dfrac{1}{4}|x|$

9. $y = |2x|$ **10.** $y = \left|\dfrac{x}{3}\right|$ **11.** $y = |x - 2| + 3$

12. $y = 2|x + 1| - 3$ **13.** $y = -3|x - 2|$ **14.** $y = \dfrac{1}{3}|x + 2| + 1$

Here is a graph of $y = f(x)$. No formula will be given for this function. In Exercises 15–33, sketch graphs by transforming this one.

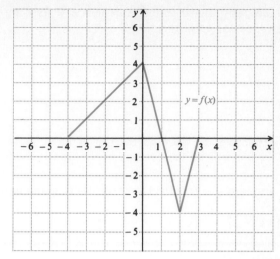

15. $y = 2 + f(x)$ **16.** $y + 1 = f(x)$ **17.** $y = f(x - 1)$ **18.** $y = f(x + 2)$

19. $\dfrac{y}{-2} = f(x)$ **20.** $y = 3f(x)$ **21.** $y = \dfrac{1}{3}f(x)$ **22.** $y = -\dfrac{1}{2}f(x)$

23. $y = f(2x)$ **24.** $y = f(3x)$ **25.** $y = f(-2x)$ **26.** $y = f(-3x)$

27. $y = f\left(\dfrac{x}{-2}\right)$ **28.** $y = f\left(\dfrac{1}{3}x\right)$ **29.** $y = f(x - 2) + 3$ **30.** $y = -3f(x - 2)$

31. $y = 2 \cdot f(x + 1) - 2$ **32.** $y = \dfrac{1}{2}f(x + 2) - 1$ **33.** $y = -\dfrac{1}{2}f(x - 3) + 2$

Here is a graph of $y = f(x)$. No formula will be given for this function. In Exercises 34–38, sketch graphs by transforming this one.

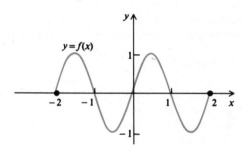

34. $y = -2f(x + 1) - 1$ **35.** $y = 3f(x + 2) + 1$ **36.** $y = \dfrac{5}{2}f(x - 3) - 2$

☆ _____

37. $y = -\sqrt{2}f(x + 1.8)$ **38.** $y = \dfrac{\sqrt{3}}{2} \cdot f(x - 2.5) - 5.3$

★ _____

39. A *linear* transformation of a coordinatized line is one that takes any point x to the point x', where $x' = ax + b$ (a and b constants). For example, the transformation $x' = 3x + 5$ takes the point 2 to the point $3 \cdot 2 + 5$, or 11. Suppose for a particular linear transformation the point 2 goes to 5 and the point 3 goes to 7.

a) Describe the transformation as $x' = ax + b$.

b) This transformation leaves one point fixed. What point is it?

c) The set $\{x \mid -2 \leq x \leq 1\}$ is mapped to what set under this transformation?

40. How does the linear transformation $x' = x - 3$ transform each of the following sets?

a) $\{x \mid 0 < x < 1\}$

b) $\{x \mid -4 \leq x < -1\}$

c) $\{x \mid 8 \leq x \leq 20\}$

41. Show that any linear transformation $x' = ax + b$, $a \neq 1$, leaves one point fixed.

42. A professor gives a test with 80 possible points. He decides that a score of 55 should be passing. He performs a linear transformation on the scores so that a perfect paper gets 100 and 70 is passing.

a) Describe the linear transformation that the professor used.

b) What grade remains unchanged under the transformation?

43. Suppose the average score on the test of Exercise 42 was 60. What will be the average of the transformed scores?

Given the graph of $y = f(x)$ for Exercises 34–38, graph each of the following.

44. $\dfrac{y}{3} = f\left(2x + \dfrac{1}{2}\right)$

45. $y = -4 \cdot f(5x + 10)$

OBJECTIVES

You should be able to:

[i] Given the graph of a function or a formula, determine whether the function is even, odd, or neither even nor odd.

[ii] Given the graph of a function, determine whether it is periodic, and if it is periodic, determine its period.

[iii] Write interval notation for certain sets.

[iv] Given the graph of a function, determine whether it is continuous over a specified interval, and indicate discontinuities.

[v] Given the graph of a function, determine whether it is increasing, decreasing, or neither increasing nor decreasing.

[vi] Graph functions defined piecewise.

1.6 SOME SPECIAL CLASSES OF FUNCTIONS

[i] Even and Odd Functions

If the graph of a function is symmetric with respect to the y-axis, it is an *even* function. Recall (Section 1.4) that a function will be symmetric to the y-axis if in its equation we can replace x by $-x$ and obtain an equivalent equation. Thus if we have a function given by $y = f(x)$, then $y = f(-x)$ will give the same function if the function is even. In other words, an even function is one for which $f(x) = f(-x)$ for all x in its domain. This is the definition of even function.

DEFINITION

A function f is an *even* function in case $f(x) = f(-x)$ for all x in the domain of f.

Example 1 Determine whether the function $f(x) = x^2 + 1$ is even.

a) Find $f(-x)$ and simplify.

$$f(-x) = (-x)^2 + 1 = x^2 + 1$$

b) Compare $f(x)$ and $f(-x)$.

Since $f(x) = f(-x)$ for all x in the domain, f is an even function.

Note that the graph is symmetric with respect to the y-axis.

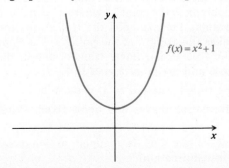

Example 2 Determine whether the function $f(x) = x^2 + 8x^3$ is even.

a) Find $f(-x)$ and simplify.

$$f(-x) = (-x)^2 + 8(-x)^3 = x^2 - 8x^3$$

b) Compare $f(x)$ and $f(-x)$.

Since $f(x)$ and $f(-x)$ are *not* the same for all x in the domain, f is *not* an even function.

DO EXERCISES 1 AND 2.

If the graph of a function is symmetric with respect to the origin, it is an *odd* function. Recall that a function will be symmetric with respect to the origin if in its equation we can replace x by $-x$ and y by $-y$ and obtain an equivalent equation. Thus if we have a function given by $y = f(x)$, then $-y = f(-x)$ will be equivalent if f is an odd function. In other words, an odd function is one for which $f(-x) = -f(x)$ for all x in the domain. Let us make this our definition.

DEFINITION

A function f is an *odd* function when $f(-x) = -f(x)$ for all x in the domain of f.

Example 3 Determine whether $f(x) = x^3$ is an odd function.

a) Find $f(-x)$ and $-f(x)$ and simplify.

$$f(-x) = (-x)^3 = -x^3,$$
$$-f(x) = -x^3$$

b) Compare $f(-x)$ and $-f(x)$.

Since $f(-x) = -f(x)$ for all x in the domain, f is odd.

Note that the graph is symmetric with respect to the origin.

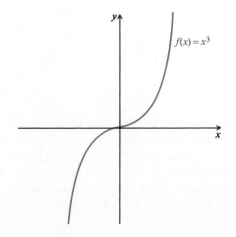

1. Determine whether each function is even.

a)
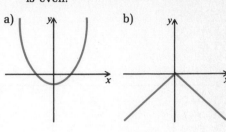

b)

c)
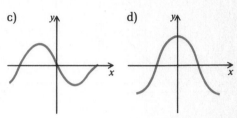

d)

2. Determine whether each function is even.

a) $f(x) = x^2 + 3x$

b) $f(x) = |x|$

c) $f(x) = 3x^2 - x^4$

d) $f(x) = 2x^2 + 1$

3. Determine whether each function is even, odd, or neither even nor odd.

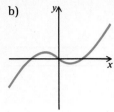

a)

b)

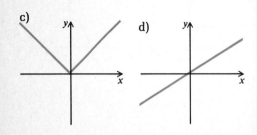

c)

d)

Example 4 Determine whether $f(x) = x^2 - 4x^3$ is even, odd, or neither even nor odd.

a) Find $f(-x)$ and $-f(x)$ and simplify.

$$f(-x) = (-x)^2 - 4(-x)^3 = x^2 + 4x^3,$$
$$-f(x) = -x^2 + 4x^3$$

b) Compare $f(x)$ and $f(-x)$ to determine whether f is even.

Since $f(x)$ and $f(-x)$ are *not* the same for all x in the domain, f is *not* even.

c) Compare $f(-x)$ and $-f(x)$ to determine whether f is odd.

Since $f(-x)$ and $-f(x)$ are *not* the same for all x in the domain, f is *not* odd. Thus f is *neither* even nor odd.

DO EXERCISES 3 AND 4.

[ii] Periodic Functions

Certain functions with a repeating pattern are called *periodic*. Here are some examples.

4. Determine whether each function is even, odd, or neither even nor odd.

a) $f(x) = x^3 + 2$

b) $f(x) = x^4 - x^6$

c) $f(x) = x^3 + x$

d) $f(x) = 3x^2 + 3x^5$

e) $f(x) = x^2 - \dfrac{1}{x}$

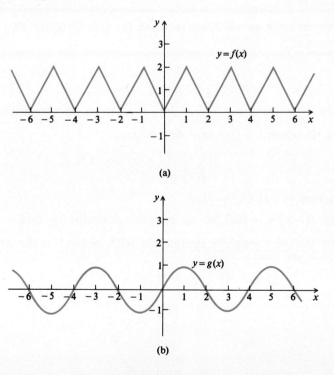

(a)

(b)

The function values of the function f repeat themselves every two units as we move from left to right. In other words, for any x, we have $f(x) = f(x + 2)$. To see this another way, think of the part of the graph between 0 and 2 and note that the rest of the graph consists of copies of it. In terms of translations, if we translate the graph two units to the left or right, the original graph will be obtained.

In the function g, the function values repeat themselves every four units. Hence $g(x) = g(x + 4)$ for any x, and if the graph is translated four units to the left, or right, it will coincide with itself. Or, think of the part of the graph between 0 and 4. The rest of the graph consists of copies of it.

We say that f has a *period* of 2, and that g has a period of 4.

DEFINITION

If a function f has the property that $f(x + p) = f(x)$ whenever x and $x + p$ are in the domain, where p is a positive constant, then f is said to be *periodic*. The smallest positive number p (if there is one) for which $f(x + p) = f(x)$ for all x is called the *period* of the function.

DO EXERCISES 5-7.

[iii] **Interval Notation**

For a and b real numbers such that $a < b$, we define the *open interval* (a, b) as follows:

(a, b) is the set of all numbers x such that $a < x < b$,

or

$$\{x \mid a < x < b\}.$$

Its graph is as follows:

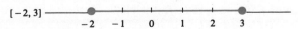

Note that the endpoints are not included. Be careful not to confuse this notation with that of an ordered pair. The context of the writing should make the meaning clear. If not, we might say "the interval $(-2, 3)$." When we mean an ordered pair, we might say "the pair $(-2, 3)$."

DO EXERCISES 8 AND 9.

The *closed interval* $[a, b]$ is defined as follows:

$[a, b]$ is the set of all x such that $a \leq x \leq b$,

or

$$\{x \mid a \leq x \leq b\}.$$

Its graph is as follows:

Note that endpoints are included. For example, the graph of $[-2, 3]$ is as follows:

5. For the function f whose graph was just considered, how does $f(x)$ compare with $f(x + 4)$? with $f(x + 6)$?

6. a) Determine whether this function is periodic.

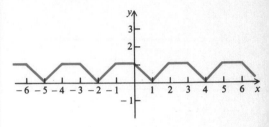

b) If so, what is the period?

7. (*Optional*). For the function t, where $t(x) = 3$, how does $t(x)$ compare with $t(x + 1)$? By the definition, is t periodic? If so, does it have a period?

8. Write interval notation for each set pictured.

a)

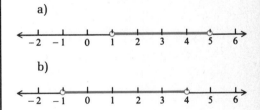

b)

9. Write interval notation for each set.

a) $\{x \mid -2 < x < 3\}$

b) $\{x \mid 0 < x < 1\}$

c) $\left\{ x \mid -\dfrac{1}{4} < x < \sqrt{2} \right\}$

10. Write interval notation for each set pictured.

a)

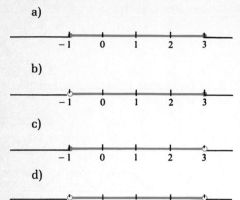

b)

c)

d)

11. Write interval notation for each set.

a) $\left\{x \mid 4 \leq x \leq 5\frac{1}{2}\right\}$

b) $\{x \mid -3 < x \leq 0\}$

c) $\left\{x \mid -\frac{1}{2} \leq x < \frac{1}{2}\right\}$

d) $\{x \mid -\pi < x < \pi\}$

12. Is this function continuous

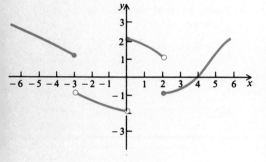

a) in the interval $(-1, 1)$?

b) in the interval $[-2, -1]$?

c) in the interval $(1, 2]$?

d) in the interval $(1, 2)$?

e) in the interval $(-5, 0)$?

13. Where are the discontinuities of the above function?

There are two kinds of *half-open intervals* defined as follows:

a) $(a, b] = \{x \mid a < x \leq b\}.$

This is open on the left. Its graph is as follows:

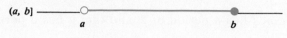

b) $[a, b) = \{x \mid a \leq x < b\}.$

This is open on the right. Its graph is as follows:

DO EXERCISES 10 AND 11.

[iv] Continuous Functions

Some functions have graphs that are continuous curves, without breaks or holes in them. Such functions are called *continuous functions*.

Example 5 The function f below is continuous because it has no breaks, jumps, or holes in it. The function g has *discontinuities* where $x = -2$ and $x = 5$. The function g is continuous on the interval $[-2, 5)$ or on any interval contained therein. It is also continuous on other intervals, but it is not continuous on an interval such as $[-3, 1]$ or $(3, 7)$.

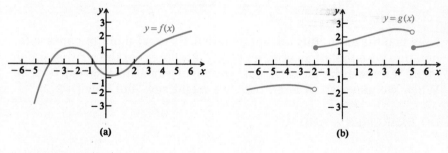

(a)   (b)

DO EXERCISES 12 AND 13.

[v] Increasing and Decreasing Functions

If the graph of a function rises from left to right it is said to be an *increasing* function. If the graph of a function drops from left to right, it is said to be a *decreasing* function. This can be stated more formally.

DEFINITION

1. A function f is an *increasing* function when for all a and b in the domain of f, if $a < b$, then $f(a) < f(b)$.

2. A function f is a *decreasing* function when for all a and b in the domain of f, if $a < b$, then $f(a) > f(b)$.

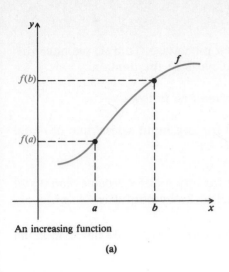

An increasing function

(a)

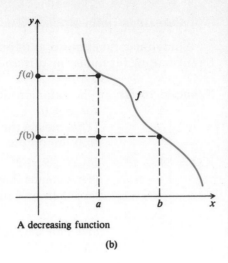

A decreasing function

(b)

Examples 6–9 Determine whether each function is increasing, decreasing, or neither.

6.

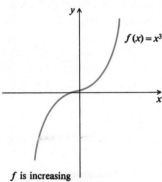

f is increasing

7.

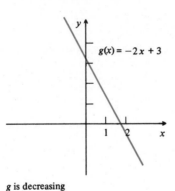

g is decreasing

8.

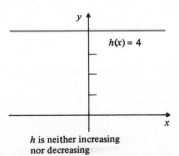

h is neither increasing nor decreasing

9.

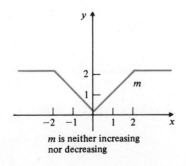

m is neither increasing nor decreasing

Note that while the function *m* is neither increasing nor decreasing, it is increasing on the interval [0, 2] and decreasing on the interval [−2, 0].

DO EXERCISES 14 AND 15.

14. Determine whether each function is increasing, decreasing, or neither increasing nor decreasing.

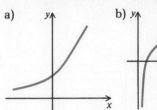

a)

b)

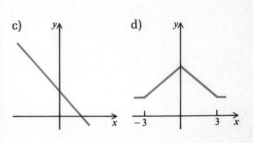

c)

d)

15. For the function in Exercise 14(d), find an interval on which the function is (a) increasing; (b) decreasing.

16. Graph the function defined as follows:

$$f(x) = \begin{cases} x + 3 & \text{for } x \le -2, \\ 1 & \text{for } -2 < x \le 3, \\ x^2 - 10 & \text{for } 3 < x. \end{cases}$$

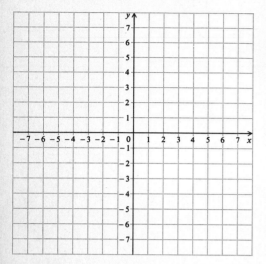

[vi] **Functions Defined Piecewise**

Sometimes functions are defined piecewise. That is, we have different output formulas for different parts of the domain.

Example 10 Graph the function defined as follows.

$$f(x) = \begin{cases} 4 & \text{for } x \le 0 \\ & \text{(This means that for any input } x \text{ less than or equal to 0 the output is 4.)} \\ 4 - x^2 & \text{for } 0 < x \le 2 \\ & \text{(This means that for any input } x \text{ greater than 0 and less than or equal to 2, the output is } 4 - x^2.) \\ 2x - 6 & \text{for } x > 2 \\ & \text{(This means that for any input } x \text{ greater than 2, the output is } 2x - 6.) \end{cases}$$

See the graph below.

a) We graph $f(x) = 4$ for inputs less than or equal to 0 (that is, $x \le 0$).

b) We graph $f(x) = 4 - x^2$ for inputs greater than 0 and less than or equal to 2 (that is, $0 < x \le 2$).

c) We graph $f(x) = 2x - 6$ for inputs greater than 2 (that is, $x > 2$).

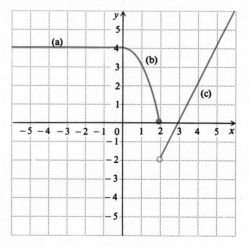

DO EXERCISE 16.

EXERCISE SET 1.6

[i]

1. Determine whether each function is even, odd, or neither even nor odd.

a)

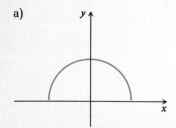

b)

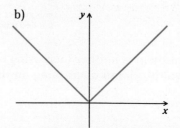

c)

d)

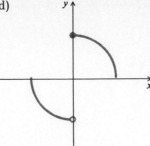

2. Determine whether each function is even, odd, or neither even nor odd.

a)

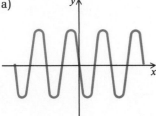

b)

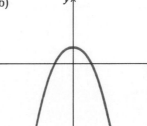

c)

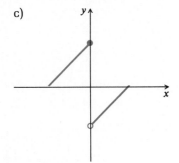

d)

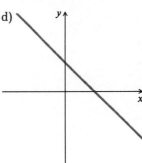

Determine whether each function is even, odd, or neither even nor odd.

3. $f(x) = 2x^2 + 4x$ **4.** $f(x) = -3x^3 + 2x$ **5.** $f(x) = 3x^4 - 4x^2$

6. $f(x) = 5x^2 + 2x^4 - 1$ **7.** $f(x) = 7x^3 + 4x - 2$ **8.** $f(x) = 4x$

9. $f(x) = |3x|$ **10.** $f(x) = x^{24}$ **11.** $f(x) = x^{17}$

12. $f(x) = x + \dfrac{1}{x}$ **13.** $f(x) = x - |x|$ **14.** $f(x) = \sqrt{x}$

15. $f(x) = \sqrt[3]{x}$ **16.** $f(x) = 7$ **17.** $f(x) = 0$

18. $f(x) = \sqrt[3]{x - 2}$ **19.** $f(x) = \sqrt{x^2 + 1}$ **20.** $f(x) = \dfrac{x^2 + 1}{x^3 - x}$

[ii]

21. Determine whether each function is periodic.

a)

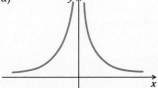

b)

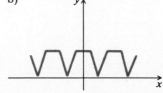

c)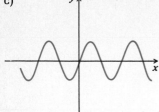

d)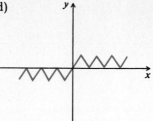

22. Determine whether each function is periodic.

a)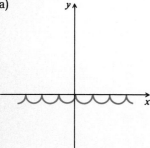

b)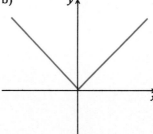

c)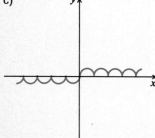

d)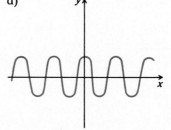

23. What is the period of this function?

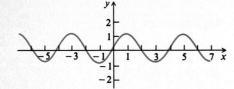

24. What is the period of this function?

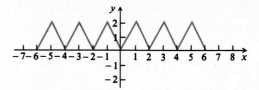

[iii]

25. Write interval notation for each set pictured.

a)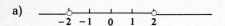

c)

b)

d)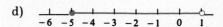

26. Write interval notation for each set pictured.

a)
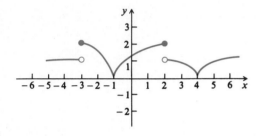

b)

c)

d)

27. Write interval notation for each set.

a) $\{x \mid -2 < x < 4\}$

b) $\left\{x \mid -\dfrac{1}{4} < x \le \dfrac{1}{4}\right\}$

c) $\{x \mid 7 \le x < 10\pi\}$

d) $\{x \mid -9 \le x \le -6\}$

28. Write interval notation for each set.

a) $\{x \mid -5 < x < 0\}$

b) $\{x \mid -\sqrt{2} \le x < \sqrt{2}\}$

c) $\left\{x \mid -\dfrac{\pi}{2} < x \le \dfrac{\pi}{2}\right\}$

d) $\left\{x \mid -12 \le x \le -\dfrac{1}{2}\right\}$

[iv]

29. Is this function continuous

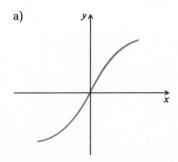

a) in the interval $[0, 2]$?

b) in the interval $(-2, 0)$?

c) in the interval $[0, 3]$?

d) in the interval $[-3, -1]$?

e) in the interval $(-3, -1]$?

30. Is this function continuous

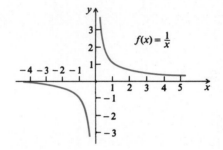

a) in the interval $[-3, -1]$?

b) in the interval $(1, 4)$?

c) in the interval $[-1, 1]$?

d) in the interval $[-2, 4]$?

e) in the interval $(0, 1]$?

31. Where are the discontinuities of the function in Exercise 29?

32. Where are the discontinuities of the function in Exercise 30?

[v]

33. Determine whether each function is increasing, decreasing, or neither increasing nor decreasing.

a)
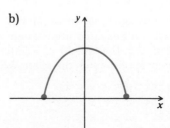

b)

c)

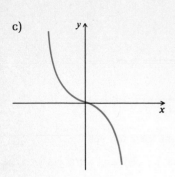

d)

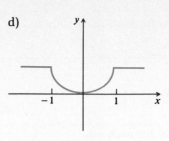

34. Determine whether each function is increasing, decreasing, or neither increasing nor decreasing.

a)

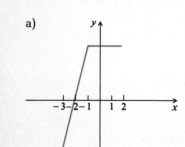

b)

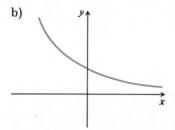

c)

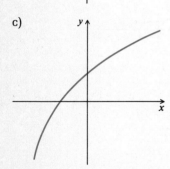

d)

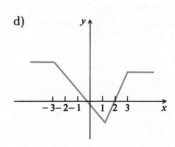

[vi] Graph.

35. $f(x) = \begin{cases} 1 & \text{for } x < 0, \\ -1 & \text{for } x \geq 0 \end{cases}$

36. $f(x) = \begin{cases} 2 & \text{for } x \text{ an integer}, \\ -2 & \text{for } x \text{ not an integer} \end{cases}$

37. $f(x) = \begin{cases} 3 & \text{for } x \leq -3, \\ |x| & \text{for } -3 < x \leq 3, \\ -3 & \text{for } x > 3 \end{cases}$

38. $f(x) = \begin{cases} -2x - 6 & \text{for } x \leq -2, \\ 2 - x^2 & \text{for } -2 < x < 2, \\ 2x - 6 & \text{for } x \geq 2 \end{cases}$

39. (*The postage function.*) Postage rates are as follows: 18¢ for the first ounce plus 17¢ for each additional ounce or fraction thereof. Thus if x is the weight of a letter in ounces, then $p(x)$ is the cost of mailing the letter, where

$$p(x) = \begin{cases} 18¢ & \text{if } 0 < x \leq 1, \\ 35¢ & \text{if } 1 < x \leq 2, \\ 52¢ & \text{if } 2 < x \leq 3, \end{cases}$$

and so on, up to 12 ounces, after which postal cost also depends on distance. Graph this function for x such that $0 < x \leq 12$.

40. Graph.

$$f(x) = \begin{cases} 3 + x & \text{for } x \leq 0, \\ \sqrt{x} & \text{for } 0 < x < 4, \\ x^2 - 4x - 1 & \text{for } x \geq 4 \end{cases}$$

☆ ———

41. For the function in Exercise 33(d), find an interval on which the function is (a) increasing; (b) decreasing.

42. For the function in Exercise 34(d), find an interval on which the function is (a) increasing; (b) decreasing.

43. Graph each function. Then determine whether it is increasing, decreasing, or neither increasing nor decreasing.

a) $f(x) = 3x + 4$ b) $f(x) = -3x + 4$

c) $f(x) = x^2 + 1$ d) $f(x) = -2$

e) $f(x) = x^3 + 1$ f) $f(x) = |x|$

44. Graph each function. Then determine whether it is increasing, decreasing, or neither increasing nor decreasing.

a) $f(x) = -2x - 3$ b) $f(x) = 2x - 3$

c) $f(x) = 3x^2$ d) $f(x) = \sqrt{3}$

e) $f(x) = |x| + 2$ f) $f(x) = x^3 - 2$

45. To what interval is each of the following mapped under the linear transformation $x' = x + 2$?

a) $[0, 1]$ b) $(-2, 7]$ c) $(-8, -1)$

46. To what interval is each of the following mapped under the linear transformation $x' = 0.4x - 3$?

a) $[0, 1]$ b) $(-4, -20)$ c) $[6, 11)$

═══

1.7 INVERSES

[i] Inverses of Relations

If in a relation we interchange first and second members in each ordered pair, we obtain a new relation. The new relation is called the *inverse* of the original relation. A relation is shown here in color; its inverse is shaded.

(1, 4)	(2, 4)	(3, 4)	(4, 4)
(1, 3)	(2, 3)	(3, 3)	(4, 3)
(1, 2)	(2, 2)	(3, 2)	(4, 2)
(1, 1)	(2, 1)	(3, 1)	(4, 1)

DO EXERCISE 1.

═══

When a relation is defined by an equation, interchanging x and y produces an equation of the inverse relation.

═══

Example 1 Find an equation of the inverse of $y = x^2 - 5$.

We interchange x and y, and obtain $x = y^2 - 5$. This is an equation of the inverse relation.

DO EXERCISE 2 ON THE FOLLOWING PAGE.

[ii] Graphs of Inverse Relations

Interchanging first and second coordinates in each ordered pair of a relation has the effect of interchanging the x-axis and the y-axis.

DO EXERCISE 3 ON THE FOLLOWING PAGE.

Interchanging the x-axis and the y-axis has the effect of reflecting across the diagonal line whose equation is $y = x$, as shown below. Thus the graphs of a relation and its inverse are always reflections of

OBJECTIVES

You should be able to:

[i] Given an equation defining a relation, write an equation of the inverse relation.

[ii] Given a graph of a relation, sketch a graph of its inverse; or given an equation defining a relation, graph it and then graph its inverse.

[iii] Given an equation defining a relation, determine whether the graph is symmetric with respect to the line $y = x$.

[iv] Given the graph of a relation or an equation for a relation, determine whether its inverse is a function.

[v] Given a function defined by a simple formula, find a formula for its inverse.

[vi] For a function f whose inverse is also a function, quickly find $f^{-1}(f(x))$ and $f(f^{-1}(x))$ for any number x in the domains of the functions.

1. a) Below, graph the relation containing (4, 1), (4, 2), (3, 2), (2, 3), (2, 4), and (1, 4).

b) Interchange the members in each ordered pair to obtain the inverse.

c) Graph the inverse and compare the graphs.

(1, 4)	(2, 4)	(3, 4)	(4, 4)
(1, 3)	(2, 3)	(3, 3)	(4, 3)
(1, 2)	(2, 2)	(3, 2)	(4, 2)
(1, 1)	(2, 1)	(3, 1)	(4, 1)

2. Write an equation of the inverse of each relation.

a) $y = 3x + 2$ b) $y = x$

c) $x^2 + 3y^2 = 4$ d) $y = 5x^2 + 2$

e) $y^2 = 4x - 5$ f) $xy = 5$

3. a) On a piece of thin paper, draw coordinate axes.

b) Draw the line $y = x$.

c) Draw a relation as shown here.

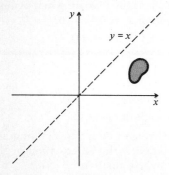

d) Flip the paper over (to interchange the x-axis and y-axis). Look through the paper. How do the graphs of the relation and its inverse compare?

4. Graph the inverse of each relation by reflecting across the line $y = x$.

a)

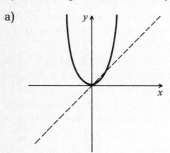

b)

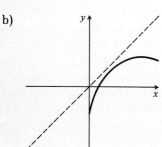

c) Graph $y = 4 - x^2$. Then by reflecting across the line $y = x$, graph its inverse.

each other across the line $y = x$. (This assumes, of course, that the same scale is used on both axes.)

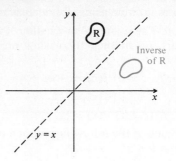

Examples 2–4 In each case a relation is shown in black. The graph of its inverse is shown in color.

2.

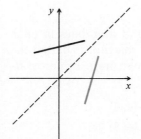

3.

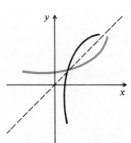

4.

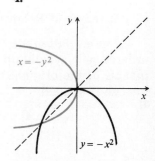

DO EXERCISE 4.

[iii] Symmetry with Respect to the Line $y = x$

It can happen that a relation is its own inverse; that is, when x and y are interchanged or the relation is reflected across the line $y = x$, there is no change. Such a relation is symmetric with respect to the line $y = x$. The following are two examples of relations symmetric with respect to the line $y = x$.

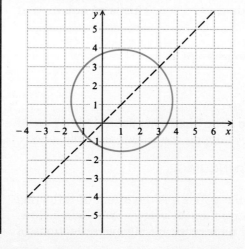

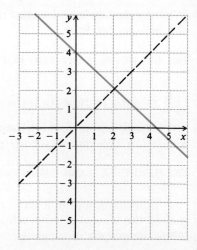

Each of the relations shown in black in Examples 2–4 is *not* symmetric with respect to the line $y = x$.

For a relation defined by an equation: If interchanging x and y produces an equivalent equation, the relation is its own inverse, and the graph of the equation is symmetric with respect to the line $y = x$.

Example 5 Test $3x + 3y = 5$ for symmetry with respect to the line $y = x$.

a) Interchange x and y. This amounts to replacing each occurrence of x by a y and each y by an x.

$$3x + 3y = 5$$
$$3y + 3x = 5$$

b) Is the resulting equation equivalent to the original?

The commutative law of addition guarantees that the resulting equation is equivalent to the original. Thus the graph is symmetric with respect to the line $y = x$.

Example 6 Test $y = x^2$ for symmetry with respect to the line $y = x$.

a) Interchange x and y.

$$y = x^2$$
$$x = y^2$$

b) Is the resulting equation equivalent to the original?

Note that $(2, 4)$ is a solution of the original equation $y = x^2$, but it is not a solution of the resulting equation $x = y^2$.

y	$= x^2$		x	$= y^2$
4	2^2		2	4^2
	4			16

Thus the equations are not equivalent, so the graph of $y = x^2$ is *not* symmetric with respect to the line $y = x$.

DO EXERCISES 5–12.

[iv] **Inverses of Functions**

Every function has an inverse, but that inverse may not be a function, as the following examples show.

Example 7 The relation g given by

$$g = \{(5, 3), (3, 1), (-7, 2), (2, 3)\}$$

is a function. Find the inverse of g and determine whether it is a function.

Test the following for symmetry with respect to the line $y = x$.

5. $y = -x$

6. $x + y = 4$

7. $xy = 3$

8. $y = |x|$

9. $3x^2 + 3y^2 = 4$

10. $|x| = |y|$

11. $y = x^3$

12. $x - y = 4$

13. Which of the following have inverses that are functions?

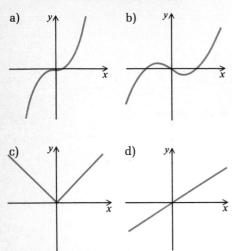

a)   b)

c)   d)

14. Graph the relation $y = x^2 - 1$ and decide whether it is a function. Then graph the inverse and decide whether it is a function.

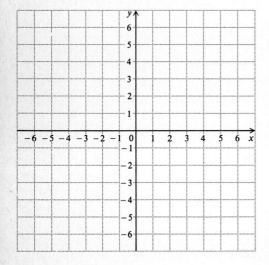

The inverse of g is found by interchanging first and second members in each ordered pair, and is $\{(3, 5), (1, 3), (2, -7), (3, 2)\}$. It is *not* a function because the pairs $(3, 5)$ and $(3, 2)$ have the same first coordinates but different second coordinates.

Example 8 Graph the relation $y = x^2$ and determine whether it is a function. Then graph its inverse and determine whether it is a function.

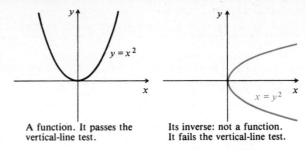

A function. It passes the vertical-line test.

Its inverse: not a function. It fails the vertical-line test.

DO EXERCISES 13 AND 14.

[v] Finding Formulas for Inverses of Functions

All functions have inverses, but in only some cases is the inverse also a function. If the inverse of a function f is also a function, we denote it by f^{-1} (read "f inverse"). [*Caution:* This is *not* exponential notation!] Recall that we obtain the inverse of any relation by reversing the coordinates in each ordered pair. In terms of mapping, let us see what this means. A function f *maps* the set of first coordinates (the domain) D to the set of second coordinates (the range) R. The inverse mapping f^{-1} maps the range of f onto the domain of f. Each y in R is mapped onto just one x in D, provided f^{-1} is a function. Note that the domain of f is the range of f^{-1} and the range of f is the domain of f^{-1}.

Let us consider inverses of functions in terms of function machines. Suppose that the function f programmed into a machine has an inverse that is also a function. Suppose then that the function machine has a reverse switch. When the switch is thrown the machine is then programmed to do the inverse mapping f^{-1}. Inputs then enter at the opposite end and the entire process is reversed.

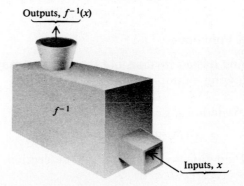

When a function is defined by a formula, we can sometimes find a formula for its inverse by thinking of interchanging x and y.

Example 9 Given $f(x) = x + 1$, find a formula for $f^{-1}(x)$.

a) Let us think of this as $y = x + 1$.

b) To find the inverse we interchange x and y: $x = y + 1$.

c) Now we solve for y: $y = x - 1$.

d) Thus $f^{-1}(x) = x - 1$.

Note in Example 9 that f maps any x onto $x + 1$ (this function adds 1 to each number of the domain). Its inverse, f^{-1}, maps any number x onto $x - 1$ (this inverse function subtracts 1 from each member of its domain). Thus the function and its inverse do opposite things.

Example 10 Let $g(x) = 2x + 3$. Find a formula for $g^{-1}(x)$.

a) Let us think of this as $y = 2x + 3$.

b) To find the inverse we interchange x and y: $x = 2y + 3$.

c) Now we solve for y: $y = (x - 3)/2$.

d) Thus $g^{-1}(x) = (x - 3)/2$.

Note in Example 10 that g maps any x onto $2x + 3$ (this function doubles each input and then adds 3). Its inverse, g^{-1}, maps any input onto $(x - 3)/2$ (this inverse function subtracts 3 from each input and then divides it by 2). Thus the function and its inverse do opposite things.

Example 11 Consider $f(x) = x^2$. Let us think of this as $y = x^2$.

To find the inverse we interchange x and y: $x = y^2$. Now we solve for y: $y = \pm\sqrt{x}$. Note that for each positive x we get two y's. For example, the pairs $(4, 2)$ and $(4, -2)$ belong to the relation. Look at the following graphs. Note also that the inverse fails the vertical-line test.

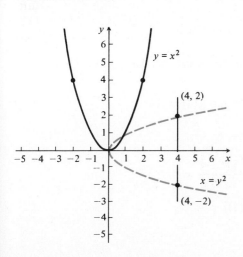
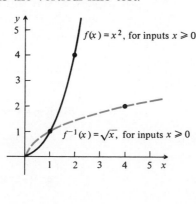

15. $g(x) = 3x - 2$. Find a formula for $g^{-1}(x)$.

If we restrict the domain of $f(x) = x^2$ to nonnegative numbers, then its inverse is a function, $f^{-1}(x) = \sqrt{x}$.

DO EXERCISES 15 AND 16.

[vi] Finding $f^{-1}(f(x))$ and $f(f^{-1}(x))$

Suppose the inverse of a function f is also a function. Let us suppose that we do the mapping f and then do the inverse mapping f^{-1}. We will be back where we started. In other words, if we find $f(x)$ for some x and then find f^{-1} for this number, we will be back at x. In function notation the statement looks like this:

$$f^{-1}(f(x)) = x.$$

This is read "f inverse of f of x equals x." It means, working from the inside out, to take x, then find $f(x)$, and then find the value of f^{-1} for that number. When we do, we will be back where we started, at x. For similar reasons, the following is also true:

$$f(f^{-1}(x)) = x.$$

For the above statements to be true, x must of course be in the domain of the function being considered.

16. $f(x) = x^2 - 1$ and the domain of f is the set of all nonnegative real numbers. Find a formula for $f^{-1}(x)$.

Let us consider function machines. To find $f^{-1}(f(x))$ we use x as an input, when the machine is set on f. We then use the output for an input, running the machine backward. The last output is the same as the first input. The first output must of course be an acceptable input when the machine is set on f^{-1}. Following is a precise statement of the property.

THEOREM 5

For any function f whose inverse is a function, $f^{-1}(f(a)) = a$ for any a in the domain of f. Also, $f(f^{-1}(a)) = a$ for any a in the domain of f^{-1}.

17. Let $g(x) = 3x - 2$ as in Exercise 15. Find $g^{-1}(g(5))$. Find $g(g^{-1}(5))$ as in Example 12.

Proof. Suppose a is in the domain of f. Then $f(a) = b$, for some b in the range of f. Then the ordered pair (a, b) is in f, and by definition of f^{-1}, (b, a) is in f^{-1}. It follows that $f^{-1}(b) = a$. Then substituting $f(a)$ for b, we get $f^{-1}(f(a)) = a$. A similar proof shows that $f(f^{-1}(a)) = a$, for any a in the domain of f^{-1}.

Example 12 For the function g of Example 10, find $g(4)$. Then find $g^{-1}(g(4))$.

$$g(x) = 2x + 3, \quad \text{so } g(4) = 11.$$

Now

$$g^{-1}(x) = \frac{x - 3}{2},$$

so

$$g^{-1}(11) = \frac{11 - 3}{2} = 4.$$

Thus

$$g^{-1}(g(4)) = 4.$$

DO EXERCISE 17.

Example 13 For the function g of Example 12, find $g^{-1}(g(283))$. Find also $g(g^{-1}(-12{,}045))$.

We note that every real number is in the domain of both g and g^{-1}. Thus we may immediately write the answers, without calculating:

$$g^{-1}(g(283)) = 283,$$
$$g(g^{-1}(-12{,}045)) = -12{,}045.$$

DO EXERCISE 18.

18. Let $g(x) = 3x - 2$ as in Exercise 17. For any number a, find $g^{-1}(g(a))$. Find $g(g^{-1}(a))$.

EXERCISE SET 1.7

[i] Write an equation of the inverse relation.

1. $y = 4x - 5$

2. $y = 3x + 5$

3. $x^2 - 3y^2 = 3$

4. $2x^2 + 5y^2 = 4$

5. $y = 3x^2 + 2$

6. $y = 5x^2 - 4$

7. $xy = 7$

8. $xy = -5$

[ii]

9. Graph $y = x^2 + 1$. Then by reflection across the line $y = x$, graph its inverse.

10. Graph $y = x^2 - 3$. Then by reflection across the line $y = x$, graph its inverse.

11. Graph $y = |x|$. Then by reflection across the line $y = x$, graph its inverse.

12. Graph $x = |y|$. Then by reflection across the line $y = x$, graph its inverse.

[iii] Test for symmetry with respect to the line $y = x$.

13. $3x + 2y = 4$

14. $5x - 2y = 7$

15. $4x + 4y = 3$

16. $5x + 5y = -1$

17. $xy = 10$

18. $xy = 12$

19. $3x = \dfrac{4}{y}$

20. $4y = \dfrac{5}{x}$

21. $y = |2x|$

22. $3x = |2y|$

23. $4x^2 + 4y^2 = 3$

24. $3x^2 + 3y^2 = 5$

[iv]

25. Which of the following have inverses that are functions?

a)
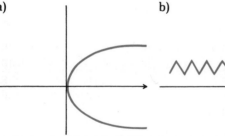

b)
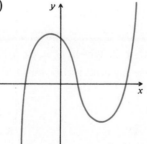

c)
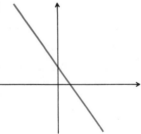

d)
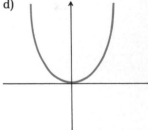

26. Which of the following have inverses that are functions?

a)
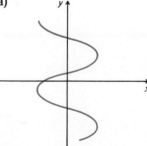

b)
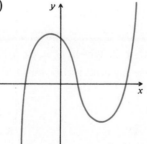

c)
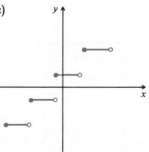

d)
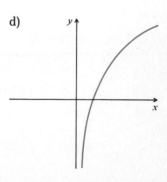

[v]

27. $f(x) = 2x + 5$. Find a formula for $f^{-1}(x)$.

29. $f(x) = \sqrt{x + 1}$. Find a formula for $f^{-1}(x)$.

28. $g(x) = 3x - 1$. Find a formula for $g^{-1}(x)$.

30. $g(x) = \sqrt{x - 1}$. Find a formula for $g^{-1}(x)$.

[vi]

31. $f(x) = 35x - 173$. Find $f^{-1}(f(3))$.
  Find $f(f^{-1}(-125))$.

33. $f(x) = x^3 + 2$. Find $f^{-1}(f(12{,}053))$.
  Find $f(f^{-1}(-17{,}243))$.

32. $g(x) = \dfrac{-173x + 15}{3}$. Find $g^{-1}(g(5))$.
  Find $g(g^{-1}(-12))$.

34. $g(x) = x^3 - 486$. Find $g^{-1}(g(489))$.
  Find $g(g^{-1}(-17{,}422))$.

☆ _____

35. Carefully graph $y = x^2$, using a large scale. Then use the graph to approximate $\sqrt{3.1}$.

37. Graph this equation and its inverse. Then test for symmetry with respect to the x-axis, the y-axis, the origin, and the line $y = x$.

$$y = \frac{1}{x^2}$$

36. Carefully graph $y = x^3$, using a large scale. Then use the graph to approximate $\sqrt[3]{-5.2}$.

★ _____

38. Ice melts at 0° Celsius or 32° Fahrenheit. Water boils at 100° Celsius or 212° Fahrenheit.

  a) What linear transformation converts Celsius temperature to Fahrenheit?

  b) Find the inverse of your answer to (a). Is it a function? What kind of conversion does it accomplish?

  c) At what temperature are the Celsius and Fahrenheit scales the same?

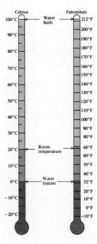

Graph each equation and its inverse. Then test for symmetry with respect to the x-axis, the y-axis, the origin, and the line $y = x$.

39. $|x| - |y| = 1$

40. $y = \dfrac{|x|}{x}$

_____

CHAPTER 1 REVIEW

[1.1, **i**]   **1.** List all the ordered pairs in the Cartesian square $G \times G$, where $G = \{1, 3, 5, 7\}$.

[1.1, **iii**]   **2.** List the domain and the range of the relation whose ordered pairs are (3, 1), (5, 3), (7, 7), and (3, 5).

[1.2, **ii**]   Graph.

[1.5, **i**]   **3.** $x = |y|$ **4.** $y = (x + 1)^2$ **5.** $y = |x| - 2$
        6. $f(x) = \sqrt{x}$ **7.** $f(x) = \sqrt{x - 2}$ **8.** $f(x) = 2\sqrt{x + 3}$

9. $f(x) = \frac{1}{4}\sqrt{x-1} + 2$ **10.** $|x - y| = 1$

Consider the following relations for Exercises 11–14.

a) $y = 7$ b) $x^2 + y^2 = 4$ c) $x^3 = y^3 - y$ d) $y^2 = x^2 + 3$
e) $x + y = 3$ f) $x = 3$ g) $y = x^2$ h) $y = x^3$

[1.4, i]   **11.** Which are symmetric with respect to the x-axis?

[1.4, i]   **12.** Which are symmetric with respect to the y-axis?

[1.4, ii]   **13.** Which are symmetric with respect to the origin?

[1.7, iii]   **14.** Which are symmetric with respect to the line $y = x$?

[1.7, i]   Write an equation of the inverse.

15. $y = 3x^2 + 2x - 1$ **16.** $y = \sqrt{x + 2}$

[1.3, i]   **17.** Which of the following are graphs of functions?

a)
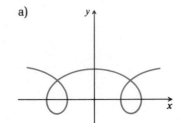

b)
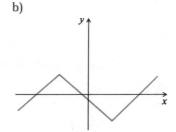

c)
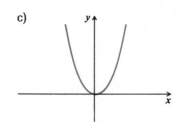

d)
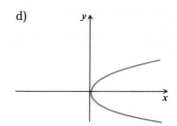

[1.7, iv]   **18.** Which of the relations in Exercise 17 have inverses that are functions?

[1.3, ii]   Use $f(x) = x^2 - x - 3$ to answer Exercises 19–21. Find:

19. $f(0)$.   **20.** $f(-3)$.   **21.** $f(a + h)$.

[1.3, ii]   Use $g(x) = 2\sqrt{x - 1}$ to answer Exercises 22–24. Find:

22. $g(1)$.   **23.** $g(5)$.   **24.** $g(a + 2)$.

[1.7, v]   **25.** $f(x) = \frac{\sqrt{x}}{2} + 2$. Find a formula for $f^{-1}(x)$.

26. $f(x) = x^2 + 2$. Find a formula for $f^{-1}(x)$.

[1.3, iii]   Find the domain of each function.

27. $f(x) = \sqrt{7 - 3x}$ **28.** $f(x) = \dfrac{1}{x^2 - 6x + 5}$

29. $f(x) = (x - 9x^{-1})^{-1}$ **30.** $f(x) = \dfrac{\sqrt{1 - x}}{x - |x|}$

[1.3, iv]   Find $f \circ g(x)$ and $g \circ f(x)$ in Exercises 31 and 32.

31. $f(x) = \dfrac{4}{x^2}$; $g(x) = 3 - 2x$ **32.** $f(x) = 3x^2 + 4x$; $g(x) = 2x - 1$

[1.7, vi]   **33.** $f(x) = x^3 + 2$. Find $f(f^{-1}(a))$.
              34. $h(x) = x^{17} + x^{65}$. Find $h^{-1}(h(t))$.

[1.5, i]   **35.** Here is a graph of $y = f(x)$.

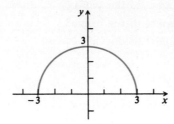

Sketch graphs of the following.

  a) $y = 1 + f(x)$ b) $y = \dfrac{1}{2}f(x)$ c) $y = f(x + 1)$

[1.6, i]   Use the following to answer Exercises 36–38.

  a)   b)

   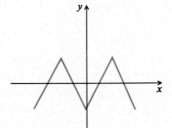

  c) $f(x) = 3x^2 - 2$ d) $f(x) = x + 3$
  e) $f(x) = 3x^3$ f) $f(x) = x^5 - x^3$

36. Which of the previous are even?   **37.** Which of the previous are odd?

38. Which of the previous are neither even nor odd?

[1.6, ii]   **39.** Which of the following functions are periodic?

  a)   b)

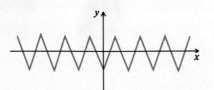   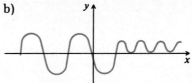

  c)   d)

    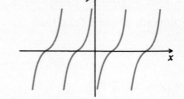

[1.6, ii]   **40.** What is the period of this function?

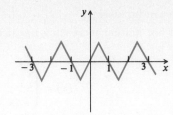

[1.6, iv]   **41.** Is this function continuous

           a) in the interval $[-3, -1]$?

           b) in the interval $[-1, 1]$?

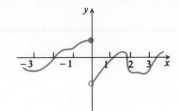

[1.6, iii]   Write interval notation for the following.

42. $\{x \mid -\pi \le x \le 2\pi\}$ **43.** $\{x \mid 0 < x \le 1\}$

[1.6, v]   Use the following for Exercises 44–46.

a)
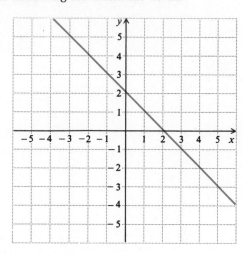

b)
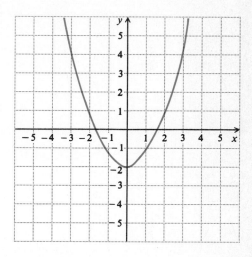

c)
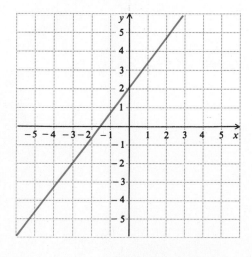

d)
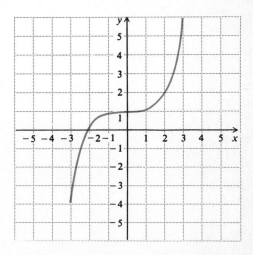

44. Which of the above are increasing?   **45.** Which of the above are decreasing?

46. Which of the above are neither increasing nor decreasing?

[1.6, vi   **47.** Graph

$$f(x) = \begin{cases} x^2 + 2 & \text{for } x < 0, \\ x^3 & \text{for } 0 \le x < 2, \\ -4x + 5 & \text{for } x \ge 2. \end{cases}$$

48. Graph several functions of the type $y = |f(x)|$. Describe a procedure, involving transformations, for graphing such functions.

2
THE CIRCULAR FUNCTIONS

OBJECTIVES

You should be able to:

[i] Define and determine the six trigonometric ratios, or functions,
and use them to find measures of
parts of triangles.

[ii] Use these functions to solve simple problems involving right triangles.

1. Find $m \angle A$.

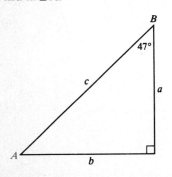

2.1 TRIANGLES*

In this chapter we consider an important class of functions called *trigonometric* functions, or *circular* functions. They are based, historically, on certain properties of triangles. In this section we consider some of those properties. We focus our attention on right triangles.

The angles and sides of a right triangle are often labeled as shown, with $\angle C$ a right angle and c the length of the hypotenuse. Side a is opposite $\angle A$ and side b is opposite $\angle B$.

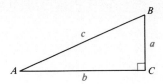

In any triangle the sum of the angle measures is 180°. Thus in a right triangle the measures of the two acute angles add up to 90°.

Example 1 In the triangle, find $m \angle B$ (the measure of $\angle B$).

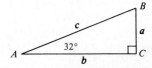

We know that $m \angle A + m \angle B = 90°$ and that $m \angle A = 32°$, so

$$m \angle B = 90° - 32° = 58°.$$

DO EXERCISE 1.

Similar Triangles

Triangles are similar if their corresponding angles have the same measure. In the following triangles, if we know that $\angle A$ and $\angle A'$ are the same size, then we know that *all* corresponding angles have the same measure; hence the triangles are similar.

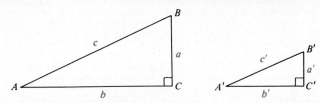

---

*The reader who wishes to approach trigonometry from the functional point of view first, and then consider triangles, can postpone this section. The reader who wishes a brief introduction to triangles before studying the functions can read this section and then proceed.

There are several paths through the trigonometry. For example, it is possible to treat triangles thoroughly before delving deeply into analytic trigonometry. For a thorough discussion of this, see the preface.

In similar triangles corresponding sides are in the same ratio (are proportional). For the triangles above, this means the following:

$$\frac{a}{a'} = \frac{b}{b'} = \frac{c}{c'}.$$

Similar triangles can be used to determine distances without measuring directly.

Example 2 Find the height of the flagpole using a rod and a measuring device as shown.

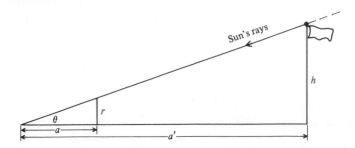

Place the rod of length r so that the top of its shadow coincides with the shadow of the top of the flagpole. We know that we have similar triangles, because the acute angle θ at the ground is the same in both triangles. We measure and find that $a = 3$ ft, that $a' = 30$ ft, and that $r = 2$ ft.

Now $h/r = a'/a$ by similar triangles, so

$$h = r\frac{a'}{a} = 2\left(\frac{30}{3}\right)$$
$$= 20 \text{ ft.}$$

DO EXERCISES 2 AND 3.

[i] Trigonometric Ratios

In a right triangle as shown below, the ratio a/c of the side *opposite* θ to the hypotenuse depends on the size of θ. In other words, this ratio is a *function* of θ.

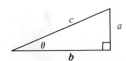

This function is called the *sine* function. Another such function, called the *cosine* function, is the ratio b/c, the ratio of the side *adjacent* to θ to the hypotenuse. There are other such ratios, or trigonometric functions, defined as follows.

2. Find b.

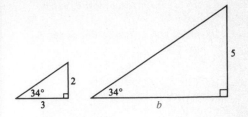

3. Find the height of this building.

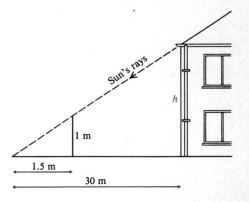

4. Find the following.

a) sin θ, cos θ, tan θ, cot θ, sec θ, csc θ

b) sin ϕ, cos ϕ, tan ϕ, cot ϕ, sec ϕ, csc ϕ

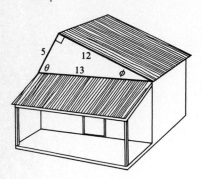

FUNCTION	ABBREVIATION FOR FUNCTION VALUE	DEFINING RATIO
sine	sin (θ)* or sin θ	$\dfrac{\text{side opposite } \theta}{\text{hypotenuse}}$
cosine	cos (θ) or cos θ	$\dfrac{\text{side adjacent } \theta}{\text{hypotenuse}}$
tangent	tan (θ) or tan θ	$\dfrac{\text{side opposite } \theta}{\text{side adjacent } \theta}$
cotangent	cot (θ) or cot θ	$\dfrac{\text{side adjacent } \theta}{\text{side opposite } \theta}$
secant	sec (θ) or sec θ	$\dfrac{\text{hypotenuse}}{\text{side adjacent } \theta}$
cosecant	csc (θ) or csc θ	$\dfrac{\text{hypotenuse}}{\text{side opposite } \theta}$

Note that the function values do not depend on the size of a triangle, but only on the size of the angle θ.

Example 3 In this triangle find:

a) the trigonometric function values for θ.

b) the trigonometric function values for ϕ.

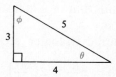

$$\sin \theta = \frac{\text{side opposite } \theta}{\text{hypotenuse}} = \frac{3}{5} \qquad \sin \phi = \frac{\text{side opposite } \phi}{\text{hypotenuse}} = \frac{4}{5}$$

$$\cos \theta = \frac{\text{side adjacent } \theta}{\text{hypotenuse}} = \frac{4}{5} \qquad \cos \phi = \frac{\text{side adjacent } \phi}{\text{hypotenuse}} = \frac{3}{5}$$

$$\tan \theta = \frac{\text{side opposite } \theta}{\text{side adjacent } \theta} = \frac{3}{4} \qquad \tan \phi = \frac{\text{side opposite } \phi}{\text{side adjacent } \phi} = \frac{4}{3}$$

$$\cot \theta = \frac{\text{side adjacent } \theta}{\text{side opposite } \theta} = \frac{4}{3} \qquad \cot \phi = \frac{\text{side adjacent } \phi}{\text{side opposite } \phi} = \frac{3}{4}$$

$$\sec \theta = \frac{\text{hypotenuse}}{\text{side adjacent } \theta} = \frac{5}{4} \qquad \sec \phi = \frac{\text{hypotenuse}}{\text{side adjacent } \phi} = \frac{5}{3}$$

$$\csc \theta = \frac{\text{hypotenuse}}{\text{side opposite } \theta} = \frac{5}{3} \qquad \csc \phi = \frac{\text{hypotenuse}}{\text{side opposite } \phi} = \frac{5}{4}$$

DO EXERCISE 4.

Values for the trigonometric functions have been worked out in great detail and are given in tables and by some calculators. When

---

*For function values we generally write $f(x)$. For these functions we often omit parentheses, as in sin θ.

these values are available we can use them to help determine distances or angles without actually measuring them directly.

[ii] Using the Trigonometric Functions

Example 4 An observer stands on level ground, 200 meters from the base of a TV tower, and looks up at an angle of 26.5° to see the top of the tower. How high is the tower above the observer's eye level?

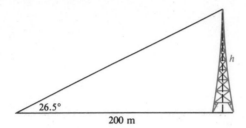

26.5°

200 m

We draw a diagram and see that a right triangle is formed. We use one of the trigonometric functions. This time the tangent function is most convenient. From the definition of the tangent function, we have

$$\frac{h}{200} = \tan 26.5°.$$

Then $h = 200 \tan 26.5°$. We find, from a table, that $\tan 26.5° = 0.499$, approximately. Thus $h = 200 \times 0.499 = 99.8\,\text{m}$.

DO EXERCISE 5.

Calculators

Scientific calculators contain tables of trigonometric functions. It is recommended that students not use them until after the trigonometric tables have been studied (Section 2.5). In the meantime, calculators will be most useful for arithmetic calculations.

Example 5 A kite flies at a height of 60 ft when 130 ft of string is out. Assuming that the string is in a straight line, what is the angle that it makes with the ground?

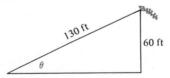

130 ft

60 ft

θ

We draw a diagram and then use the most convenient trigonometric function. A portion of a sine table is given below.

$\sin 26.5° = 0.4462$	$\sin 28.5° = 0.4772$
$\sin 27° = 0.4540$	$\sin 29° = 0.4848$
$\sin 27.5° = 0.4617$	$\sin 29.5° = 0.4924$
$\sin 28° = 0.4695$	$\sin 30° = 0.5000$

5. An observer stands 120 meters from a tree, and finds that the line of sight to the top of the tree is 32.3° above the horizontal. The tangent of 32.3° is 0.632. Find the height of the tree above eye level.

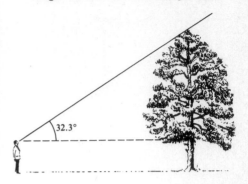

32.3°

6. A guy wire is 13.6 meters long, and is fastened from the ground to a pole 6.5 meters above the ground. What angle does the wire make with the ground?

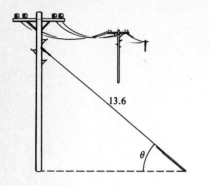

From the definition of the sine function, we have

$$\frac{60}{130} = \sin \theta$$

$$= 0.4615.$$

From the table, we find that θ is about 27.5°.

DO EXERCISE 6.

EXERCISE SET 2.1

[i]

1. Find $m \angle B$.

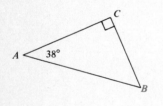

2. Find $m \angle A$.

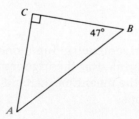

3. Find b.

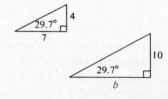

4. Find c.

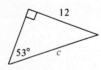

5. Find the trigonometric function values for θ.

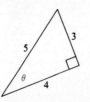

6. Find the trigonometric function values for ϕ.

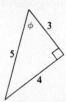

7. ▦ Find the trigonometric function values for θ. Round to four decimal places.

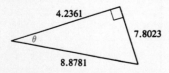

8. ▦ Find the trigonometric function values for ϕ. Round to four decimal places.

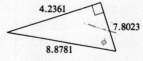

[ii] For Exercises 9–14, use the following values.

sin 36° = 0.5878	cos 36° = 0.8090	tan 36° = 0.7265	cot 36° = 1.376
sin 36.5° = 0.5948	cos 36.5° = 0.8039	tan 36.5° = 0.7400	cot 36.5° = 1.351
sin 37° = 0.6018	cos 37° = 0.7986	tan 37° = 0.7536	cot 37° = 1.327
sin 37.5° = 0.6088	cos 37.5° = 0.7934	tan 37.5° = 0.7673	cot 37.5° = 1.303
sin 38° = 0.6157	cos 38° = 0.7880	tan 38° = 0.7813	cot 38° = 1.280

9. a) Find ∠B.*
   b) Find a.
   c) Find b.

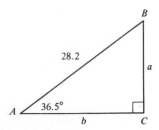

10. a) Find ∠B.
   b) Find a.
   c) Find b.

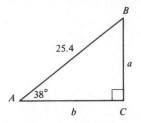

11. A guy wire is attached to a pole, and makes an angle of 37.5° with the ground. Find:

   a) the angle B that the wire makes with the pole;

   b) the distance b from A to the pole;

   c) the length of the wire.

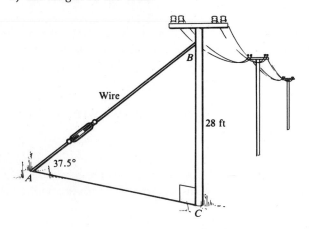

12. A cable across a river is attached at points B and C. A picnic grill at A, together with B and C, form the vertices of a right triangle. Find:

   a) the angle at A;

   b) the width of the river, a;

   c) the distance c.

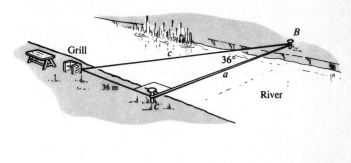

☆ ───

13. Find h.

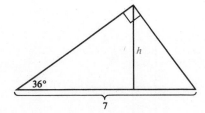

14. Find a.

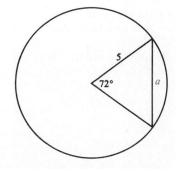

─────────────
*Instead of saying "find $m \angle B$," we often say "find ∠B."

15. A 45° right triangle is half of a square. Use the Pythagorean theorem to find the length of a diagonal and then find:

a) sin 45°

b) cos 45°;

c) tan 45°.

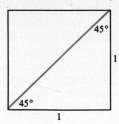

16. A 30°–60° right triangle is half of an equilateral triangle. Using this fact, find:

a) sin 30°;

b) cos 30°;

c) sin 60°;

d) cos 60°.

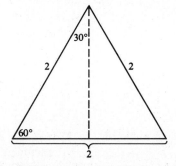

OBJECTIVES

You should be able to:

[i] Given the coordinates of a point on a circle centered at the origin, find its reflection across the u-axis, the v-axis, and the origin.

[ii] For the unit circle, state the coordinates of the intercepts and the midpoints of the arcs in each quadrant.

[iii] Given a real number that is an integral number of 4ths or 6ths of π, find on the unit circle the point determined by it. Given a point on the unit circle that is an integral number of 8ths or 12ths of the distance around the circle, find the real numbers between −2π and 2π that determine that point.

[iv] Given a real number n and the coordinates of the point it determines on the unit circle, state the coordinates of the point determined by −n.

2.2 THE UNIT CIRCLE

To develop the important circular functions we will use a circle with a radius of length 1. Such a circle is called a *unit circle*. When its center is at the origin of a coordinate system, it has an equation

$$u^2 + v^2 = 1.$$

(We use u and v for the variables here because we wish to reserve x and y for other purposes later.)

[i] Reflections

Any circle centered at the origin is symmetric with respect to the u-axis, the v-axis, and the origin.

Example 1 Suppose the point (1, 2) is on a circle centered at the origin. Find its reflection across the u-axis, the v-axis, and the origin.

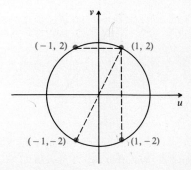

The reflection across the u-axis is (1, −2).
The reflection across the v-axis is (−1, 2).
The reflection across the origin is (−1, −2).
By symmetry, all these points are on the circle.

DO EXERCISES 1 AND 2.

[ii] Special Points

The intercepts (points of intersection with the axes) of the unit circle A, B, C, and D have coordinates as shown. Let us find the coordinates of point E, which is the intersection of the circle with the line $u = v$. Note that this point divides the arc in the first quadrant in half.

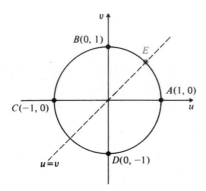

Since for any point (u, v) on the circle $u^2 + v^2 = 1$, and for point E we have $u = v$, we know that the coordinates of E satisfy $u^2 + u^2 = 1$ (substituting u for v in the equation of the circle). Let us solve this equation:

$$u^2 + u^2 = 1, \quad \text{or} \quad 2u^2 = 1$$

$$u^2 = \frac{1}{2}$$

$$u = \sqrt{\frac{1}{2}}, \quad \text{or} \quad \frac{\sqrt{1}}{\sqrt{2}}, \quad \text{or} \quad \frac{1}{\sqrt{2}}.$$

Now

$$\frac{1}{\sqrt{2}} = \frac{1}{\sqrt{2}} \cdot \frac{\sqrt{2}}{\sqrt{2}} = \frac{\sqrt{2}}{2}.$$

Thus the coordinates of point E are $\left(\dfrac{\sqrt{2}}{2}, \dfrac{\sqrt{2}}{2}\right)$.

DO EXERCISE 3.

[iii] Distances on the Unit Circle

The circumference of a circle of radius r is $2\pi r$. Thus for the unit circle the circumference is 2π. If a point starts at A and travels counter-

---

1. The point (3, −4) is on a circle centered at the origin. Find the coordinates of its reflection across

   a) the u-axis,
   b) the v-axis, and
   c) the origin.
   d) Are these points on the circle? Why?

2. The point (−5, −2) is on a circle centered at the origin. Find the coordinates of its reflection across

   a) the u-axis,
   b) the v-axis, and
   c) the origin.
   d) Are these points on the circle? Why?

3. Find the coordinates of the reflection of point E across

   a) the u-axis,
   b) the v-axis, and
   c) the origin.
   d) Are these points on the circle? Why?

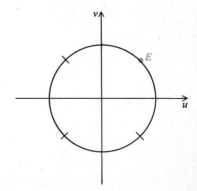

4. How far will a point travel if it goes

   a) $\frac{1}{4}$ of the way around the unit circle?

   b) $\frac{3}{8}$ of the way around the unit circle?

   c) $\frac{3}{4}$ of the way around the unit circle?

clockwise around the circle, it will travel a distance of 2π. If it travels halfway around the circle, it will travel a distance of π.

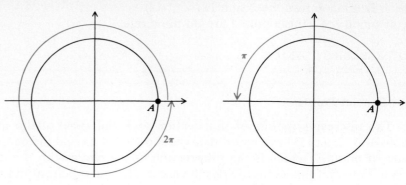

DO EXERCISE 4.

If a point travels $\frac{1}{8}$ of the way around the circle, it will travel a distance of $(\frac{1}{8}) \cdot 2\pi$, or $\pi/4$. Note that the stopping point is $\frac{1}{4}$ of the way from A to C and $\frac{1}{2}$ of the way from A to B.

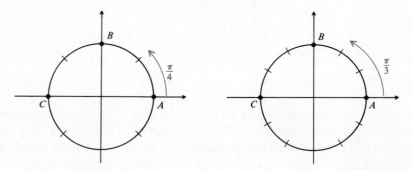

5. How far will a point move on the unit circle, counterclockwise, going from point A to

   a) M?   b) N?
   c) P?   d) Q?

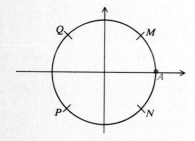

If a point travels $\frac{1}{6}$ of the way around the circle ($\frac{2}{12}$ of the way, as shown), it will travel a distance of $(\frac{1}{6}) \cdot 2\pi$, or $\pi/3$. Note that the stopping point is $\frac{1}{3}$ of the way from A to C and $\frac{2}{3}$ of the way from A to B.

DO EXERCISES 5 AND 6.

A point may travel completely around the circle and then continue. For example, if it goes around once and then continues $\frac{1}{4}$ of the way around, it will have traveled a distance of $2\pi + (\pi/2)$, or $(\frac{5}{2})\pi$.

6. How far will a point move on the unit circle, counterclockwise, in going from A to

   a) F?   b) G?
   c) H?   d) J?

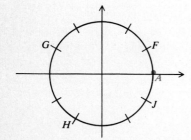

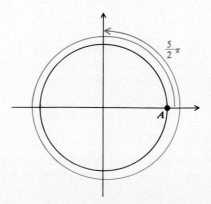

Any positive number thus determines a point on the unit circle. For the number 35, for example, we start at A and travel counterclockwise a distance of 35. The point at which we stop is the point "determined" by the number 35.

DO EXERCISES 7 AND 8.

[iv] Negative Inputs

Negative numbers also determine points on the unit circle. For a negative number, we move clockwise around the circle. Points for $-\pi/4$ and $-3\pi/2$ are shown. Any negative number thus determines a point on the unit circle. The same is true for any positive number. The number 0 determines the point A. Hence there is a point on the unit circle for every real number.

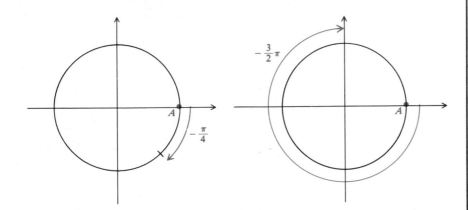

Points on the unit circle determined by two numbers that are additive inverses are situated as shown. The points are reflections of each other across the u-axis. Thus their first coordinates are the same and their second coordinates are additive inverses of each other.

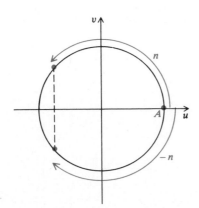

DO EXERCISE 9.

7. On this unit circle, mark the points determined by the following numbers:

a) $\frac{\pi}{4}$; b) $\frac{7}{4}\pi$; c) $\frac{9}{4}\pi$; d) $\frac{13}{4}\pi$.

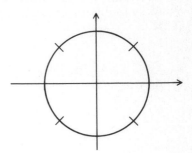

8. On this unit circle, mark the points determined by the following numbers:

a) $\frac{\pi}{6}$; b) $\frac{7}{6}\pi$; c) $\frac{13}{6}\pi$; d) $\frac{25}{6}\pi$.

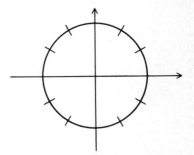

9. On this unit circle, mark the points determined by the following numbers:

a) $-\frac{\pi}{2}$; b) $-\frac{3}{4}\pi$; c) $-\frac{7}{4}\pi$;

d) -2π.

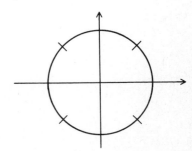

EXERCISE SET 2.2

[i]

1. The point $(-5, 2)$ is on a circle centered at the origin. Find the coordinates of its reflection across (a) the u-axis, (b) the v-axis, and (c) the origin. (d) Are these points on the circle? Why?

2. The point $(3, -5)$ is on a circle centered at the origin. Find the coordinates of its reflection across (a) the u-axis, (b) the v-axis, and (c) the origin. (d) Are these points on the circle? Why?

[ii]

3. Find the coordinates of these points on a unit circle.

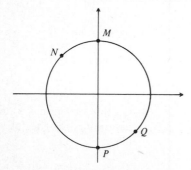

4. Find the coordinates of these points on a unit circle.

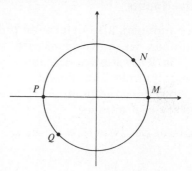

[iii] For each of Exercises 5–10, sketch a unit circle such as the one shown below.

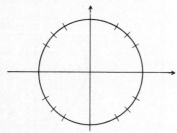

5. Mark the points determined by (a) $\frac{\pi}{4}$, (b) $\frac{3}{2}\pi$, (c) $\frac{3}{4}\pi$, (d) π, (e) $\frac{11}{4}\pi$, and (f) $\frac{17}{4}\pi$.

6. Mark the points determined by (a) $\frac{\pi}{2}$, (b) $\frac{5}{4}\pi$, (c) 2π, (d) $\frac{9}{4}\pi$, (e) $\frac{13}{4}\pi$, and (f) $\frac{23}{4}\pi$.

7. Mark the points determined by (a) $\frac{\pi}{6}$, (b) $\frac{2}{3}\pi$, (c) $\frac{7}{6}\pi$, (d) $\frac{10}{6}\pi$, (e) $\frac{14}{6}\pi$, and (f) $\frac{23}{4}\pi$.

8. Mark the points determined by (a) $\frac{\pi}{3}$, (b) $\frac{5}{6}\pi$, (c) $\frac{11}{6}\pi$,

9. Mark the points determined by (a) $-\frac{\pi}{2}$, (b) $-\frac{3}{4}\pi$, (c) $-\frac{5}{6}\pi$, (d) $-\frac{5}{2}\pi$, (e) $-\frac{17}{6}\pi$, and (f) $-\frac{9}{4}\pi$.

10. Mark the points determined by (a) $-\frac{3}{2}\pi$, (b) $-\frac{\pi}{3}$, (c) $-\frac{7}{6}\pi$, (d) $-\frac{13}{6}\pi$, (e) $-\frac{19}{6}\pi$, and (f) $-\frac{11}{4}\pi$.

(d) $\frac{13}{6}\pi$, (e) $\frac{23}{6}\pi$, and (f) $\frac{33}{6}\pi$.

11. Find two real numbers between -2π and 2π that determine each of these points.

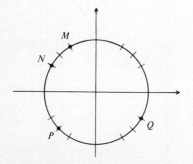

12. Find two real numbers between -2π and 2π that determine each of these points.

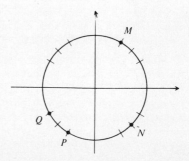

[iv]

13. The number $\frac{\pi}{6}$ determines a point on the unit circle with coordinates

$$\left(\frac{\sqrt{3}}{2}, \frac{1}{2}\right).$$

What are the coordinates of the point determined by $-\frac{\pi}{6}$?

14. The number $\frac{\pi}{3}$ determines a point on the unit circle with coordinates

$$\left(\frac{1}{2}, \frac{\sqrt{3}}{2}\right).$$

What are the coordinates of the point determined by $-\frac{\pi}{3}$?

☆

---

15. For the points given in Exercise 11, find a real number between 2π and 4π that determines each point.

16. For the points given in Exercise 12, find a real number between -2π and -4π that determines each point.

17. A point on the unit circle has u-coordinate $-\frac{1}{2}$. What is its v-coordinate?

18. A point on the unit circle has u-coordinate $-\frac{\sqrt{3}}{2}$. What is its v-coordinate?

19. ▦ A point on the unit circle has u-coordinate 0.25671. What is its v-coordinate?

20. ▦ A point on the unit circle has v-coordinate -0.88041. What is its u-coordinate?

2.3 THE SINE AND COSINE FUNCTIONS

[i] The Sine Function

Every real number s determines a point T on the unit circle, as shown. To the real number s we assign the second coordinate of point T as a function value. The function thus defined is called the *sine function*. Function values are denoted by "sine s" or, more briefly, "sin s".* By considering the unit circle, we see that

$$\sin 0 = 0, \qquad \sin \frac{\pi}{2} = 1, \qquad \sin \frac{\pi}{4} = \sin \frac{3}{4}\pi = \frac{\sqrt{2}}{2}.$$

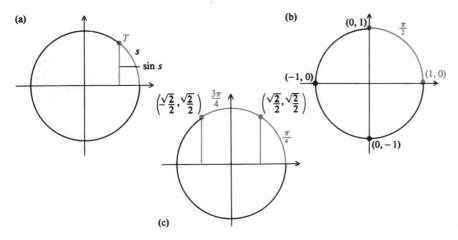

*The sine function is a generalization of what was done in Section 2.1. To help visualize this, we could draw a triangle as shown here. Its hypotenuse has length 1, so $\sin \theta = v/1$, or v. Note also that the arc length s determines the angle θ, so we think of sin s rather than sin θ.

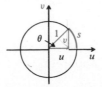

OBJECTIVES

You should be able to:

[i] Determine sin x for any real number x that is a multiple of $\pi/4$, $\pi/6$, or $\pi/3$ and sketch a graph of the sine function.

[ii] State the properties of the sine function and state or complete the following identities.

$\sin x \equiv \sin (x + 2k\pi)$, k any integer
$\sin (-x) \equiv -\sin x$
$\sin (x \pm \pi) \equiv -\sin x$
$\sin (\pi - x) \equiv \sin x$

[iii] Determine cos x for any real number x that is a multiple of $\pi/4$, $\pi/6$, or $\pi/3$ and sketch a graph of the cosine function.

[iv] State the properties of the cosine function and state or complete the following identities.

$\cos x \equiv \cos (x + 2k\pi)$, k any integer
$\cos (-x) \equiv \cos x$
$\cos (x \pm \pi) \equiv -\cos x$
$\cos (\pi - x) \equiv -\cos x$

1. Using a unit circle, find:

   a) $\sin \pi$; b) $\sin \frac{3}{2}\pi$;

   c) $\sin \frac{\pi}{4}$; d) $\sin\left(-\frac{\pi}{4}\right)$;

   e) $\sin \frac{2\pi}{3}$; f) $\sin\left(-\frac{\pi}{6}\right)$.

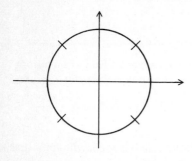

2. Is the sine function periodic? If so, what is its period?

3. Is the sine function even? Is it odd?

4. Does the sine function appear to be continuous?

5. What is the domain of the sine function?

6. What is the range of the sine function?

DO EXERCISE 1.

The sine function also has the following values, as we shall later verify:

$$\sin \frac{\pi}{6} = \frac{1}{2},$$

$$\sin \frac{\pi}{3} = \frac{\sqrt{3}}{2}.$$

[ii] Properties of the Sine Function

Function values increase to a maximum of 1 at $\pi/2$, then decrease to 0 at π, decrease further to -1 at $(\frac{3}{2})\pi$, then increase to 0 at 2π, and so on. The graph of the sine function is shown below. Note that on the s-axis, π appears at about 3.14.

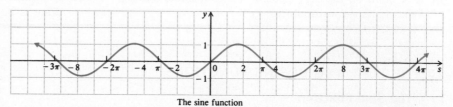

The sine function

The following is a device that could be constructed to draw a graph of the sine function. A light shines horizontally, casting a shadow of the ball fastened to the wheel at P. The wheel rotates at a constant speed and the paper on which the shadow falls moves at a constant speed. The shadow of the ball traces a sine graph.

DO EXERCISES 2–6.

From the graph of the sine function, certain properties are apparent.

1. **The sine function is periodic, with period 2π.**

2. **It is an odd function, because it is symmetric with respect to the origin. Thus we know that $\sin(-s) = -\sin s$ for all real numbers s.**

3. **It is continuous everywhere.**

4. **The domain is the set of all real numbers. The range is the set of all real numbers from -1 to 1, inclusive.**

The unit circle can be used to verify some of the above, as well as to determine other important properties of the sine function.

Let us verify that the period is 2π. Consider any real number s and its point T on the unit circle. Now we increase s by 2π. The point for $s + 2\pi$ is again the point T. Hence for any real number s, $\sin s = \sin(s + 2\pi)$. There is no number smaller than 2π for which this is true. Hence the period is 2π. Actually, we know that $\sin s = \sin(s + 2k\pi)$, where k is any integer.*

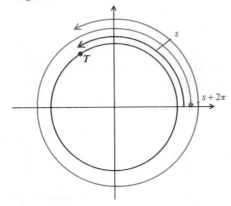

Some Identities

Consider any real number s and its additive inverse $-s$. These numbers determine points on the unit circle that are symmetric with respect to the u-axis. Hence their second coordinates are additive inverses of each other. Thus we know for any number s that $\sin(-s) = -\sin s$, so the sine function is odd.

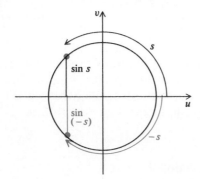

DO EXERCISE 7.

7. A real number s determines the point T as shown.

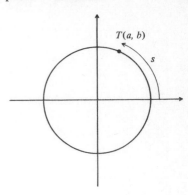

a) Find the point for $s + \pi$. What are its coordinates?

b) Find the point for $s - \pi$. What are its coordinates?

---

*If $k = 1$, we have $\sin s = \sin(s + 2\pi)$. If $k = 3$, we have $\sin s = \sin(s + 6\pi)$. If $k = -5$, we have $\sin s = \sin(s - 10\pi)$, and so on.

Consider any real number s and its point T on the unit circle. Then $s + \pi$ and $s - \pi$ both determine a point T_1 symmetric with T with respect to the origin. Thus the sines of $s + \pi$ and $s - \pi$ are the same, and are the additive inverse of $\sin s$. This gives us another important property.

For any real number s, $\sin (s + \pi) = -\sin s$, and $\sin (s - \pi) = -\sin s$.

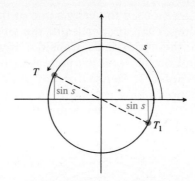

Let us add to our list a property concerning $\sin (\pi - s)$. Since $\pi - s$ is the additive inverse of $s - \pi$, $\sin (\pi - s) = \sin [-(s - \pi)]$. Since the sine function is odd, the latter is equal to $-\sin (s - \pi)$. By the second property in the preceding paragraph, it is equal to $-(-\sin s)$ or $\sin s$. Thus our last property is

$$\sin (\pi - s) = \sin s.$$

Equations that hold for all sensible replacements for the variables are known as *identities*. We shall use the sign $\equiv$ instead of $=$ to indicate that an equation is an identity. We now have the following identities concerning the sine function.

$$\sin s \equiv \sin (s + 2k\pi), \text{ } k \text{ any integer}$$
$$\sin (-s) \equiv -\sin s$$
$$\sin (s \pm \pi) \equiv -\sin s$$
$$\sin (\pi - s) \equiv \sin s$$

The *amplitude* of a periodic function is half the difference between its maximum and minimum function values. It is always positive. The maximum value of the sine function can be seen to be 1, either from the graph or the unit circle, whereas the minimum value is -1. Thus the amplitude of the sine function is 1.

[iii] The Cosine Function

Every real number s determines a point T on the unit circle. To the real number s we assign the first coordinate of point T as a function value. This defines the function known as the *cosine function*. Function values are denoted by "cosine s" or, more briefly, "cos s." By considering the unit circle, we see that*

$$\cos 0 = 1,$$

$$\cos \frac{\pi}{2} = 0,$$

$$\cos \frac{\pi}{4} = \frac{\sqrt{2}}{2},$$

$$\cos \frac{3}{4}\pi = -\frac{\sqrt{2}}{2}.$$

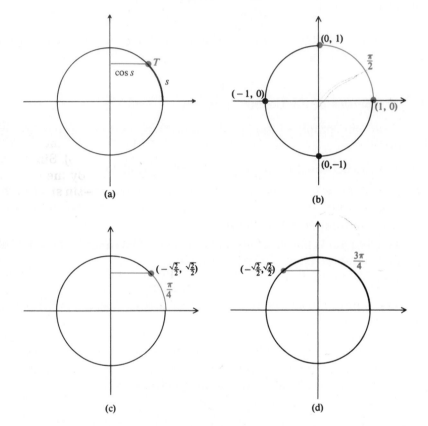

(a)   (b)

(c)   (d)

The cosine function also has the following values, as we shall later verify:

$$\cos \frac{\pi}{6} = \frac{\sqrt{3}}{2}, \cos \frac{\pi}{3} = \frac{1}{2}.$$

DO EXERCISE 8.

---

*The cosine function is also a generalization of what was done in Section 2.1. The triangle here has a hypotenuse of length 1, so $\cos \theta = u/1$, or u. Since the arc length s determines the angle θ, we think of $\cos s$ rather than $\cos \theta$.

8. Using a unit circle, find:

a) $\cos \pi$;   b) $\cos \frac{3}{2}\pi$;

c) $\cos \frac{\pi}{4}$;   d) $\cos \left(-\frac{\pi}{4}\right)$;

e) $\cos \frac{5\pi}{6}$;   f) $\cos \left(-\frac{\pi}{3}\right)$.

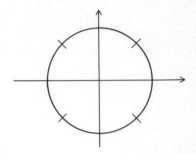

9. Is the cosine function periodic? If so, what is its period?

10. Is the cosine function even? Is it odd?

11. Does the cosine function appear to be continuous?

12. What is the domain of the cosine function?

13. What is the range of the cosine function?

The graph of the cosine function is as follows. Note that on the s-axis, π appears at about 3.14.

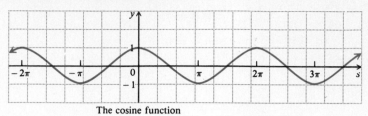

The cosine function

DO EXERCISES 9–13.

It should be noted that for any point on the unit circle, the coordinates are (cos s, sin s).

[iv] Properties of the Cosine Function

From the graph of the cosine function, certain properties are apparent.

1. The cosine function is periodic, with period 2π.

2. It is an even function, because it is symmetric with respect to the y-axis. Thus we know that cos $(-s) = \cos s$ for all real numbers s.

3. It is continuous everywhere.

4. The domain is the set of all real numbers. The range is the set of all real numbers from -1 to 1 inclusive.

5. The amplitude is 1.

The unit circle can be used to verify that the period is 2π and, in fact, that cos $s = \cos (s + 2k\pi)$ for any integer k.

Consider any real number s and its additive inverse. They determine points symmetric with respect to the u-axis, hence they have the same first coordinates. Thus we know that for any number s, cos $(-s) = \cos s$, and that the cosine function is even.

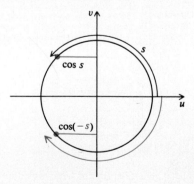

Let us consider cos s and cos $(s \pm \pi)$. Recall that s and $s \pm \pi$ determine points that are symmetric with respect to the origin. Thus their coordinates are additive inverses of each other. This verifies that cos $(s \pm \pi) = -\cos s$ for all real numbers s.

Let us consider $\cos(\pi - s)$. This is equal to $\cos[-(s-\pi)]$. Since the cosine function is even, this is equal to $\cos(s-\pi)$, and by the above definition this is in turn equal to $-\cos s$. Hence we have $\cos(\pi - s) = -\cos s$. We now list the identities that we have developed for the cosine function.

$$\cos s \equiv \cos(s + 2k\pi), \; k \text{ any integer}$$
$$\cos(-s) \equiv \cos s$$
$$\cos(s \pm \pi) \equiv -\cos s$$
$$\cos(\pi - s) \equiv -\cos s$$

Instead of the letter s, we can use any letter. We usually use x, as is the custom with most functions.

EXERCISE SET 2.3

[i], [iii]

1. Construct a graph of the sine function. The lower set of axes is shown here at half size, so on other paper, copy the axes at twice the size shown below. Then from the unit circle (shown full size here) transfer vertical distances with a compass.

2. Construct a graph of the cosine function. Follow the instructions for Exercise 1, except that horizontal distances on the unit circle will be transferred with a compass.

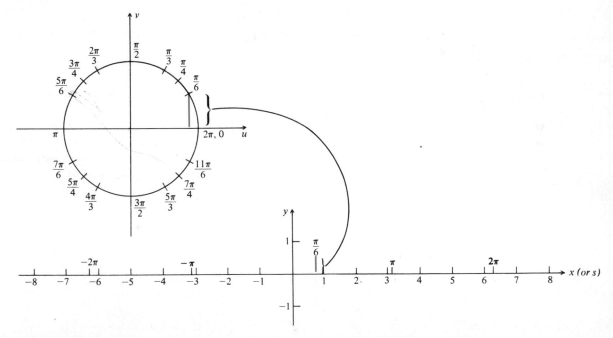

3. a) Sketch a graph of $y = \sin x$.

   b) By reflecting the graph in (a), sketch a graph of $y = \sin(-x)$.

   c) By reflecting the graph in (a), sketch a graph of $y = -\sin x$.

   d) How do the graphs in (b) and (c) compare?

4. a) Sketch a graph of $y = \cos x$.

   b) By reflecting the graph in (a), sketch a graph of $y = \cos(-x)$.

   c) By reflecting the graph in (a), sketch a graph of $y = -\cos x$.

   d) How do the graphs in (a) and (b) compare?

5. a) Sketch a graph of $y = \sin x$.

   b) By translating, sketch a graph of $y = \sin(x + \pi)$.

   c) By reflecting the graph of (a), sketch a graph of $y = -\sin x$.

   d) How do the graphs of (b) and (c) compare?

7. a) Sketch a graph of $y = \cos x$.

   b) By translating, sketch a graph of $y = \cos(x + \pi)$.

   c) By reflecting the graph of (a), sketch a graph of $y = -\cos x$.

   d) How do the graphs in (b) and (c) compare?

6. a) Sketch a graph of $y = \sin x$.

   b) By translating, sketch a graph of $y = \sin(x - \pi)$.

   c) By reflecting the graph of (a), sketch a graph of $y = -\sin x$.

   d) How do the graphs of (b) and (c) compare?

8. a) Sketch a graph of $y = \cos x$.

   b) By translating, sketch a graph of $y = \cos(x - \pi)$.

   c) By reflecting the graph of (a), sketch a graph of $y = -\cos x$.

   d) How do the graphs in (b) and (c) compare?

[ii], [iv] Complete.

9. $\cos(-x) \equiv$ _____

11. $\sin(x + \pi) \equiv$ _____

13. $\cos(\pi - x) \equiv$ _____

15. $\cos(x + 2k\pi) \equiv$ _____

17. $\cos(x - \pi) \equiv$ _____

10. $\sin(-x) \equiv$ _____

12. $\sin(x - \pi) \equiv$ _____

14. $\sin(\pi - x) \equiv$ _____

16. $\sin(x + 2k\pi) \equiv$ _____

18. $\cos(x + \pi) \equiv$ _____

[i]

19. Find:

  a) $\sin \dfrac{\pi}{4}$;   b) $\sin \pi$;   c) $\sin \dfrac{\pi}{6}$;

  d) $\sin \dfrac{5}{4}\pi$;   e) $\sin \dfrac{3}{2}\pi$;

  f) $\sin \dfrac{5}{6}\pi$;   g) $\sin\left(-\dfrac{\pi}{4}\right)$;

  h) $\sin(-\pi)$;   j) $\sin\left(-\dfrac{5}{4}\pi\right)$;

  k) $\sin\left(-\dfrac{3}{2}\pi\right)$;   m) $\sin\left(-\dfrac{\pi}{3}\right)$.

20. Find:

  a) $\sin \dfrac{\pi}{2}$;   b) $\sin \dfrac{\pi}{3}$;   c) $\sin \dfrac{3}{4}\pi$;

  d) $\sin \dfrac{7}{4}\pi$;   e) $\sin 2\pi$;   f) $\sin\left(-\dfrac{\pi}{2}\right)$;

  g) $\sin\left(-\dfrac{3}{4}\pi\right)$;   h) $\sin\left(-\dfrac{7}{4}\pi\right)$;

  j) $\sin(-2\pi)$;   k) $\sin\left(-\dfrac{2}{3}\pi\right)$;

  m) $\sin\left(-\dfrac{\pi}{6}\right)$.

[iii]

21. Find:

  a) $\cos \dfrac{\pi}{4}$;   b) $\cos \pi$;   c) $\cos \dfrac{5}{4}\pi$;

  d) $\cos \dfrac{\pi}{6}$;   e) $\cos \dfrac{3}{2}\pi$;   f) $\cos \dfrac{5}{6}\pi$;

  g) $\cos\left(-\dfrac{\pi}{4}\right)$;   h) $\cos(-\pi)$;

  j) $\cos\left(-\dfrac{5}{4}\pi\right)$;   k) $\cos\left(-\dfrac{3}{2}\pi\right)$;

  m) $\cos\left(-\dfrac{\pi}{3}\right)$.

22. Find:

  a) $\cos \dfrac{\pi}{2}$;   b) $\cos \dfrac{\pi}{3}$;   c) $\cos \dfrac{3}{4}\pi$;

  d) $\cos \dfrac{7}{4}\pi$;   e) $\cos 2\pi$;   f) $\cos\left(-\dfrac{\pi}{2}\right)$;

  g) $\cos\left(-\dfrac{3}{4}\pi\right)$;   h) $\cos\left(-\dfrac{\pi}{6}\right)$;

  j) $\cos\left(-\dfrac{7}{4}\pi\right)$;   k) $\cos\left(-\dfrac{2}{3}\pi\right)$.

23. For which numbers is
    a) $\sin x = 1$?
    b) $\sin x = -1$?

24. For which numbers is
    a) $\cos x = 1$?
    b) $\cos x = -1$?

25. Solve for x.
$$\sin x = 0$$

26. Solve for x.
$$\cos x = 0$$

27. Find $f \circ g$ and $g \circ f$, where $f(x) = x^2 + 2x$ and $g(x) = \cos x$.

28. ▦ Calculate the following. Do not use the trigonometric function keys.

    a) $\sin \dfrac{\pi}{4}$ b) $\sin \dfrac{\pi}{3}$

29. ▦ Calculate the following. Do not use the trigonometric function keys.

    a) $\cos \dfrac{\pi}{6}$ b) $\cos \dfrac{\pi}{4}$

30. Show that the sine function does not have the *linearity* property, i.e., that it is not true that $\sin(x + y) = \sin x + \sin y$ for all numbers x and y.

2.4 THE OTHER CIRCULAR FUNCTIONS

[i] Definitions

There are four other circular functions, all defined in terms of the sine and cosine functions. Their definitions are as follows.

DEFINITIONS

The tangent function:	$\tan s \equiv \dfrac{\sin s}{\cos s}$
The cotangent function:	$\cot s \equiv \dfrac{\cos s}{\sin s}$
The secant function:	$\sec s \equiv \dfrac{1}{\cos s}$
The cosecant function:	$\csc s \equiv \dfrac{1}{\sin s}$

The Tangent Function

The tangent of a number is the quotient of its sine by its cosine.

Example 1 Find $\tan 0$.

$$\tan 0 = \frac{\sin 0}{\cos 0} = \frac{0}{1} = 0$$

Example 2 Find $\tan \dfrac{\pi}{4}$.

$$\tan \frac{\pi}{4} = \frac{\sin \dfrac{\pi}{4}}{\cos \dfrac{\pi}{4}} = \frac{\dfrac{\sqrt{2}}{2}}{\dfrac{\sqrt{2}}{2}} = 1$$

OBJECTIVES

You should be able to:

[i] Define the tangent, cotangent, secant, and cosecant functions; state their properties; sketch their graphs; and find function values (when they exist) for multiples of $\pi/6$, $\pi/4$, and $\pi/3$.

[ii] Given values of the sine and cosine, calculate the values of the other functions.

[iii] Verify simple identities.

1. Find $\tan \pi$.

2. Find $\tan \frac{3}{4}\pi$.

3. Find $\tan \frac{\pi}{2}$.

4. Find three other real numbers at which the tangent function is undefined.

Example 3 Find $\tan \frac{\pi}{3}$.

$$\tan \frac{\pi}{3} = \frac{\sin \frac{\pi}{3}}{\cos \frac{\pi}{3}} = \frac{\frac{\sqrt{3}}{2}}{\frac{1}{2}} = \sqrt{3}$$

Not every real number has a tangent.

Example 4 Show that $\tan \frac{\pi}{2}$ does not exist.

$$\tan \frac{\pi}{2} = \frac{\sin \frac{\pi}{2}}{\cos \frac{\pi}{2}} = \frac{1}{0},$$

but division by 0 is undefined. Thus $\pi/2$ is not in the domain of the tangent function, and $\tan(\pi/2)$ does not exist.

DO EXERCISES 1–4.

The tangent function is undefined for any number whose cosine is 0. Thus it is undefined for $(\pi/2) + k\pi$, where k is any integer. In the first quadrant the function values are positive. In the second quadrant, however, the sine and cosine values have opposite signs, so the tangent values are negative. Tangent values are also negative in the fourth quadrant, and they are positive in quadrant three.

A graph of the tangent function is shown below. Note that the function value is 0 when $x = 0$, and the values increase as x increases toward $\pi/2$. As we approach $\pi/2$, the denominator becomes very small, so the tangent values become very large. In fact, they increase without bound. The dashed vertical line at $\pi/2$ is not a part of the graph. The graph approaches this vertical line closely, however. The vertical dashed lines are known as *asymptotes* to the curve.

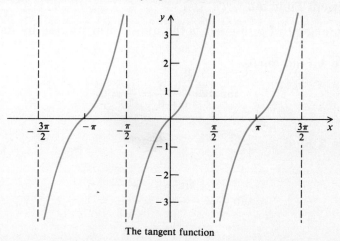

The tangent function

The tangent function is periodic, with a period of π. Its domain is the set of all real numbers except $(\pi/2) + k\pi$, k any integer. Its range is the set of all real numbers. It is continuous except where it is not defined.

DO EXERCISES 5–9.

The Cotangent Function

The cotangent of a number is the quotient of its cosine by its sine. Thus the cotangent function is undefined for any number whose sine is 0. The graph of this function is shown here.

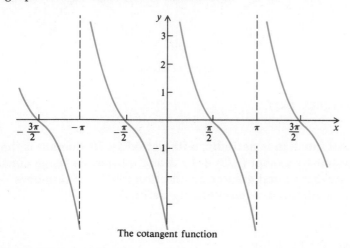

The cotangent function

DO EXERCISES 10–14.

The cotangent function is periodic, with a period of π. Its domain is the set of all real numbers except $k\pi$, k any integer. Its range is the set of all real numbers. It is continuous except where it is not defined.

Like the tangent function, the cotangent has positive function values in quadrants I and III and negative function values in quadrants II and IV. The tangent and cotangent are reciprocals, hence the following identities hold.

$$\tan x \equiv \frac{1}{\cot x}, \qquad \cot x \equiv \frac{1}{\tan x}$$

The Secant Function

The secant and cosine functions are reciprocals.

Example 5 Find $\sec \dfrac{\pi}{4}$.

$$\sec \frac{\pi}{4} = \frac{1}{\cos \dfrac{\pi}{4}} = \frac{1}{\dfrac{\sqrt{2}}{2}} = \sqrt{2}$$

5. What is the period of the tangent function?

6. What is the domain of the tangent function?

7. What is the range of the tangent function?

8. Does the tangent function appear to be even? odd?

9. Find three numbers at which the cotangent function is undefined.

10. What is the period of the cotangent function?

11. What is the domain of the cotangent function?

12. What is the range of the cotangent function?

13. In which quadrants are the function values of the cotangent function positive? negative?

14. Does the cotangent function appear to be odd? even?

15. Find $\sec \frac{\pi}{4}$.

16. Find $\sec \pi$.

17. Find three numbers at which the secant function is undefined.

18. What is the period of the cosecant function?

19. What is the domain of the cosecant function?

20. What is the range of the cosecant function?

21. In which quadrants are the function values of the cosecant function positive? negative?

22. Does the cosecant function appear to be even? odd?

This function is undefined for those numbers x for which $\cos x = 0$. A graph of the secant function is shown below. The cosine graph is also shown for reference, since these functions are reciprocals.

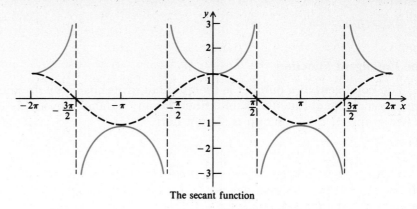

The secant function

DO EXERCISES 15–17.

The secant function is periodic, with period 2π. Its domain is the set of all real numbers except $(\pi/2) + k\pi$, k any integer. Its range consists of all real numbers 1 and greater, in addition to all real numbers -1 and less. It is continuous wherever it is defined.

The Cosecant Function

Since the cosecant and sine functions are reciprocals, the cosecant will be undefined wherever the sine function is 0. The functions will be positive together and negative together. A graph of the cosecant function is shown below. The sine graph is shown for reference, since these functions are reciprocals.

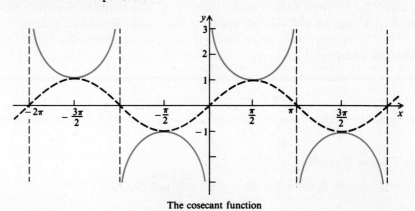

The cosecant function

DO EXERCISES 18–22.

The cosecant function is periodic, with period 2π. Its domain is the set of all real numbers except $k\pi$, k any integer. Its range is the same as for the secant function. It is continuous wherever it is defined.

Signs of the Functions

Suppose a real number s determines a point on the unit circle in the first quadrant. Since both coordinates are positive there, the function values for all the circular functions will be positive. In the second quadrant the first coordinate is negative and the second coordinate is positive. Thus the sine function has positive values and the cosine has negative values. The secant, being the reciprocal of the cosine, also has negative values, and the cosecant, being the reciprocal of the sine, has positive values. The tangent and cotangent both have negative values in the second quadrant.

DO EXERCISE 23.

The following diagram summarizes, showing the quadrants in which the circular function values are positive. This diagram need not be memorized, because the signs can readily be determined by reference to a unit circle.

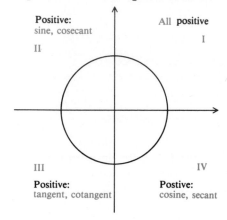

[ii] Calculating Function Values

Given values for sine and cosine, we can calculate the other function values.

▦ **Example 6** Given that $\sin(\pi/9) \approx 0.34202$ and $\cos(\pi/9) \approx 0.93969$, calculate the other function values for $\pi/9$.

$$\tan\frac{\pi}{9} = \frac{\sin\dfrac{\pi}{9}}{\cos\dfrac{\pi}{9}} \approx \frac{0.34202}{0.93969} \approx 0.36397$$

$$\cot\frac{\pi}{9} = \frac{\cos\dfrac{\pi}{9}}{\sin\dfrac{\pi}{9}} \approx \frac{0.93969}{0.34202} \approx 2.74747$$

$$\sec\frac{\pi}{9} = \frac{1}{\cos\dfrac{\pi}{9}} \approx \frac{1}{0.93969} \approx 1.06418$$

$$\csc\frac{\pi}{9} = \frac{1}{\sin\dfrac{\pi}{9}} \approx \frac{1}{0.34202} \approx 2.92381$$

23. In the third quadrant, what are the signs of the six circular functions?

▦ **24.** Given that $\sin(\pi/11) = 0.28173$ and $\cos(\pi/11) = 0.95949$, calculate the other function values for $\pi/11$.

▦ **Example 7** Given that $\sin(2/3)\pi \approx 0.86603$ and $\cos(2/3)\pi \approx -0.50000$, calculate the other function values for $(2/3)\pi$.

$$\tan\frac{2}{3}\pi = \frac{\sin\dfrac{2}{3}\pi}{\cos\dfrac{2}{3}\pi} \approx \frac{0.86603}{-0.50000} \approx -1.73206$$

$$\cot\frac{2}{3}\pi = \frac{\cos\dfrac{2}{3}\pi}{\sin\dfrac{2}{3}\pi} \approx \frac{-0.50000}{0.86603} \approx -0.57735$$

$$\sec\frac{2}{3}\pi = \frac{1}{\cos\dfrac{2}{3}\pi} \approx \frac{1}{-0.50000} \approx -2.0000$$

$$\csc\frac{2}{3}\pi = \frac{1}{\sin\dfrac{2}{3}\pi} \approx \frac{1}{0.86603} \approx 1.15469$$

DO EXERCISE 24.

[iii] **Some Identities**

The tangent function appears to be odd. Let us investigate by finding $\tan(-x)$. By definition of the tangent function,

$$\tan(-x) \equiv \frac{\sin(-x)}{\cos(-x)}.$$

But we know that $\sin(-x) \equiv -\sin x$ and $\cos(-x) \equiv \cos x$. Hence

$$\tan(-x) \equiv \frac{\sin(-x)}{\cos(-x)} \equiv \frac{-\sin x}{\cos x} \equiv -\tan x.$$

Thus the tangent function is indeed odd. Similarly, it can be shown that the cotangent function is odd, the secant function is even, and the cosecant function is odd. Each such demonstration gives us an identity.

By use of the previously developed identities involving the sine and cosine functions we can develop a number of other identities.

Example 8 Show that $\tan(x + \pi) \equiv \tan x$.

$$\tan(x + \pi) \equiv \frac{\sin(x + \pi)}{\cos(x + \pi)} \qquad \text{By definition of the tangent function}$$

$$\equiv \frac{-\sin x}{-\cos x} \qquad \text{Using } \sin(x + \pi) \equiv -\sin x \text{ and } \cos(x + \pi) \equiv -\cos x$$

$$\equiv \frac{\sin x}{\cos x}$$

$$\equiv \tan x$$

Example 9 Show that $\sec(\pi - x) \equiv -\sec x$.

$$\sec(\pi - x) \equiv \frac{1}{\cos(\pi - x)} \qquad \text{By definition of the secant function}$$

$$\equiv \frac{1}{-\cos x} \qquad \text{Using } \cos(\pi - x) \equiv -\cos x$$

$$\equiv -\frac{1}{\cos x}$$

$$\equiv -\sec x$$

Example 10 Show that $\csc(x - \pi) \equiv -\csc x$.

$$\csc(x - \pi) \equiv \frac{1}{\sin(x - \pi)} \qquad \text{By definition of the cosecant function}$$

$$\equiv \frac{1}{-\sin x} \qquad \text{Using } \sin(x - \pi) \equiv -\sin x$$

$$\equiv -\csc x$$

DO EXERCISES 25–27.

25. Show that $\cot(-x) \equiv -\cot x$.

26. Show that $\tan(x - \pi) \equiv \tan x$.

27. Show that $\csc(\pi - x) \equiv \csc x$.

EXERCISE SET 2.4

[i] Find the following function values.

1. $\cot \dfrac{\pi}{4}$

2. $\tan\left(-\dfrac{\pi}{4}\right)$

3. $\tan \dfrac{\pi}{6}$

4. $\cot\left(\dfrac{5\pi}{6}\right)$

5. $\sec \dfrac{\pi}{4}$

6. $\csc \dfrac{3\pi}{4}$

7. $\tan \dfrac{3\pi}{2}$

8. $\cot \pi$

9. $\tan \dfrac{2\pi}{3}$

10. $\cot\left(-\dfrac{2\pi}{3}\right)$

[ii]

11. ▦ Complete this table of approximate function values. Do not use trigonometric function keys. Round to five decimal places.

	$\pi/16$	$\pi/8$	$\pi/6$	$\pi/4$	$3\pi/8$	$7\pi/16$
sin	0.19509	0.38268			0.92388	0.98079
cos	0.98079	0.92388			0.38268	0.19509
tan						
cot						
sec						
csc						

12. ▦ Complete this table of approximate function values. Do not use trigonometric function keys. Round to five decimal places.

	$-\pi/16$	$-\pi/8$	$-\pi/6$	$-\pi/4$	$-\pi/3$
sin	-0.19509	-0.38268	-0.50000	-0.70711	-0.86603
cos	0.98079	0.92388	0.86603	0.70711	0.50000
tan					
cot					
sec					
csc					

In Exercises 13–16, use graph paper. Use the tables above, plus your other knowledge of the functions, to make graphs.

13. Graph the tangent function between -2π and 2π.

14. Graph the cotangent function between -2π and 2π.

15. Graph the secant function between -2π and 2π.

16. Graph the cosecant function between -2π and 2π.

17. Of the six circular functions, which are even?

18. Of the six circular functions, which are odd?

19. Which of the six circular functions have period 2π?

20. Which of the six circular functions have period π?

21. In which quadrants is the tangent function positive? negative?

22. In which quadrants is the cotangent function positive? negative?

23. In which quadrants is the secant function positive? negative?

24. In which quadrants is the cosecant function positive? negative?

[iii] Verify the following identities.

25. $\sec(-x) \equiv \sec x$

26. $\csc(-x) \equiv -\csc x$

27. $\cot(x + \pi) \equiv \cot x$

28. $\cot(x - \pi) \equiv \cot x$

29. $\sec(x + \pi) \equiv -\sec x$

30. $\tan(\pi - x) \equiv -\tan x$

☆ ——————————————————————————————

31. Verify the identity $\sec(x - \pi) \equiv -\sec x$ graphically.

32. Verify the identity $\tan(x + \pi) \equiv \tan x$ graphically.

33. Describe how the graphs of the tangent and cotangent functions are related.

34. Describe how the graphs of the secant and cosecant functions are related.

35. Which pairs of circular functions have the same zeros? (A "zero" of a function is an input that produces an output of 0.)

36. Describe how the asymptotes of the tangent, cotangent, secant, and cosecant functions are related to the inputs that produce outputs of 0.

37. Graph $f(x) = |\tan x|$.

★ ——————————————————————————————

38. Solve $\cos x \le \sec x$.

39. Solve $\sin x > \csc x$.

OBJECTIVES

You should be able to:

[i] State the Pythagorean identities and derive variations of them.

[ii] State the cofunction identities for the sine and cosine, and derive cofunction identities for the other functions.

2.5 SOME RELATIONS AMONG THE CIRCULAR FUNCTIONS

We have already seen that there are some relations among the six circular functions. In fact, the sine and cosine functions are used to define the others. There are certain other important relations among these functions. These are given by identities.

[i] The Pythagorean Identities

Recall that the equation of the unit circle in a *u-v* plane is

$$u^2 + v^2 = 1.$$

For any point on the unit circle, the coordinates *u* and *v* satisfy this equation. Suppose a real number *s* determines a point *T* on the unit circle, with coordinates (*u*, *v*). Then $u = \cos s$ and $v = \sin s$. Substituting these in the equation of the unit circle gives us the identity

$$\sin^2 s + \cos^2 s \equiv 1.$$

This is true because the coordinates of a point on the unit circle are (cos *s*, sin *s*). When exponents are used with the circular functions, we write $\sin^2 s$ instead of $(\sin s)^2$. This identity relates the sine and cosine of any real number *s*. It is an important identity, known as one of the *Pythagorean* identities. We now develop another. We will divide the above identity by $\sin^2 s$:

$$\frac{\sin^2 s}{\sin^2 s} + \frac{\cos^2 s}{\sin^2 s} \equiv \frac{1}{\sin^2 s}.$$

Simplifying, we get

$$1 + \cot^2 s \equiv \csc^2 s.$$

This relation is valid for any number *s* for which $\sin^2 s \neq 0$, since we divided by $\sin^2 s$. But the numbers for which $\sin^2 s = 0$ (or $\sin s = 0$) are exactly the ones for which the cotangent and cosecant functions are undefined. Hence our new equation holds for all numbers *s* for which cot *s* and csc *s* are defined, and is thus an identity.

DO EXERCISE 1.

The third Pythagorean identity, obtained by dividing the first by $\cos^2 s$, is

$$1 + \tan^2 s \equiv \sec^2 s.$$

The Pythagorean identities should be memorized. Certain variations of them, although most useful, need not be memorized, because they are so easily obtained from these three.

Example 1 Derive identities that give $\cos^2 x$ and $\cos x$ in terms of $\sin x$.

We begin with $\sin^2 x + \cos^2 x \equiv 1$, and solve for $\cos^2 x$. We have $\cos^2 x \equiv 1 - \sin^2 x$. This gives $\cos^2 x$ in terms of $\sin x$. If we now take the principal square root, we obtain

$$|\cos x| \equiv \sqrt{1 - \sin^2 x}.$$

1. Consider $\sin^2 s + \cos^2 s \equiv 1$.

a) Divide by $\cos^2 s$ and simplify.

b) For what values of cos *s* is the new equation meaningless?

c) Does the new equation hold for all values of *s* for which the functions are defined? (Is it an identity?)

2. Derive an identity for $\sin^2 x$ in terms of $\cos x$.

We can also express this as

$$\cos x \equiv \pm \sqrt{1 - \sin^2 x},$$

with the understanding that the sign must be determined by the quadrant in which the point for x lies.

DO EXERCISES 2 AND 3.

3. Derive an identity for $\sin x$ in terms of $\cos x$.

[ii] The Cofunction Identities

The sine and cosine are called *cofunctions* of each other. Similarly, the tangent and cotangent are cofunctions and the secant and cosecant are cofunctions. Another class of identities gives functions in terms of their cofunctions at numbers differing by $\pi/2$. We consider first the sine and cosine. Consider this graph.

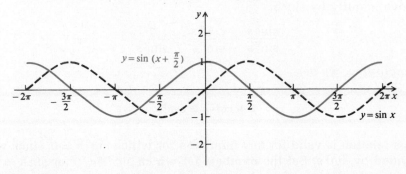

4. a) Graph $y = \cos x$.
   b) Translate, to obtain a graph of

$$y = \cos \left(x - \frac{\pi}{2} \right).$$

   c) Graph $y = \sin x$.
   d) How do the graphs of (b) and (c) compare?
   e) Write the identity thus established.

The graph of $y = \sin x$ has been translated to the left a distance of $\pi/2$, to obtain the graph of $y = \sin (x + \pi/2)$. The latter is also a graph of the cosine function. Thus we obtain the identity

$$\sin \left(x + \frac{\pi}{2} \right) \equiv \cos x.$$

DO EXERCISE 4.

By means similar to that above, we obtain the identity

$$\cos \left(x - \frac{\pi}{2} \right) \equiv \sin x.$$

Consider the following graph. The graph of $y = \sin x$ has been translated to the right a distance of $\pi/2$, to obtain the graph of $y = \sin (x - \pi/2)$. The latter is a reflection of the cosine function across the x-axis. In other words, it is a graph of $y = -\cos x$. We thus obtain the following identity:

$$\sin \left(x - \frac{\pi}{2} \right) \equiv -\cos x.$$

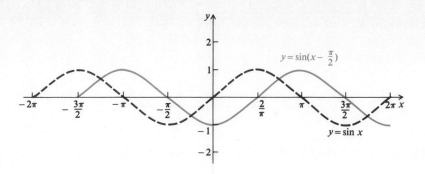

DO EXERCISE 5.

By means similar to that above, we obtain the identity

$$\cos\left(x + \frac{\pi}{2}\right) \equiv -\sin x.$$

We have now established the following cofunction identities.*

$$\sin\left(x \pm \frac{\pi}{2}\right) \equiv \pm\cos x$$

$$\cos\left(x \pm \frac{\pi}{2}\right) \equiv \mp\sin x$$

These should be learned. The other cofunction identities can be obtained easily, using these and the definitions of the other circular functions.

Example 2 Derive an identity relating $\sin(\pi/2 - x)$ to $\cos x$.

Since the sine function is odd, we know that

$$\sin\left(\frac{\pi}{2} - x\right) \equiv \sin\left[-\left(x - \frac{\pi}{2}\right)\right] \equiv -\sin\left(x - \frac{\pi}{2}\right).$$

Now consider the identity already established,

$$\sin\left(x - \frac{\pi}{2}\right) \equiv -\cos x.$$

This is equivalent to

$$-\sin\left(x - \frac{\pi}{2}\right) \equiv \cos x,$$

and we now have

$$\sin\left(\frac{\pi}{2} - x\right) \equiv \cos x.$$

DO EXERCISE 6.

_____
*There are four identities in this list. Two of them are obtained by taking the top signs, the other two by taking the bottom signs.

5. a) Graph $y = \cos x$.
   b) Translate, to obtain a graph of
   $$y = \cos\left(x + \frac{\pi}{2}\right).$$
   c) Graph $y = \sin x$.
   d) Reflect the graph of c, to obtain a graph of $y = -\sin x$.
   e) How do the graphs of (b) and (d) compare?
   f) Write the identity thus established.

6. Prove that $\cos\left(\frac{\pi}{2} - x\right) \equiv \sin x.$

7. Find an identity for $\cot\left(x - \dfrac{\pi}{2}\right)$.

Example 3 Find an identity for $\tan\left(x + \dfrac{\pi}{2}\right)$.

By definition of the tangent function,

$$\tan\left(x + \frac{\pi}{2}\right) \equiv \frac{\sin\left(x + \dfrac{\pi}{2}\right)}{\cos\left(x + \dfrac{\pi}{2}\right)}.$$

Using the cofunction identities above, we obtain

$$\frac{\sin\left(x + \dfrac{\pi}{2}\right)}{\cos\left(x + \dfrac{\pi}{2}\right)} \equiv \frac{\cos x}{-\sin x} \equiv -\frac{\cos x}{\sin x} \equiv -\cot x.$$

Thus the identity we seek is

$$\tan\left(x + \frac{\pi}{2}\right) \equiv -\cot x.$$

Example 4 Find an identity for $\sec\left(\dfrac{\pi}{2} - x\right)$.

By definition of the secant function,

$$\sec\left(\frac{\pi}{2} - x\right) \equiv \frac{1}{\cos\left(\dfrac{\pi}{2} - x\right)}.$$

If we proceed as in Example 2, the following can be established: $\cos\left(\pi/2 - x\right) \equiv \cos\left(x - \pi/2\right)$. Then we obtain

$$\frac{1}{\cos\left(\dfrac{\pi}{2} - x\right)} \equiv \frac{1}{\sin x} \equiv \csc x.$$

Thus the identity we seek is

$$\sec\left(\frac{\pi}{2} - x\right) \equiv \csc x.$$

DO EXERCISE 7.

EXERCISE SET 2.5

[i]

1. From the identity $1 + \cot^2 x \equiv \csc^2 x$, obtain two other identities.

2. From the identity $1 + \tan^2 x \equiv \sec^2 x$, obtain two other identities.

3. a) Derive an identity for $\csc x$ in terms of $\cot x$.
   b) Derive an identity for $\cot x$ in terms of $\csc x$.

4. a) Derive an identity for $\tan x$ in terms of $\sec x$.
   b) Derive an identity for $\sec x$ in terms of $\tan x$.

[ii]

5. Find an identity for $\tan\left(x - \dfrac{\pi}{2}\right)$.

6. Find an identity for $\cot\left(x + \dfrac{\pi}{2}\right)$.

7. Find an identity for $\sec\left(\dfrac{\pi}{2} - x\right)$.

8. Find an identity for $\csc\left(\dfrac{\pi}{2} - x\right)$.

9. For the six circular functions, list the cofunction identities involving $x \pm \dfrac{\pi}{2}$.

☆ ───

10. In which quadrant is $\sin x = \sqrt{1 - \cos^2 x}$?

11. ▦ Given that $\sin x = 0.1425$ in quadrant II, find the other circular function values.

12. ▦ Given that $\tan x = 0.8012$ in quadrant III, find the other circular function values.

13. ▦ Given that $\cot x = 0.7534$ and $\sin x > 0$, find the other circular function values.

14. ▦ Given that $\sin\dfrac{\pi}{8} = 0.38268$, use identities to find:

a) $\cos\dfrac{\pi}{8}$; b) $\cos\dfrac{5\pi}{8}$; c) $\sin\dfrac{5\pi}{8}$; d) $\sin\dfrac{-3\pi}{8}$; e) $\cos\dfrac{-3\pi}{8}$.

15. Using sines, cosines, and a determinant, write an equation of the unit circle.

16. The coordinates of a point on the unit circle are $(\cos s, \sin s)$, for some suitable s. Show that for a circle with center at the origin of radius r, the coordinates of any point on the circle are $(r \cos s, r \sin s)$.

2.6 GRAPHS

[i], [ii] Variations of Basic Graphs

We will consider graphs of some variations of the sine and cosine functions. In particular, we are interested in the following:

$$y = A + B \sin (Cx - D),$$

and

$$y = A + B \cos (Cx - D),$$

where A, B, C, and D are constants, some of which may be 0. These constants have the effect of translating, stretching, reflecting, or shrinking the basic graphs. Let us examine an equation to see what each of these constants does.* We first change the form a bit.

$$y - A = B \sin C\left(x - \dfrac{D}{C}\right).$$

| Translates, upward if positive. | Stretches, vertically if greater than 1. | Stretches, horizontally if between 0 and 1. | Translates, to the right if positive. |

The amplitude of the sine or cosine function is multiplied by B, when that constant is present. The period will be affected by the constant C. If $C = 2$, for example, the period will be divided by 2. If $C = \frac{1}{2}$, the period will be multiplied by 2. In general, the new period will be $2\pi/C$.

OBJECTIVES

You should be able to:

[i] Sketch graphs of
$y = A + B \sin (Cx + D)$ and
$y = A + B \cos (Cx + D)$, for various values of the constants A, B, C, and D.

[ii] For functions like these, determine the amplitude, period, and phase shift.

[iii] By addition of ordinates, graph sums of functions.

─────────────
*You may wish to review pp. 26–34.

1. Sketch a graph of $y = -2 + \cos x$.

Example 1 The graph of $y = 3 + \sin x$ is a translation of $y = \sin x$ upward 3 units ($A = 3$, $B = C = 1$, and $D = 0$).

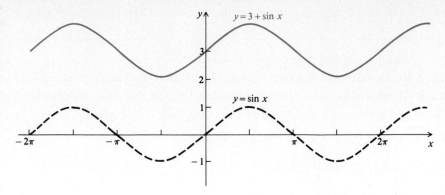

DO EXERCISE 1.

Example 2 The graph of $y = 2 \sin x$ is a vertical stretching of the graph of $y = \sin x$. The amplitude of this function is 2 ($A = 0$, $B = 2$, $C = 1$, and $D = 0$).

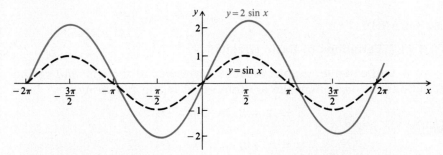

2. Sketch a graph of $y = -2 \cos x$. What is the amplitude?

If the constant B in $y = B \sin x$ is negative, there will also be a reflection across the x-axis. If the absolute value of B is less than 1, then there will be a vertical shrinking. $|B|$ will be the amplitude.

Example 3 The graph of $y = -(\frac{1}{2}) \sin x$ is a vertical shrinking and a reflection of the graph of $y = \sin x$. The amplitude is $\frac{1}{2}$.

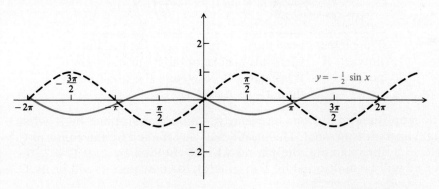

DO EXERCISE 2.

Example 4 The graph of $y = \sin 2x$ is a horizontal shrinking of the graph of $y = \sin x$. The period of this function is π (half the period of $y = \sin x$).

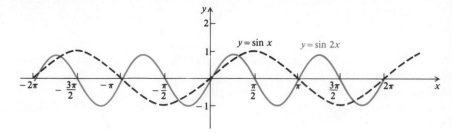

DO EXERCISES 3 AND 4.

Example 5 The graph of $y = \sin(2x - \pi)$ is the same as that of $y = \sin 2(x - \pi/2)$. We merely factored the 2 out of the parentheses. The $\pi/2$ translates the graph of $y = \sin 2x$ a distance of $\pi/2$ to the right. The 2 in front of the parentheses shrinks by half, making the period π.

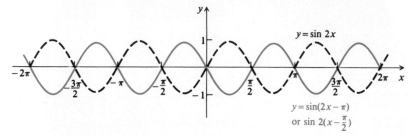

DO EXERCISE 5.

The quantity D/C translates to the right or left, and is called the *phase shift*. In Example 5, the phase shift is $\pi/2$. This means that the graph of $\sin 2x$ has been shifted forward a distance of $\pi/2$.

Example 6 The graph of $y = 3 \sin [2x + (\pi/2)]$ has an amplitude of 3.

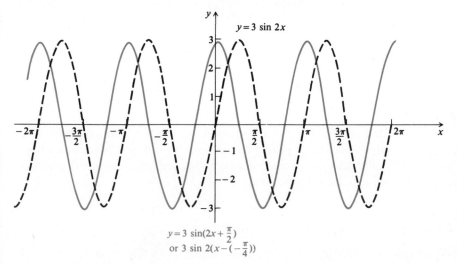

3. Sketch a graph of $y = \cos \frac{1}{2}x$. What is the period?

4. Sketch a graph of $y = \sin(-2x)$. What is the period?

5. Sketch a graph of $y = \cos(2x + \pi)$. What is the period? What is the phase shift?

6. Sketch a graph of
$$y = 3 \cos\left(2x - \frac{\pi}{2}\right).$$

What is the amplitude? What is the period? What is the phase shift?

The period is π and the phase shift is $-\pi/4$. This means that the graph of $3 \sin 2x$ has been shifted back a distance of $\pi/4$.

DO EXERCISE 6.

The oscilloscope is an electronic device that draws graphs on a cathode-ray tube. By manipulating the controls, one can change such things as the amplitude, period, and phase. The oscilloscope has many applications, and the trigonometric functions play a major role in many of them.

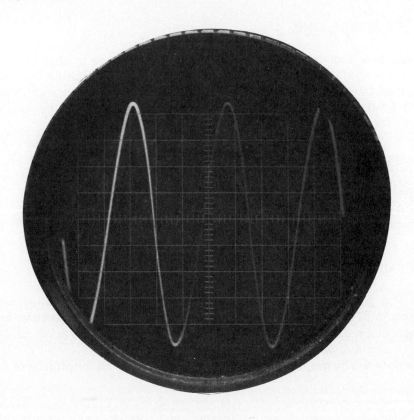

[iii] Graphs of Sums

A function that is a sum of two functions can often be graphed by a method called *addition of ordinates*. We graph the two functions separately and then add the second coordinates (called *ordinates*) graphically. A compass may be helpful.

Example 7 Graph $y = 2 \sin x + \sin 2x$.

We graph $y = 2 \sin x$ and $y = \sin 2x$ using the same axes.

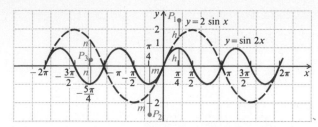

Now we graphically add some ordinates to obtain points on the graph we seek. At $x = \pi/4$ we transfer the distance h, which is the value of $\sin 2x$, upward to add it to the value of $2 \sin x$. Point P_1 is on the graph we seek. At $x = -\pi/4$ we do a similar thing, but the distance m is negative. Point P_2 is on our graph. At $x = -(\frac{5}{4})\pi$ the distance n is negative, so we in effect subtract it from the value of $2 \sin x$. Point P_3 is on the graph we seek. We continue to plot points in this fashion and then connect them to get the desired graph, shown here.

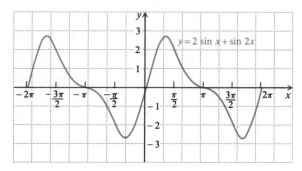

DO EXERCISE 7.

A "sawtooth" function, such as the one shown on this oscilloscope, has numerous applications—for example, in television circuits. This sawtooth function can be approximated extremely well by adding, electronically, several sine and cosine functions.

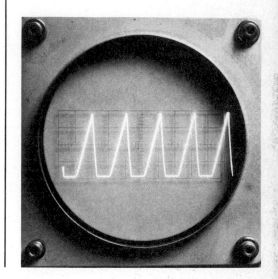

EXERCISE SET 2.6

[i] In Exercises 1–28, use graph paper. Sketch graphs of the functions.

1. $y = 2 + \sin x$

2. $y = -3 + \cos x$

3. $y = \frac{1}{2} \sin x$

4. $y = \frac{1}{3} \cos x$

5. $y = 2 \cos x$

6. $y = 3 \cos x$

7. $y = -\frac{1}{2} \cos x$

8. $y = -2 \sin x$

9. $y = \cos 2x$

10. $y = \cos 3x$

11. $y = \cos (-2x)$

12. $y = \cos (-3x)$

13. $y = \sin \frac{1}{2} x$

14. $y = \cos \frac{1}{3} x$

15. $y = \sin \left(-\frac{1}{2} x\right)$

16. $y = \cos \left(-\frac{1}{3} x\right)$

17. $y = \cos(2x - \pi)$

18. $y = \sin(2x + \pi)$

19. $y = 2\cos\left(\frac{1}{2}x - \frac{\pi}{2}\right)$

20. $y = 4\sin\left(\frac{1}{4}x + \frac{\pi}{8}\right)$

21. $y = -3\cos(4x - \pi)$

22. $y = -3\sin\left(2x + \frac{\pi}{2}\right)$

[iii]

23. $y = 2\cos x + \cos 2x$

24. $y = 3\cos x + \cos 3x$

25. $y = \sin x + \cos 2x$

26. $y = 2\sin x + \cos 2x$

27. $y = 3\cos x - \sin 2x$

28. $y = 3\sin x - \cos 2x$

[ii] In Exercises 29–34, determine the amplitude, period, and phase shift.

29. $y = 3\cos\left(3x - \frac{\pi}{2}\right)$

30. $y = 4\sin\left(4x - \frac{\pi}{3}\right)$

31. $y = -5\cos\left(4x + \frac{\pi}{3}\right)$

32. $y = -4\sin\left(5x + \frac{\pi}{2}\right)$

33. $y = \frac{1}{2}\sin(2\pi x + \pi)$

34. $y = -\frac{1}{4}\cos(\pi x - 4)$

☆

35. Graph $y = \sec^2 x$.

36. Graph $y = \tan x \csc x$.

37. Graph $y = 3\cos 2(x + \pi/4)$ by first converting to a sine.

★

38. Graph $y = x + \sin x$. (*Hint:* Use addition of ordinates. You might also wish to use a calculator to find some function values to plot.)

39. Graph $y = 2^{-x}\sin x$, for $0 \le x \le 4\pi$. Functions like this have applications in the theory of damped oscillations. (*Hint:* First graph $y = 2^{-x}$ and also $y = -(2^{-x})$. Then graph $y = \sin x$ and consider multiplying the ordinates. You may wish to use a calculator to find some function values to plot.)

OBJECTIVES

You should be able to:

[i] Multiply and factor expressions containing trigonometric expressions.

[ii] Compute with and simplify expressions containing trigonometric expressions.

[iii] Compute with radical expressions containing trigonometric expressions, including rationalizing numerators or denominators.

[iv] Solve equations containing trigonometric expressions for the values of those expressions.

2.7 ALGEBRAIC AND TRIGONOMETRIC MANIPULATIONS

The circular functions and variations of them are called *trigonometric functions*. Trigonometric expressions such as sin 2x or tan (x − π) represent numbers, just as algebraic expressions such as 3x or $3x^2 - 2$ represent numbers. Thus we can manipulate expressions containing trigonometric expressions in much the same way as we manipulate purely algebraic expressions.

[i] **Multiplying and Factoring**

Example 1 Multiply and simplify $\cos y (\tan y - \sec y)$.

$$\cos y (\tan y - \sec y) = \cos y \tan y - \cos y \sec y$$

The multiplication has now been accomplished, but we can simplify:

$$\cos y \tan y - \cos y \sec y = \cos y \frac{\sin y}{\cos y} - \cos y \frac{1}{\cos y}$$

$$= \sin y - 1.$$

In Example 1 it has been illustrated how we can use certain identities to accomplish simplification. There is no rule for doing such simplifications, but it is often helpful to put everything in terms of sines and cosines, as we did in Example 1.

Example 2 Factor and simplify $\sin^2 x \cos^2 x + \cos^4 x$.

$$\sin^2 x \cos^2 x + \cos^4 x = \cos^2 x (\sin^2 x + \cos^2 x)$$

The factoring has been accomplished, but we can now simplify, using the Pythagorean identity $\sin^2 x + \cos^2 x \equiv 1$:

$$\cos^2 x (\sin^2 x + \cos^2 x) = \cos^2 x.$$

Example 3 Factor and simplify $\tan x + \cos \left(x - \dfrac{\pi}{2} \right)$.

$$\tan x + \cos \left(x - \frac{\pi}{2} \right) = \sin x \, \frac{1}{\cos x} + \sin x.$$

Here we have used the identity $\cos (x - \pi/2) \equiv \sin x$, as well as the definition of the tangent function. Now we factor as follows:

$$\sin x \left(\frac{1}{\cos x} + 1 \right) \quad \text{or} \quad \sin x \, (\sec x + 1).$$

DO EXERCISES 1–3.

[ii] **Simplifying**

Example 4 Simplify $\dfrac{\sin x - \sin x \cos x}{\sin x + \sin x \tan x}$.

$$\frac{\sin x - \sin x \cos x}{\sin x + \sin x \tan x} = \frac{\sin x \, (1 - \cos x)}{\sin x \, (1 + \tan x)} \qquad \text{Factoring}$$

$$= \frac{1 - \cos x}{1 + \tan x} \qquad \text{Simplifying}$$

Example 5 Subtract and simplify $\dfrac{2}{\sin x - \cos x} - \dfrac{3}{\sin x + \cos x}$.

$$\frac{2}{\sin x - \cos x} - \frac{3}{\sin x + \cos x} = \frac{2}{\sin x - \cos x} \cdot \frac{\sin x + \cos x}{\sin x + \cos x}$$

$$- \frac{3}{\sin x + \cos x} \cdot \frac{\sin x - \cos x}{\sin x - \cos x}$$

$$= \frac{2(\sin x + \cos x) - 3(\sin x - \cos x)}{\sin^2 x - \cos^2 x}$$

$$= \frac{-\sin x + 5 \cos x}{\sin^2 x - \cos^2 x}$$

1. Multiply and simplify.

$$\sin x (\cot x + \csc x)$$

2. Factor and simplify.

$$\sin^3 x + \sin x \cos^2 x$$

3. Factor and simplify.

$$\cot x - \sin \left(\frac{\pi}{2} - x \right)$$

4. Simplify.

$$\frac{\cos x + \sin x \cos x}{\cos x - \cos x \cot x}$$

5. Add and simplify.

$$\frac{2}{\sin x - \cos x} + \frac{3}{\sin x + \cos x}$$

6. Add and simplify.

$$\frac{2}{\cot^3 x - 2 \cot^2 x} + \frac{2}{\cot x - 2}$$

7. Multiply and simplify.

$$\sqrt{\tan x \sin^2 x} \cdot \sqrt{\tan x \sin x}$$

8. Rationalize the numerator.

$$\sqrt{\frac{\cos x}{3}}$$

Example 6 Add and simplify $\dfrac{2}{\tan^3 x - 2 \tan^2 x} + \dfrac{2}{\tan x - 2}$.

$$\frac{2}{\tan^3 x - 2 \tan^2 x} + \frac{2}{\tan x - 2} = \frac{2}{\tan^2 x (\tan x - 2)} + \frac{2}{\tan x - 2} \cdot \frac{\tan^2 x}{\tan^2 x}$$

$$= \frac{2 + 2\tan^2 x}{\tan^2 x (\tan x - 2)} = \frac{2(1 + \tan^2 x)}{\tan^2 x (\tan x - 2)}$$

Now we use the Pythagorean identity $1 + \tan^2 x \equiv \sec^2 x$ in the numerator, and obtain for that numerator,

$$2 \sec^2 x, \quad \text{or} \quad 2\frac{1}{\cos^2 x}.$$

Thus we have

$$\frac{2}{\cos^2 x \dfrac{\sin^2 x}{\cos^2 x} (\tan x - 2)},$$

and finally

$$\frac{2}{\sin^2 x (\tan x - 2)}.$$

DO EXERCISES 4–6.

[iii] Radical Expressions

When radicals occur, the use of absolute value is sometimes necessary, but its consideration can be complex. In the examples and exercises that follow, we shall assume that all quantities are nonnegative.

Example 7 Multiply and simplify $\sqrt{\sin^3 x \cos x} \cdot \sqrt{\cos x}$.

$$\sqrt{\sin^3 x \cos x} \cdot \sqrt{\cos x} = \sqrt{\sin^3 x \cos^2 x}$$
$$= \sin x \cos x \sqrt{\sin x}$$

Example 8 Rationalize the denominator: $\sqrt{\dfrac{2}{\tan x}}$.

$$\sqrt{\frac{2}{\tan x}} = \sqrt{\frac{2}{\tan x} \cdot \frac{\tan x}{\tan x}} = \sqrt{\frac{2 \tan x}{\tan^2 x}} = \frac{\sqrt{2 \tan x}}{\tan x}$$

DO EXERCISES 7 AND 8.

[iv] Equations

Example 9 Solve for tan x: $\tan^2 x + \tan x = 56$.

$$\tan^2 x + \tan x - 56 = 0 \qquad \text{Reducible to quadratic; let } u = \tan x$$
$$(\tan x + 8)(\tan x - 7) = 0 \qquad \text{Factoring}$$
$$\tan x + 8 = 0 \quad \text{or} \quad \tan x - 7 = 0$$
$$\tan x = -8 \quad \text{or} \qquad \tan x = 7$$

Example 10 Solve for sec x: $\sec^2 x - \frac{3}{4}\sec x = \frac{1}{2}$.

$$\sec^2 x - \frac{3}{4}\sec x - \frac{1}{2} = 0$$

We now use the quadratic formula, and obtain

$$\sec x = \frac{\frac{3}{4} \pm \sqrt{\frac{9}{16} - 4 \cdot 1 \cdot \left(-\frac{1}{2}\right)}}{2} = \frac{\frac{3}{4} \pm \sqrt{\frac{41}{16}}}{2}$$

$$\sec x = \frac{\frac{3}{4} \pm \frac{\sqrt{41}}{4}}{2} = \frac{3 \pm \sqrt{41}}{8}.$$

Example 11 Solve for sin x:

$$8\sin^2 x + 2\cos\left(x + \frac{\pi}{2}\right) - 3 = 0.$$

We first use the identity $\cos(x + \pi/2) \equiv -\sin x$, to get everything in terms of sin x:

$$8\sin^2 x - 2\sin x - 3 = 0.$$

Now we factor:

$$(4\sin x - 3)(2\sin x + 1) = 0$$
$$4\sin x - 3 = 0 \quad \text{or} \quad 2\sin x + 1 = 0$$
$$\sin x = \frac{3}{4} \quad \text{or} \quad \sin x = -\frac{1}{2}.$$

Note that for the manipulations in Examples 9–11 no knowledge of trigonometry is necessary. A symbol such as sin x or tan x plays the same role as a single letter such as *u* or *v*. Later, when trigonometric equations are considered, we shall solve for x rather than sin x or tan x.

DO EXERCISES 9–11.

9. Solve for cot x.

$$\cot^2 x + \cot x = 12$$

10. Solve for cos x.

$$\cos^2 x - \frac{1}{4}\cos x = \frac{3}{2}$$

11. Solve for cos x.

$$6\cos^2 x + \sin\left(\frac{\pi}{2} - x\right) - 2 = 0$$

EXERCISE SET 2.7A

[i] In Exercises 1–20, multiply and simplify.

1. $(\sin x - \cos x)(\sin x + \cos x)$

2. $(\tan y - \cot y)(\tan y + \cot y)$

3. $\tan x(\cos x - \csc x)$

4. $\cot x(\sin x + \sec x)$

5. $\cos y \sin y(\sec y + \csc y)$

6. $\tan y \sin y(\cot y - \csc y)$

7. $(\sin x + \cos x)(\csc x - \sec x)$

8. $(\sin x + \cos x)(\sec x + \csc x)$

9. $(\sin y - \cos y)^2$

10. $(\sin y + \cos y)^2$

11. $(1 + \tan x)^2$

12. $(1 + \cot x)^2$

13. $(\sin y - \csc y)^2$

14. $(\cos y + \sec y)^2$

15. $(\cos x - \sec x)(\cos^2 x + \sec^2 x + 1)$

16. $(\sin x + \csc x)(\sin^2 x + \csc^2 x - 1)$

17. $(\cot x - \tan x)(\cot^2 x + 1 + \tan^2 x)$

18. $(\cot y + \tan y)(\cot^2 y - 1 + \tan^2 y)$

19. $\sin\left(\frac{\pi}{2} - x\right)[\sec x - \cos x]$

20. $\cos\left(\frac{\pi}{2} - x\right)[\csc x - \sin x]$

In Exercises 21–40, factor and simplify.

21. $\sin x \cos x + \cos^2 x$

22. $\sec x \csc x - \csc^2 x$

23. $\sin^2 y - \cos^2 y$

24. $\tan^2 y - \cot^2 y$

25. $\tan x + \sin(\pi - x)$

26. $\cot x - \cos(\pi - x)$

27. $\sin^4 x - \cos^4 x$

28. $\tan^4 x - \sec^4 x$

29. $3\cot^2 y + 6\cot y + 3$

30. $4\sin^2 y + 8\sin y + 4$

31. $\csc^4 x + 4\csc^2 x - 5$

32. $-8 + \tan^4 x - 2\tan^2 x$

33. $\sin^3 y + 27$

34. $1 - 125\tan^3 y$

35. $\sin^3 y - \csc^3 y$

36. $\cos^3 x - \sec^3 x$

37. $\sin x \cos y - \cos\left(x + \frac{\pi}{2}\right)\tan y$

38. $\cos\left(x - \frac{\pi}{2}\right)\tan y + \sin x \cot y$

39. $\cos(\pi - x) + \cot x \sin\left(x - \frac{\pi}{2}\right)$

40. $\sin(\pi - x) - \tan x \cos\left(\frac{\pi}{2} - x\right)$

EXERCISE SET 2.7B

[ii] Simplify.

1. $\dfrac{\sin^2 x \cos x}{\cos^2 x \sin x}$

2. $\dfrac{\cos^2 x \sin x}{\sin^2 x \cos x}$

3. $\dfrac{4\sin x \cos^3 x}{18\sin^2 x \cos x}$

4. $\dfrac{30\sin^3 x \cos x}{6\cos^2 x \sin x}$

5. $\dfrac{\cos^2 x - 2\cos x + 1}{\cos x - 1}$

6. $\dfrac{\sin^2 x + 2\sin x + 1}{\sin x + 1}$

7. $\dfrac{\cos^2 x - 1}{\cos x - 1}$

8. $\dfrac{\sin^2 x - 1}{\sin x + 1}$

9. $\dfrac{4\tan x \sec x + 2\sec x}{6\sin x \sec x + 2\sec x}$

10. $\dfrac{6\tan x \sin x - 3\sin x}{9\sin^2 x + 3\sin x}$

11. $\dfrac{\cos^2 x - 1}{\sin\left(\frac{\pi}{2} - x\right) - 1}$

12. $\dfrac{\sin^2 x - 1}{\cos\left(\frac{\pi}{2} - x\right) + 1}$

13. $\dfrac{\sin x - \cos\left(x - \frac{\pi}{2}\right)\cos x}{-\sin x - \cos\left(x - \frac{\pi}{2}\right)\tan x}$

14. $\dfrac{\cos x - \sin\left(\frac{\pi}{2} - x\right)\sin x}{\cos x - \cos(\pi - x)\tan x}$

15. $\dfrac{\sin^4 x - \cos^4 x}{\sin^2 x - \cos^2 x}$

16. $\dfrac{\sec^4 x - \tan^4 x}{\sec^2 x + \tan^2 x}$

17. $\dfrac{\cos^2 x + 2\sin\left(x - \frac{\pi}{2}\right) + 1}{\sin\left(\frac{\pi}{2} - x\right) - 1}$

18. $\dfrac{\sin^2 x - 2\cos\left(x - \frac{\pi}{2}\right) + 1}{\cos\left(\frac{\pi}{2} - x\right) - 1}$

19. $\dfrac{\sin^2 y \cos\left(y + \frac{\pi}{2}\right)}{\cos^2 y \cos\left(\frac{\pi}{2} - y\right)}$

20. $\dfrac{\cos^2 y \sin\left(y + \frac{\pi}{2}\right)}{\sin^2 y \sin\left(\frac{\pi}{2} - y\right)}$

21. $\dfrac{2\sin^2 x}{\cos^3 x}\cdot\left(\dfrac{\cos x}{2\sin x}\right)^2$

22. $\dfrac{4\cos^3 x}{\sin^2 x}\cdot\left(\dfrac{\sin x}{4\cos x}\right)^2$

23. $\dfrac{3\sin x}{\cos^2 x} \cdot \dfrac{\cos^2 x + \cos x \sin x}{\cos^2 x - \sin^2 x}$

24. $\dfrac{5\cos x}{\sin^2 x} \cdot \dfrac{\sin^2 x - \sin x \cos x}{\sin^2 x - \cos^2 x}$

25. $\dfrac{\tan^2 y}{\sec y} \div \dfrac{3\tan^3 y}{\sec y}$

26. $\dfrac{\cot^3 y}{\csc y} \div \dfrac{4\cot^2 y}{\csc y}$

27. $\dfrac{1}{\sin^2 y - \cos^2 y} - \dfrac{2}{\cos y + \sin y}$

28. $\dfrac{3}{\cos y - \sin y} - \dfrac{2}{\sin^2 y - \cos^2 y}$

29. $\left(\dfrac{\sin x}{\cos x}\right)^2 - \dfrac{1}{\cos^2 x}$

30. $\left(\dfrac{\cot x}{\csc x}\right)^2 + \dfrac{1}{\csc^2 x}$

31. $\dfrac{\sin^2 x - 9}{2\cos x + 1} \cdot \dfrac{10\cos x + 5}{3\sin x + 9}$

32. $\dfrac{9\cos^2 x - 25}{2\cos x - 2} \cdot \dfrac{\cos^2 x - 1}{6\cos x - 10}$

EXERCISE SET 2.7C

[iii] In Exercises 1–10, simplify.

1. $\sqrt{\sin^2 x \cos x} \cdot \sqrt{\cos x}$

2. $\sqrt{\cos^2 x \sin x} \cdot \sqrt{\sin x}$

3. $\sqrt{\sin^3 y} + \sqrt{\sin y \cos^2 y}$

4. $\sqrt{\cos y \sin^2 y} - \sqrt{\cos^3 y}$

5. $\sqrt{\sin^2 x + 2\cos x \sin x + \cos^2 x}$

6. $\sqrt{\tan^2 x - 2\tan x \sin x + \sin^2 x}$

7. $(1 - \sqrt{\sin y})(\sqrt{\sin y} + 1)$

8. $(2 - \sqrt{\tan y})(\sqrt{\tan y} + 2)$

9. $\sqrt{\sin x}\,(\sqrt{2\sin x} + \sqrt{\sin x \cos x})$

10. $\sqrt{\cos x}\,(\sqrt{3\cos x} - \sqrt{\sin x \cos x})$

In Exercises 11–18, rationalize the denominator.

11. $\sqrt{\dfrac{\sin x}{\cos x}}$

12. $\sqrt{\dfrac{\cos x}{\sin x}}$

13. $\sqrt{\dfrac{\sin x}{\cot x}}$

14. $\sqrt{\dfrac{\cos x}{\tan x}}$

15. $\sqrt{\dfrac{\cos^2 x}{2\sin^2 x}}$

16. $\sqrt{\dfrac{\sin^2 x}{3\cos^2 x}}$

17. $\sqrt{\dfrac{1 + \sin x}{1 - \sin x}}$

18. $\sqrt{\dfrac{1 - \cos x}{1 + \cos x}}$

[iv]

19. Solve for $\tan x$.
$$\tan^2 x + 4\tan x = 21$$

20. Solve for $\sec x$.
$$\sec^2 x - 7\sec x = -10$$

21. Solve for $\sin x$.
$$8\sin^2 x - 2\sin x = 3$$

22. Solve for $\cos x$.
$$6\cos^2 x + 17\cos x = -5$$

23. Solve for $\cot x$.
$$\cot^2 x + 9\cot x - 10 = 0$$

24. Solve for $\csc x$.
$$\csc^2 x + 3\csc x - 10 = 0$$

25. Solve for $\sin x$.
$$\sin^2 x - \cos\left(x + \dfrac{\pi}{2}\right) = 6$$

26. Solve for $\sin x$.
$$2\cos^2\left(x - \dfrac{\pi}{2}\right) - 3\cos\left(x + \dfrac{\pi}{2}\right) - 2 = 0$$

27. Solve for $\tan x$.
$$\tan^2 x - 6\tan x = 4$$

28. Solve for $\cot x$.
$$\cot^2 x + 8\cot x = 5$$

29. Solve for $\sec x$.
$$6\sec^2 x - 5\sec x - 2 = 0$$

30. Solve for $\csc x$.
$$2\csc^2 x - 3\csc x - 4 = 0$$

CHAPTER 2 REVIEW

The point $(3, -2)$ is on a circle centered at the origin. Find the coordinates of its reflections across:

[2.2, i]   **1.** The y-axis.   **2.** The origin.   **3.** The x-axis.

On a unit circle, mark the points determined by:

[2.2, iii]   **4.** $\dfrac{7\pi}{6}$.   **5.** $\dfrac{3\pi}{4}$.   **6.** $\dfrac{\pi}{6}$.   **7.** $\dfrac{9\pi}{4}$.

[2.3, i]   **8.** Sketch a graph of $y = \sin x$.

[2.3, ii]   **9.** What is the domain of the sine function?

[2.3, iii]   **10.** Sketch a graph of $y = \cos x$.

[2.3, iv]   **11.** What is the period of the cosine function?

[2.3, i]   **12.** Sketch a graph of $y = \sin\left(x + \dfrac{\pi}{2}\right)$. Use the axes of Exercise 8.

[2.3, iii]   **13.** Complete the following table.

	$\pi/6$	$\pi/4$	$\pi/3$	$\pi/2$	$3\pi/4$	$5\pi/4$
sin x						
cos x						

[2.4, i]   **14.** Sketch a graph of $y = \csc x$.

              15. What is the period of the cosecant function?

              16. What is the range of the cosecant function?

              17. In which quadrants are the signs of the sine and the tangent the same?

[2.4, iii]   **18.** Verify the following identity: $\cot(x - \pi) \equiv \cot x$.

Complete these Pythagorean identities.

[2.5, i]   **19.** $\sin^2 x + \cos^2 x \equiv$ _____

              20. $1 + \cot^2 x \equiv$ _____

[2.5, ii]   Complete these cofunction identities.

              21. $\cos\left(x + \dfrac{\pi}{2}\right) \equiv$ _____

              22. $\cos\left(\dfrac{\pi}{2} - x\right) \equiv$ _____

              23. $\sin\left(x - \dfrac{\pi}{2}\right) \equiv$ _____

[2.5, i]   **24.** Express $\tan x$ in terms of $\sec x$.

[2.6, i]   **25.** Sketch a graph of $y = 3 + \cos\left(x - \dfrac{\pi}{4}\right)$.

[2.6, ii]   **26.** What is the phase shift of the function in Exercise 25?

              27. What is the period of the function in Exercise 25?

[2.6, iii]   **28.** Sketch a graph of $y = 3\cos x + \sin x$ for values of x between 0 and 2π.

Simplify.

[2.7, i]   **29.** $\cos x\,(\tan x + \cot x)$

[2.7, ii]   **30.** $\dfrac{\csc x \, (\sin^2 x + \cos^2 x \tan x)}{\sin x + \cos x}$

[2.7, iii]   **31.** Rationalize the denominator.

$$\sqrt{\dfrac{\tan x}{\sec x}}$$

[2.7, iv]   **32.** Solve for $\tan x$: $3 \tan^2 x - 2 \tan x - 2 = 0$.

              33. Does $5 \sin x = 7$ have a solution for x? Why or why not?

              34. For what values of x in $\left(0, \dfrac{\pi}{2}\right]$ is the following true? $\sin x < x$

TRIGONOMETRIC FUNCTIONS, ROTATIONS, AND ANGLES

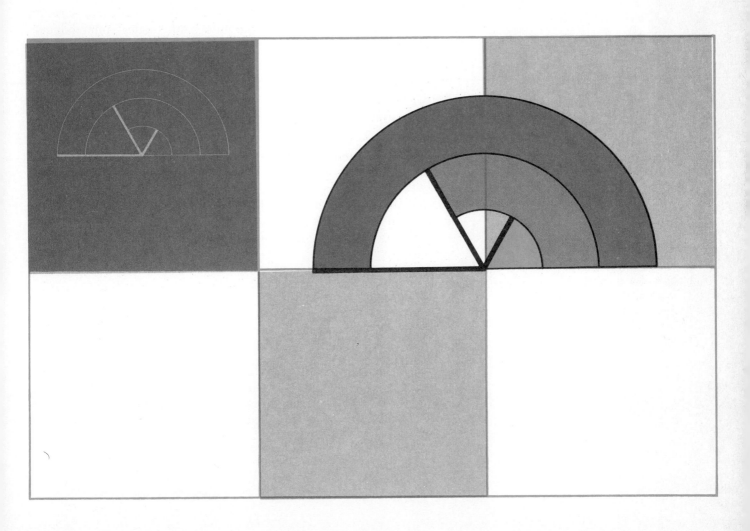

OBJECTIVES

You should be able to:

[i] Given the measure of an angle or rotation in degrees or radians, tell in which quadrant the terminal side lies.

[ii] Convert from degree measure to radian measure, and convert from radian measure to degree measure.

[iii] Find the length of an arc of a circle, given the measure of its central angle and the length of a radius; also find the measure of a central angle of a circle, given the length of its arc and the length of a radius.

1. In which quadrant does the terminal side of each angle lie?

   a) 47°
   b) 212°
   c) −43°
   d) −135°
   e) 365°
   f) −365°
   g) 740°

2. How many degrees are there in

   a) one revolution?
   b) half of a revolution?
   c) one-fourth of a revolution?
   d) one-eighth of a revolution?
   e) one-sixth of a revolution?
   f) one-twelfth of a revolution?

3.1 ROTATIONS AND ANGLES

We shall consider rotations abstractly. That is, instead of considering a specific rotating object, such as a wheel or part of a machine, we will consider a rotating ray, with its endpoint at the origin of an xy-plane. The ray starts in position along the positive half of the x-axis. A counterclockwise rotation will be called positive. Clockwise rotations will be called negative.

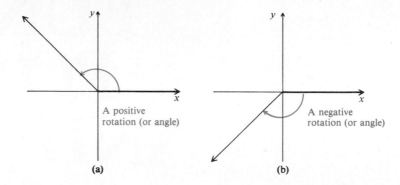

(a)   (b)

Note that the rotating ray and the positive half of the x-axis form an angle. Thus we often speak of "rotations" and "angles" interchangeably. The rotating ray is often called the *terminal side* of the angle, and the positive half of the x-axis is called the *initial side*.

[i] Measures of Rotations or Angles

The measure of an angle, or rotation, may be given in degrees. For example, a complete revolution has a measure of 360°, half a revolution has a measure of 180°, and so on.* We also speak of an *angle* of 90°, or 720°, or −240°, and so on. An angle between 0° and 90° has its terminal side in the first quadrant. An angle between 90° and 180° has its terminal side in the second quadrant. An angle between 0° and −90° has its terminal side in the fourth quadrant, and so on.

DO EXERCISES 1 AND 2.

[ii] Radian Measure

Consider a circle with its center at the origin and radius of length 1 (a *unit* circle). The distance around this circle from the initial side of an angle to the terminal side can be used as a measure of the rotation. This kind of angle measure is very useful. The unit is called a *radian*. One radian is about 57°.

---

*Subunits are *minutes* (60 minutes = 1°) and *seconds* (60 seconds = 1 minute).

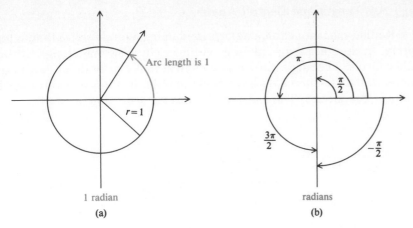

1 radian
(a)

radians
(b)

Since the circumference of a circle is $2\pi r$, and in this case $r = 1$, a rotation of 360° (1 revolution) has a measure of 2π radians. Half of a revolution is a rotation of 180° or π radians. A quarter revolution is a rotation of 90°, or $\pi/2$ radians, and so on. To convert between degrees and radians we can use the notion of "multiplying by one." Note the following:

$$\frac{1 \text{ revolution}}{1 \text{ revolution}} = 1 = \frac{2\pi \text{ radians}}{360 \text{ degrees}} = \frac{\pi \text{ radians}}{180 \text{ degrees}};$$

also

$$\frac{180 \text{ degrees}}{\pi \text{ radians}} = 1.$$

When a rotation is given in radians, the word "radians" is optional and is most often omitted. Thus if no unit is given for a rotation, it is understood to be in radians.

Example 1 Convert 60° to radians.

$$60° = 60° \cdot \frac{\pi \text{ radians}}{180°}$$

$$= \frac{60°}{180°} \pi \text{ radians}$$

$$= \frac{\pi}{3} \text{ radians, or } \frac{\pi}{3}$$

Using 3.14 for π, we find $\pi/3$ radians is about 1.047 radians.

Example 2 Convert $3\pi/4$ radians to degrees.

$$\frac{3\pi}{4} \text{ radians} = \frac{3\pi}{4} \text{ radians} \cdot \frac{180°}{\pi \text{ radians}}$$

$$= \frac{3\pi}{4\pi} \cdot 180° = 135°$$

DO EXERCISES 3–6.

3. Convert to radian measure. Leave answers in terms of π.

a) 225°
b) 315°
c) −720°

4. Convert to radian measure. Do not leave answers in terms of π (use 3.14 for π).

a) $72\frac{1}{2}°$ (a safe working angle for a certain ladder)
b) 300°
c) −315°

72½°

5. Convert to degree measure.

a) $4\pi/3$
b) $5\pi/2$
c) $-4\pi/5$

6. In which quadrant does the terminal side of each angle lie?

a) $5\pi/4$
b) $17\pi/8$
c) $-\pi/15$
d) 37.3π

7. Find the length of an arc of a circle with 10-cm radius, associated with a central angle of measure $11\pi/6$. (Use 3.14 for π.)

8. Find the radian measure of a rotation where a point 2.5 cm from the center of rotation travels 15 cm.

[iii] Arc Length and Central Angles

Radian measure can be determined using a circle other than a unit circle. In the following drawing a unit circle is shown along with another circle. The angle shown is a central angle of both circles; hence the arcs that it intercepts have their lengths in the same ratio as the radii of the circles. The radii of the circles are r and 1.

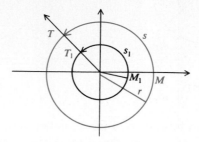

The corresponding arc lengths are MT and M_1T_1, or more simply, s and s_1. We therefore have the proportion

$$\frac{s}{r} = \frac{s_1}{1}.$$

Now s_1 is the radian measure of the rotation in question. It is more common to use a Greek letter, such as θ, for the measure of an angle or rotation. We commonly use the letter s for arc length. Adopting this convention, the above proportion becomes $\theta = s/r$. In any circle, arc length, central angle, and length of radius are related in this fashion. Or, in general, the following is true.

The radian measure θ of a rotation is the ratio of the distance s traveled by a point at a radius r from the center of rotation, to the length of the radius r:

$$\theta = \frac{s}{r}.$$

Example 3 Find the length of an arc of a circle of 5-cm radius associated with a central angle of $\pi/3$ radians.

$$\theta = \frac{s}{r}, \quad \text{or} \quad s = r\theta$$

Therefore, $s = 5 \cdot \dfrac{\pi}{3}$ cm or, using 3.14 for π, about 5.23 cm.

Example 4 Find the measure of a rotation in radians where a point 2 m from the center of rotation travels 4 m.

$$\theta = \frac{s}{r} = \frac{4 \text{ m}}{2 \text{ m}} = 2 \qquad \text{The unit is understood to be radians.}$$

A look at Examples 3 and 4 will show why the word "radian" is most often omitted. In Example 4, we have the division 4 m/2 m, which simplifies to the number 2, since m/m = 1. From this point of view it would seem preferable to omit the word "radians." In Example 3, had we used the word "radians" all the way through, our answer would have come out to be 5.23 cm-radians. It is a distance we seek; hence we know the unit should be centimeters. Thus we must omit the word "radians." A measure in radians is simply a number, so it is usually preferable to omit the word "radians."

> **CAUTION!** **In using the formula $\theta = s/r$ you must make sure that θ is in radians and that s and r are expressed in the same unit.**

DO EXERCISES 7–9. (NOTE THAT EXERCISES 7 AND 8 ARE ON THE PRECEDING PAGE.)

9. Find the radian measure of a rotation where a point 24 in. from the center of rotation travels 3 ft.

EXERCISE SET 3.1

[i] For angles of the following measures, state in which quadrant the terminal side lies.

1. 34°   **2.** 320°   **3.** −120°   **4.** 175°

5. $\dfrac{\pi}{3}$ **6.** $-\dfrac{3}{4}\pi$ **7.** $\dfrac{11}{4}\pi$ **8.** $\dfrac{19}{4}\pi$

[ii] Convert to radian measure. Leave answers in terms of π.

9. 30°   **10.** 15°   **11.** 60°

12. 200°   **13.** 75°   **14.** 300°

15. 37.71°   **16.** 12.73°   **17.** 214.6°   **18.** 73.87°

Convert to radian measure. Do not leave answers in terms of π. Use 3.14 for π.

19. 120°   **20.** 240°   **21.** 320°

22. 75°   **23.** 200°   **24.** 300°

25. ▦ 117.8°   **26.** ▦ 231.2°   **27.** ▦ 1.354°   **28.** ▦ 327.9°

Convert these radian measures to degree measure.

29. 1 radian   **30.** 2 radians   **31.** 8π

32. -12π **33.** $\dfrac{3}{4}\pi$ **34.** $\dfrac{5}{4}\pi$

35. ▦ 1.303   **36.** ▦ 2.347   **37.** ▦ 0.7532π **38.** ▦ -1.205π

[iii]

39. In a circle with 120-cm radius, an arc 132 cm long subtends a central angle of how many radians? how many degrees, to the nearest degree?

40. In a circle with 200-cm radius, an arc 65 cm long subtends a central angle of how many radians? how many degrees, to the nearest degree?

41. Through how many radians does the minute hand of a clock rotate in 50 min?

42. A wheel on a car has a 14-in. radius. Through what angle (in radians) does the wheel turn while the car travels 1 mi?

43. In a circle with 10-m radius, how long is an arc associated with a central angle of 1.6 radians?

44. In a circle with 5-m radius, how long is an arc associated with a central angle of 2.1 radians?

[ii]

45. Certain positive angles are marked here in degrees. Find the corresponding radian measures.

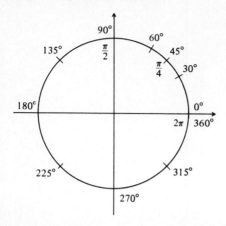

46. Certain negative angles are marked here in degrees. Find the corresponding radian measures.

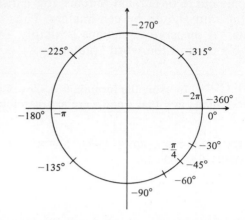

☆ ───────────────────────────────────

47. The *grad* is a unit of angle measure similar to a degree. A right angle has a measure of 100 grads. Convert the following to grads.

a) 48° b) 153° c) $\frac{\pi}{8}$ radians d) $\frac{5\pi}{7}$ radians

49. On the earth, one degree of latitude is how many kilometers? how many miles? (Assume that the radius of the earth is 6400 km, or 4000 miles, approximately.)

51. An astronaut on the moon observes the earth, about 240,000 miles away. The diameter of the earth is about 8000 miles. Find the angle α.

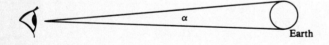

48. The *mil* is a unit of angle measure. A right angle has a measure of 1600 mils. Convert the following to degrees, minutes, and seconds.

a) 100 mils b) 350 mils

50. One minute of latitude on the earth is equivalent to one *nautical mile*. Find the circumference and radius of the earth in nautical miles.

52. The circumference of the earth was computed by Eratosthenes (276–195 B.C.). He knew the distance from Aswan to Alexandria to be about 500 miles. From each town he observed the sun at noon, finding the angular difference to be 7°12′ (7 degrees, 12 minutes). Do Eratosthenes' calculation.

★ ───────────────────────────────────

53. What is the angle between the hands of a clock at 7:45?

54. At what time between noon and 1:00 P.M. are the hands of a clock perpendicular?

─────────────────────────────────────

OBJECTIVES

You should be able to:

[i] Convert between linear and angular speed.
[ii] Find total distance or total angle, when speed, radius, and time are given.

3.2 ANGULAR SPEED

Speed is defined to be distance traveled per unit of time. Similarly, *angular speed* is defined to be amount of rotation per unit of time. For example, we might speak of the angular speed of a wheel as 150 revolutions per minute or the angular speed of the earth as 2π radians per day. The Greek letter ω (omega) is usually used for angular speed. Thus angular speed is defined as

$$\omega = \frac{\theta}{t}.$$

[i] Relating Linear and Angular Speed

For many applications it is important to know a relationship between angular speed and linear speed. For example, we might wish to find the linear speed of a point on the earth, knowing its angular speed. Or, we might wish to know the linear speed of an earth satellite, knowing its angular speed. To develop the relationship we seek, we recall the relation between angle and distance from the preceding section, $\theta = s/r$. This is equivalent to

$$s = r\theta.$$

We divide by the time, t, to obtain

$$\frac{s}{t} = r\frac{\theta}{t}.$$

Now s/t is linear speed v, and θ/t is angular speed ω. We thus have the relation we seek.

The linear speed v of a point a distance of r from the center of rotation is given by

$$v = r\omega,$$

where ω is the angular speed in radians per unit of time.

In deriving this formula we used the equation $s = r\theta$, in which the units for s and r must be the same and θ must be in radians. So, for our new formula $v = r\omega$, the units of distance for v and r must be the same, ω must be in radians per unit of time, and the units of time must be the same for v and ω.

Example 1 An earth satellite in a circular orbit 1200 km high makes one complete revolution every 90 min. What is its linear speed? Use 6400 km for the length of a radius of the earth.

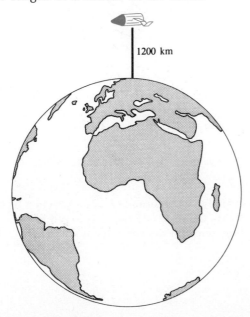

1200 km

1. A wheel with 12-cm diameter is rotating at 10 revolutions per second. What is the velocity of a point on the rim?

2. In using $v = r\omega$, if v is in cm/sec, what must be the units for r and ω?

3. In using $v = r\omega$, if ω is in radians/yr and r is in km, what must be the units for v?

4. The old oaken bucket is being raised at 3 ft/sec. The radius of the drum is 10 in. What is the angular speed of the handle?

We will use the formula $v = r\omega$; thus we shall need r and ω.

$$r = 6400 \text{ km} + 1200 \text{ km} \quad \text{Radius of earth plus height of satellite}$$

$$= 7600 \text{ km},$$

$$\omega = \frac{2\pi \text{ radians}}{90 \text{ min}}$$

$$= \frac{\pi}{45 \text{ min}} \quad \text{We have, as usual, omitted the word "radians."}$$

Now, using $v = r\omega$, we have

$$v = 7600 \text{ km} \cdot \frac{\pi}{45 \text{ min}} = \frac{7600\pi}{45} \cdot \frac{\text{km}}{\text{min}}.$$

Using 3.14 for π, we obtain $v = 530 \frac{\text{km}}{\text{min}}$.

DO EXERCISE 1.

Example 2 An anchor is being hoisted at 2 ft/sec, the chain being wound around a capstan with a 1.8-yd diameter. What is the angular speed of the capstan?

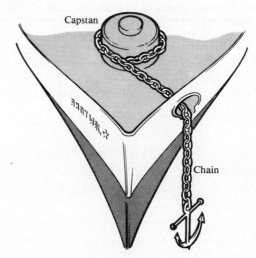

We will use the formula $\omega = v/r$, taking care to use the proper units. Since v is in ft/sec, we need r in ft. Then ω will be in radians/sec:

$$r = \frac{1.8}{2} \text{ yd} \cdot \frac{3 \text{ ft}}{\text{yd}} = 2.7 \text{ ft},$$

$$\omega = \frac{v}{r} = \frac{2}{2.7} = 1.11 \text{ radians/sec}.$$

CAUTION! **In applying the formula $v = r\omega$, we must be sure that the units for v and r are the same and that ω is in radians per unit of time. The units of time must be the same for v and ω.**

DO EXERCISES 2–4.

[ii] Total Distance and Total Angle

The formulas $\theta = s/r$ and $v = r\omega$ can be used in combination to find distances and angles in various situations involving rotational motion.

Example 3 A car is traveling at 45 mph. Its tires have a 20-in. radius. Find the angle through which a wheel turns in 5 sec.

Recall that $\omega = \theta/t$ or $\theta = \omega t$. Thus we can find θ if we know ω and t. To find ω we use $v = r\omega$. For convenience we will first convert 45 mph to ft/sec:

$$v = 45 \, \frac{\text{mi}}{\text{hr}} \cdot \frac{1 \, \text{hr}}{60 \, \text{min}} \cdot \frac{1 \, \text{min}}{60 \, \text{sec}} \cdot \frac{5280 \, \text{ft}}{1 \, \text{mi}}$$

$$= 66 \, \frac{\text{ft}}{\text{sec}}.$$

Now $r = 20$ in. We shall convert to ft, since v is in ft/sec:

$$r = 20 \, \text{in.} \cdot \frac{1 \, \text{ft}}{12 \, \text{in.}}$$

$$= \frac{20}{12} \, \text{ft} = \frac{5}{3} \, \text{ft}.$$

Using $v = r\omega$, we have

$$66 \, \frac{\text{ft}}{\text{sec}} = \frac{5}{3} \, \text{ft} \cdot \omega, \quad \text{so} \quad \omega = 39.6 \, \frac{\text{radians}}{\text{sec}}.$$

Then

$$\theta = \omega t = 39.6 \, \frac{\text{radians}}{\text{sec}} \cdot 5 \, \text{sec} = 198 \, \text{radians}.$$

DO EXERCISE 5.

5. The diameter of a wheel of a car is 30 in. When the car is traveling 60 mph (88 ft/sec), how many revolutions does the wheel make in 1 sec?

EXERCISE SET 3.2

[i]

1. A flywheel is rotating 7 radians/sec. It has a 15-cm diameter. What is the speed of a point on its rim, in cm/min?

2. A wheel is rotating at 3 radians/sec. The wheel has a 30-cm radius. What is the speed of a point on its rim, in m/min?

3. A $33\frac{1}{3}$-rpm record has a radius of 15 cm. What is the linear velocity of a point on the rim, in cm/sec?

4. A 45-rpm record has a radius of 8.7 cm. What is the linear velocity of a point on the rim, in cm/sec?

5. The earth has a 4000-mile radius and rotates one revolution every 24 hours. What is the linear speed of a point on the equator, in mph?

6. The earth is 93,000,000 miles from the sun and traverses its orbit, which is nearly circular, every 365.25 days. What is the linear velocity of the earth in its orbit, in mph?

7. A wheel has a 32-cm diameter. The speed of a point on its rim is 11 m/s. What is its angular speed?

8. A horse on a merry-go-round is 7 m from the center and travels at 10 km/h. What is its angular speed?

9. (*Determining the speed of a river*). A water wheel has a 10-ft radius. To get a good approximation to the speed of the river, you count the revolutions of the wheel and find that it makes 14 revolutions per minute. What is the speed of the river?

10. (*Determining the speed of a river*). A water wheel has a 10-ft radius. To get a good approximation to the speed of the river, you count the revolutions of the wheel and find that it makes 16 rpm. What is the speed of the river?

[ii]

11. The wheels of a bicycle have a 24-in. diameter. When the bike is being ridden so that the wheels make 12 rpm, how far will the bike travel in 1 min?

12. The wheels of a car have a 15-in. radius. When the car is being driven so that the wheels make 10 revolutions/sec, how far will the car travel in one minute?

13. A car is traveling 30 mph. Its wheels have a 14-in. radius. Find the angle through which a wheel rotates in 10 sec.

14. A car is traveling 40 mph. Its wheels have a 15-in. radius. Find the angle through which a wheel rotates in 12 sec.

☆ ————————————————————————————————————

15. Two pulleys, 50 cm and 30 cm in diameter, respectively, are connected by a belt. The larger pulley makes 12 revolutions per minute. Find the angular speed of the smaller pulley in radians/sec.

16. One gear wheel turns another, the teeth being on the rims. The wheels have 40-cm and 50-cm radii and the smaller wheel rotates at 20 rpm. Find the angular speed of the larger wheel in radians/sec.

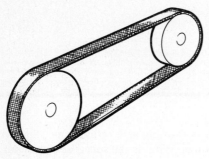

17. An airplane engine is idling at 800 rpm. When the throttle is opened, it takes 4.3 seconds for the speed to come up to 2500 rpm. What was the angular acceleration (a) in rpm per second? (b) in radians per second per second?

18. The linear speed of an airplane, flying at low level over the sea, is 175 knots (nautical miles per hour). It accelerates to 325 knots in 12 seconds.

a) What was its linear acceleration in knots per second?

b) What was its angular acceleration in radians per second per second? (*Hint:* One nautical mile is equivalent to one minute of latitude.)

3.3 TRIGONOMETRIC FUNCTIONS OF ANGLES OR ROTATIONS

[i] Function Values Defined

We now extend the domains of the functions defined on p. 60 to angles or rotations of any size. To do this, we consider a triangle with one vertex at the origin of a coordinate system and one side along the x-axis. The point R can be in any of the four quadrants, or on one of the axes. The sides of the triangle have lengths x, y, and r, as shown, and the angle θ is always measured from the positive half of the x-axis. The numbers x and y can be positive, negative, or zero, but the number r will always be considered positive.

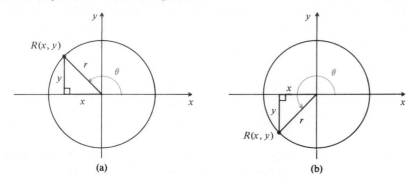

(a)   (b)

The values of the six trigonometric functions are as follows.

$$\sin \theta = \frac{y}{r} = \frac{\text{second coordinate}}{\text{radius}}$$

$$\cos \theta = \frac{x}{r} = \frac{\text{first coordinate}}{\text{radius}}$$

$$\tan \theta = \frac{y}{x} = \frac{\text{second coordinate}}{\text{first coordinate}}$$

$$\cot \theta = \frac{x}{y} = \frac{\text{first coordinate}}{\text{second coordinate}}$$

$$\sec \theta = \frac{r}{x} = \frac{\text{radius}}{\text{first coordinate}}$$

$$\csc \theta = \frac{r}{y} = \frac{\text{radius}}{\text{second coordinate}}$$

Example 1 Find the six trigonometric function values for the angle shown.

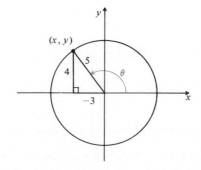

OBJECTIVES

You should be able to:

[i] Given a triangle in standard position with known sides, in any quadrant, determine the six function values.

[ii] Find the function values, without tables, for any angle whose terminal side makes an angle of 30°, 45°, or 60° with the x-axis.

[iii] Find the function values for any angle whose terminal side lies on a coordinate axis.

1. Find the six trigonometric function values for the angle shown here.

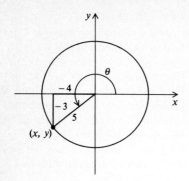

$$\sin \theta = \frac{y}{r} = \frac{4}{5} = 0.8, \qquad \cos \theta = \frac{x}{r} = \frac{-3}{5} = -0.6,$$

$$\tan \theta = \frac{y}{x} = \frac{4}{-3} \approx -1.33, \qquad \cot \theta = \frac{x}{y} = \frac{-3}{4} = -0.75,$$

$$\sec \theta = \frac{r}{x} = \frac{5}{-3} \approx -1.67, \qquad \csc \theta = \frac{r}{y} = \frac{5}{4} = 1.25.$$

DO EXERCISE 1.

[ii] Function Values for Some Special Angles

Our knowledge of triangles enables us to determine function values for certain angles. Let us first recall the Pythagorean theorem about right triangles. It says that in any right triangle, as shown, $a^2 + b^2 = c^2$, where c is the length of the hypotenuse.

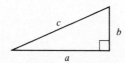

A 45° right triangle is half of a square, so its legs are the same length. Let us consider such a triangle whose legs have length 1. Then its hypotenuse has length c given by

$$1^2 + 1^2 = c^2, \quad \text{or} \quad c^2 = 2, \quad \text{or} \quad c = \sqrt{2}.$$

Such a triangle is shown below. From this diagram we can easily determine the function values for 45° or $\pi/4$:

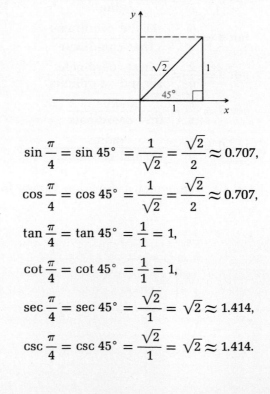

$$\sin \frac{\pi}{4} = \sin 45° = \frac{1}{\sqrt{2}} = \frac{\sqrt{2}}{2} \approx 0.707,$$

$$\cos \frac{\pi}{4} = \cos 45° = \frac{1}{\sqrt{2}} = \frac{\sqrt{2}}{2} \approx 0.707,$$

$$\tan \frac{\pi}{4} = \tan 45° = \frac{1}{1} = 1,$$

$$\cot \frac{\pi}{4} = \cot 45° = \frac{1}{1} = 1,$$

$$\sec \frac{\pi}{4} = \sec 45° = \frac{\sqrt{2}}{1} = \sqrt{2} \approx 1.414,$$

$$\csc \frac{\pi}{4} = \csc 45° = \frac{\sqrt{2}}{1} = \sqrt{2} \approx 1.414.$$

These function values should be memorized. The decimal values need not be, and it is sufficient to memorize the first three, since the others are their reciprocals.

$$\sin 45° = \frac{\sqrt{2}}{2}, \quad \cos 45° = \frac{\sqrt{2}}{2}, \quad \tan 45° = 1.$$

Next we consider an equilateral triangle, each of whose sides has length 2. If we take half of it as shown, we obtain a right triangle having a hypotenuse of length 2 and a leg of length 1. The other leg has length a, given by the Pythagorean theorem as follows:

$$a^2 + 1^2 = 2^2, \quad \text{or} \quad a^2 = 3, \quad \text{or} \quad a = \sqrt{3}.$$

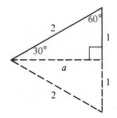

The acute angles of this triangle have measures of 30° and 60°. We can now determine function values for these angles. The function values for 30° or $\pi/6$ are as follows:

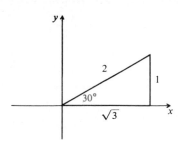

$$\sin \frac{\pi}{6} = \sin 30° = \frac{1}{2},$$

$$\cos \frac{\pi}{6} = \cos 30° = \frac{\sqrt{3}}{2},$$

$$\tan \frac{\pi}{6} = \tan 30° = \frac{1}{\sqrt{3}} = \frac{\sqrt{3}}{3},$$

$$\cot \frac{\pi}{6} = \cot 30° = \sqrt{3},$$

$$\sec \frac{\pi}{6} = \sec 30° = \frac{2}{\sqrt{3}} = \frac{2\sqrt{3}}{3},$$

$$\csc \frac{\pi}{6} = \csc 30° = 2.$$

We obtain the function values for 60° or $\pi/3$ by repositioning the triangle:

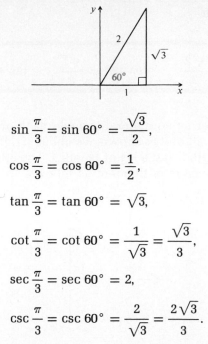

$$\sin \frac{\pi}{3} = \sin 60° = \frac{\sqrt{3}}{2},$$

$$\cos \frac{\pi}{3} = \cos 60° = \frac{1}{2},$$

$$\tan \frac{\pi}{3} = \tan 60° = \sqrt{3},$$

$$\cot \frac{\pi}{3} = \cot 60° = \frac{1}{\sqrt{3}} = \frac{\sqrt{3}}{3},$$

$$\sec \frac{\pi}{3} = \sec 60° = 2,$$

$$\csc \frac{\pi}{3} = \csc 60° = \frac{2}{\sqrt{3}} = \frac{2\sqrt{3}}{3}.$$

The function values for 30° and 60° should be memorized, but again it is sufficient to learn only those for the sine, cosine, and tangent, since the others are their reciprocals.

$$\sin 30° = \frac{1}{2}, \qquad \cos 30° = \frac{\sqrt{3}}{2}, \qquad \tan 30° = \frac{\sqrt{3}}{3},$$

$$\sin 60° = \frac{\sqrt{3}}{2}, \qquad \cos 60° = \frac{1}{2}, \qquad \tan 60° = \sqrt{3}.$$

Signs of the Functions

Function values of the generalized trigonometric functions can be positive, negative, or zero, depending on where the terminal side of the angle lies. In the first quadrant, all function values are positive. In the second quadrant, first coordinates are negative and second coordinates are positive.

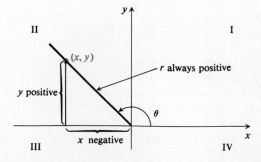

Thus if θ is in quadrant II,

$\sin \theta = \dfrac{y}{r}$ is positive, because y and r are positive;

$\cos \theta = \dfrac{x}{r}$ is negative, because x is negative and r is positive;

$\tan \theta = \dfrac{y}{x}$ is negative, because y is positive and x is negative;

and so on.

The following diagram summarizes, showing the quadrants in which the function values are positive. The diagram need not be memorized, because the signs can readily be determined as in the diagram above.

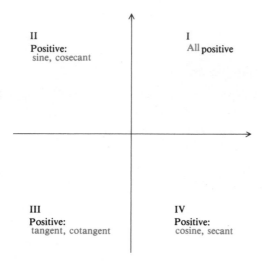

II
Positive:
  sine, cosecant

I
All positive

III
Positive:
  tangent, cotangent

IV
Positive:
  cosine, secant

Example 2 Find the trigonometric function values for 210°.

We draw a diagram showing the terminal side of a 210° angle. We now have a right triangle with a 30° angle as shown. We can read off function values.

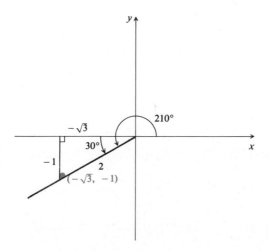

2. Find the trigonometric function values for 120°.

$$\sin 210° = \frac{-1}{2} = -\frac{1}{2}$$

$$\cos 210° = \frac{-\sqrt{3}}{2} = -\frac{\sqrt{3}}{2}$$

$$\tan 210° = \frac{-1}{-\sqrt{3}} = \frac{1}{\sqrt{3}} \text{ or } \frac{\sqrt{3}}{3}$$

$$\cot 210° = \frac{-\sqrt{3}}{-1} = \sqrt{3}$$

$$\sec 210° = \frac{2}{-\sqrt{3}} = -\frac{2}{\sqrt{3}} \text{ or } -\frac{2\sqrt{3}}{3}$$

$$\csc 210° = \frac{2}{-1} = -2$$

DO EXERCISES 2 AND 3.

[iii] Terminal Side on an Axis

Now let us suppose that the terminal side of an angle falls on one of the axes. In that case, one of the coordinates is zero. The definitions of the functions still apply, but in some cases functions will not be defined because a denominator will be 0.

3. Find the trigonometric function values for −135°.

Example 3 Find the trigonometric function values for 180°, or π radians.

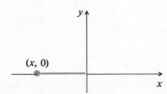

We note that the first coordinate is negative, that the second coordinate is 0, and that x and r have the same absolute value (r always being positive). Thus we have:

$$\sin 180° = \frac{0}{r} = 0;$$

$$\cos 180° = \frac{x}{r} = -1 \qquad \text{Since } |x| = |r|, \text{ but } x \text{ and } r \text{ have opposite signs}$$

$$\tan 180° = \frac{y}{x} = \frac{0}{x} = 0;$$

$$\cot 180° = \frac{x}{y} = \frac{x}{0}; \qquad \text{Thus } \cot 180° \text{ is undefined.}$$

$$\sec 180° = \frac{r}{x} = -1; \qquad \text{The reciprocal of } \cos 180°$$

$$\csc 180° = \frac{r}{0}. \qquad \text{Thus } \csc 180° \text{ is undefined.}$$

Example 4 The valve cap on a bicycle wheel is 24.5 in. from the center of the wheel. From the position shown, the wheel starts rolling. After the wheel has turned 390°, how far above the ground is the valve cap? Assume that the outer radius of the tire is 26 in.

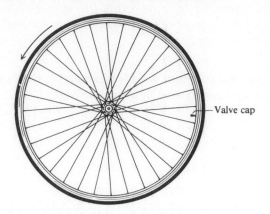

Valve cap

We draw the terminal side of an angle of 390°. Since sin 390° = $\frac{1}{2}$, the valve cap is $\frac{1}{2} \times 24.5$ in. above the center of the wheel. The distance above the ground is then 12.25 + 26 in. or 38.25 in.

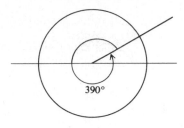

390°

DO EXERCISES 4 AND 5.

4. Find the trigonometric function values for 270°, or $3\pi/2$ radians.

5. The seats of a ferris wheel are 35 ft from the center of the wheel. When you board the wheel you are 5 ft above the ground. After you have rotated through an angle of 765°, how far above the ground are you?

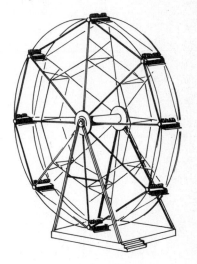

[ii], [iii] Function Values for Any Angle

We are now in a position to be able to determine the trigonometric function values for many other angles. If the terminal side of an angle falls on one of the axes, the function values are 0 or 1 or −1 or else are undefined. Thus we can determine function values for any multiple of 90°, or $\pi/2$. We can also determine the function values for any angle whose terminal side makes a 30°, 45°, or 60° angle with the x-axis. Consider, for example, an angle of 150°, or $5\pi/6$. The terminal side makes a 30° angle with the x-axis, since 180° − 150° = 30°. As the diagram shows, triangle *ONR* is congruent to triangle *ON'R'*; hence the

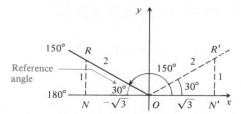

6. Given that

$$\sin 36° = 0.6283,$$
$$\cos 36° = 0.8090,$$
$$\tan 36° = 0.7265,$$

find the trigonometric function values for 324°.

ratios of the sides of the two triangles are the same except perhaps for sign. We could determine the function values directly from triangle ONR, but this is not necessary. If we remember that the sine is positive in quadrant II and that the cosine and tangent are negative, we can simply use the values for 30°, prefixing the appropriate sign. The triangle ONR is called a *reference* triangle and its acute angle at the origin is called a *reference* angle.

In general, to find the function values of an angle, we find them for the reference angle and prefix the appropriate sign.

Example 5 Find the sine, cosine, and tangent of 600°, or $10\pi/3$.

We find the multiple of 180° nearest 600°:

$$180° \times 2 = 360°,$$
$$180° \times 3 = 540°,$$

and

$$180° \times 4 = 720°.$$

The nearest multiple is 540°. The difference between 600° and 540° is 60°. This gives us the reference angle.

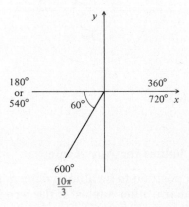

We recall that $\sin 60° = \sqrt{3}/2$, $\cos 60° = \frac{1}{2}$, and $\tan 60° = \sqrt{3}$. We also note that the sine and cosine are negative in the third quadrant, and that the tangent is positive. Hence we have

$$\sin \frac{10\pi}{3} = \sin 600° = -\frac{\sqrt{3}}{2},$$

$$\cos \frac{10\pi}{3} = \cos 600° = -\frac{1}{2},$$

$$\tan \frac{10\pi}{3} = \tan 600° = \sqrt{3}.$$

DO EXERCISE 6.

EXERCISE SET 3.3

Before beginning, make a list of the function values of 30°, 60°, and 45°. Then memorize your list. In Exercises 1–4, find the six trigonometric function values for the angle θ as shown.

[i]

1.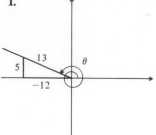

2.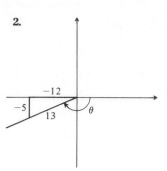

3.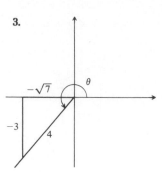

4.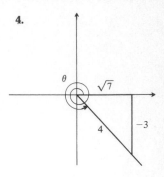

[ii] In Exercises 5 and 6, use the fact that $\sqrt{2} \approx 1.414$ and $\sqrt{3} \approx 1.732$.

5. Find decimal values for the six trigonometric functions of 30°, or $\frac{\pi}{6}$.

6. Find decimal values for the six trigonometric functions of 60°, or $\frac{\pi}{3}$.

[ii], [iii] Find the following if they exist.

7. $\cos 180°$

8. $\sin 360°$

9. $\tan \frac{\pi}{2}$

10. $\cot \pi$

11. $\sec 720°$

12. $\csc 720°$

13. $\sin \left(-\frac{3\pi}{4} \right)$

14. $\cos \frac{3\pi}{4}$

15. $\sin \frac{5\pi}{6}$

16. $\cos \frac{5\pi}{6}$

17. $\tan 240°$

18. $\cot 240°$

19. $\sec \frac{7\pi}{4}$

20. $\csc \frac{7\pi}{4}$

21. $\tan (-315°)$

22. $\cot (-315°)$

23. $\sin (-210°)$

24. $\cos (-210°)$

25. $\tan \frac{7\pi}{6}$

26. $\cot \frac{7\pi}{6}$

27. $\sin \frac{11\pi}{4}$

28. $\cos \frac{11\pi}{4}$

29. $\tan \frac{11\pi}{6}$

30. $\cot \frac{11\pi}{6}$

☆

31. Compile a list of function values for the sine function, from -2π to 2π. Then graph the function.

32. Is the sine function an even function? an odd function? periodic?

33. Compile a list of function values for the cosine function, from -2π to 2π. Then graph the function.

34. Is the cosine function an even function? an odd function? periodic?

★ ───

35. This diagram shows a piston of a steam engine, driving a drive wheel of a locomotive. The radius of the drive wheel (from the center of the wheel to the pin *P*) is *R*, and the length of the rod is *L*. Suppose that the drive wheel is rotating at a speed of ω radians per second (so that $\theta = \omega t$). Show that the distance of pin *Q* from 0, the center of the wheel, is a function of time, given by

$$x = \sqrt{L^2 - R^2 \sin^2 \omega t} + R \cos \omega t.$$

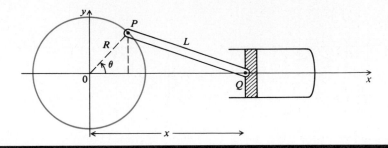

───

OBJECTIVES

You should be able to:

[i] Given a function value for an angle and the quadrant in which the terminal side lies, find the other five function values.

[ii] Given the function values for an acute angle, find the function values for its complement.

3.4 RELATIONSHIPS AMONG FUNCTION VALUES

[i] The Six Functions Related

When we know one of the trigonometric function values for an angle, we can find the other five if we know the quadrant in which the terminal side lies. The idea is to sketch a triangle in the appropriate quadrant, use the Pythagorean theorem to find the lengths of its sides as needed, and then read off the ratios of its sides.

Example 1 Given that $\tan \theta = -\frac{2}{3}$ and that θ is in the second quadrant, find the other function values.

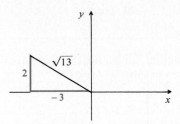

We first sketch a second quadrant triangle. Since $\tan \theta = -\frac{2}{3}$, we make the legs of lengths 2 and 3. The hypotenuse must then have length $\sqrt{13}$. Now we can read off the appropriate ratios:

$$\sin \theta = \frac{2}{\sqrt{13}}, \qquad \cos \theta = -\frac{3}{\sqrt{13}},$$

$$\tan \theta = -\frac{2}{3}, \qquad \cot \theta = -\frac{3}{2},$$

$$\sec \theta = -\frac{\sqrt{13}}{3}, \qquad \csc \theta = \frac{\sqrt{13}}{2}.$$

Example 2 Given that $\cot \theta = 2$ and θ is in the third quadrant, find $\sin \theta$, $\cos \theta$, and $\tan \theta$.

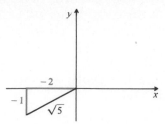

Again, we sketch a triangle and label the sides, then read off the appropriate ratios:

$$\sin \theta = -\frac{1}{\sqrt{5}},$$

$$\cos \theta = -\frac{2}{\sqrt{5}},$$

$$\tan \theta = \frac{1}{2}.$$

DO EXERCISES 1 AND 2.

[ii] Cofunctions and Complements

In a right triangle the acute angles are complementary, since the sum of all three angle measures is 180° and the right angle accounts for 90° of this total. Thus if one acute angle of a right triangle is θ, the other is $90° - \theta$, or $\pi/2 - \theta$. Note that the sine of $\angle A$ is also the cosine of $\angle B$, its complement:

$$\sin \theta = \frac{a}{c}, \qquad \text{cosine } (90° - \theta) = \frac{a}{c}.$$

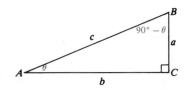

Similarly, the tangent of $\angle A$ is the cotangent of its complement and the secant of $\angle A$ is the cosecant of its complement.

These pairs of functions are called *cofunctions*. The name *cosine* originally meant the sine of the complement. The name *cotangent* meant the tangent of the complement and *cosecant* meant the secant of the complement. A complete list of the cofunction relations is as follows. Equations that hold for all sensible replacements for the variables are known as *identities*. The sign $\equiv$ is used instead of $=$ to indicate that an equation is an identity.

$$\sin \theta \equiv \cos (90° - \theta), \qquad \cos \theta \equiv \sin (90° - \theta),$$
$$\tan \theta \equiv \cot (90° - \theta), \qquad \cot \theta \equiv \tan (90° - \theta),$$
$$\sec \theta \equiv \csc (90° - \theta), \qquad \csc \theta \equiv \sec (90° - \theta).$$

1. Given $\cos \theta = \frac{3}{4}$ and that the terminal side is in quadrant IV, find the other function values.

2. Given $\cot \theta = -3$ and that the terminal side is in quadrant II, find the other function values.

3. Given that

$$\sin 75° = 0.9659, \quad \cos 75° = 0.2588,$$
$$\tan 75° = 3.732, \quad \cot 75° = 0.2679,$$
$$\sec 75° = 3.864, \quad \csc 75° = 1.035,$$

find the function values for the complement of 75°.

Example 3 Given that

$$\sin 18° = 0.3090, \quad \cos 18° = 0.9511,$$
$$\tan 18° = 0.3249, \quad \cot 18° = 3.078,$$
$$\sec 18° = 1.051, \quad \csc 18° = 3.236,$$

find the six function values for 72°.

Since 72° and 18° are complements, we have $\sin 72° = \cos 18°$, etc., and the function values are

$$\sin 72° = 0.9511, \quad \cos 72° = 0.3090,$$
$$\tan 72° = 3.078, \quad \cot 72° = 0.3249,$$
$$\sec 72° = 3.236, \quad \csc 72° = 1.051.$$

DO EXERCISE 3.

EXERCISE SET 3.4

[i] In Exercises 1–6, a function value and a quadrant are given. Find the other five function values.

1. $\sin \theta = -\frac{1}{3}$, III

2. $\sin \theta = -\frac{1}{5}$, IV

3. $\cos \theta = \frac{3}{5}$, IV

4. $\cos \theta = -\frac{4}{5}$, II

5. $\cot \theta = -2$, IV

6. $\tan \theta = 5$, III

[ii]

7. Given that
$$\sin 65° = 0.9063, \quad \cos 65° = 0.4226,$$
$$\tan 65° = 2.145, \quad \cot 65° = 0.4663,$$
$$\sec 65° = 2.366, \quad \csc 65° = 1.103,$$
find the six function values for 25°.

8. Given that
$$\sin 32° = 0.5299, \quad \cos 32° = 0.8480,$$
$$\tan 32° = 0.6249, \quad \cot 32° = 1.600,$$
$$\sec 32° = 1.179, \quad \csc 32° = 1.887,$$
find the six function values for 58°.

☆

9. Given that $\sin \theta = 1/3$, and that the terminal side is in quadrant II:

   a) find the other function values for θ;

   b) find the six function values for $\pi + \theta$;

   c) find the six function values for $\pi - \theta$;

   d) find the six function values for $2\pi - \theta$.

10. Given that $\cos \theta = 4/5$, and that the terminal side is in quadrant IV:

   a) find the other function values for θ;

   b) find the six function values for $\pi + \theta$;

   c) find the six function values for $\pi - \theta$;

   d) find the six function values for $2\pi - \theta$.

11. Given that $\sin 27° = 0.45399$:

   a) find the other function values for 27°;

   b) find the six function values for 63°.

12. Given that $\cot 54° = 0.72654$:

   a) find the other function values for 54°;

   b) find the six function values for 36°.

13. Given that $\cos 38° = 0.78801$, find the six function values for 128°.

14. Given that $\tan 73° = 3.27085$, find the six function values for 343°.

★

15. For any acute angle θ, we know that $\cos(90° - \theta) = \sin \theta$. Consider angles other than acute angles. Does this relation still hold, and if so to what extent?

16. Consider the equation $\sin(x + \pi/2) = \cos x$, where x is an angle in radians. Does this relation hold for acute angles? Does it hold for other than acute angles?

3.5 TABLES OF TRIGONOMETRIC FUNCTIONS

OBJECTIVES

You should be able to:

[i] Use Table 2 to find function values.

[ii] Convert between degrees and minutes and degrees and tenths.

[iii] Use Table 2 with interpolation.

[iv] Use Table 2 in reverse.

[v] Use Table 2 to find angles of any size, and in reverse.

[i] Finding Function Values

By use of certain formulas, theoretically determined, tables of values for the trigonometric functions have been constructed. Table 2 at the back of the book is such a table. A portion of it is shown below. This table gives the function values for angles from 0° to 90° only. Values for other angles can always be determined from these, since any angle has an acute reference angle. Note that angle measures are given in degrees and minutes (60 minutes = 1°) as well as radians. Values in the table are given for intervals of 10′ (10 minutes) and are correct to four digits.

Degrees	Radians	Sin	Cos	Tan	Cot	Sec	Csc		
36°00′	.6283	.5878	.8090	.7265	1.376	1.236	1.701	.9425	54°00′
10	312	901	073	310	368	239	695	396	50
20	341	925	056	355	360	241	688	367	40
30	.6370	.5948	.8039	.7400	1.351	1.244	1.681	.9338	30
40	400	972	021	445	343	247	675	308	20
50	429	995	004	490	335	249	668	279	10
37°00′	.6458	.6018	.7986	.7536	1.327	1.252	1.662	.9250	53°00′
10	487	041	969	581	319	255	655	221	50
20	516	065	951	627	311	258	649	192	40
30	.6545	.6088	.7934	.7673	1.303	1.260	1.643	.9163	30
40	574	111	916	720	295	263	636	134	20
50	603	134	898	766	288	266	630	105	10
44°00′	.7679	.6947	.7193	.9657	1.036	1.390	1.440	.8029	46°00′
10	709	967	173	713	030	394	435	999	50
20	738	988	153	770	024	398	431	970	40
30	.7767	.7009	.7133	.9827	1.018	1.402	1.427	.7941	30
40	796	030	112	884	012	406	423	912	20
50	825	050	092	942	006	410	418	883	10
45°00′	.7854	.7071	.7071	1.000	1.000	1.414	1.414	.7854	45°00′
		Cos	Sin	Cot	Tan	Csc	Sec	Radians	Degrees

The headings on the left of Table 2 range from 0° to 45° only. Thus the table may seem to be only half of what it should be. Function values from 45° to 90° are the cofunction values of their complements, however. Hence these can be found without further entries. For angles from 45° to 90° the headings on the right are used, together with the headings at the bottom. For example, sin 37° is found to be 0.6018 using the top and left headings. The cosine of 53° (the complement of 37°) is found also to be 0.6018, using the bottom and right headings.

Example 1 Find cos 37°20′.

We find 37°20′ in the left column and then *cos* at the top. At the intersection of this row and column we find the entry we seek:

$$\cos 37°20' = 0.7951.$$

Example 2 Find cot 0.9192.

Since no degree symbol is given, we know that the angle is in radians. We do not find 0.9192 on the left, under radians. This is because the angle is greater than 45° or $\pi/4$ radians. Therefore, we will use the right and bottom headings. At the right, under radians, we find 0.9192. Next

Use Table 2 to find the following.

1. sin 15°20′

2. cot 64°50′

3. cos 0.4451

4. tan 0.8319

Use Table 2 to answer the following.

5. How big must an angle be in order that it first differs from its sine in the fourth decimal place? the third decimal place?

we find *cot* at the bottom. At the intersection of this row and column we find the entry we seek:

$$\cot 0.9192 = 0.7627.$$

DO EXERCISES 1–4.

Function Values for Small Angles

Consider this diagram showing a unit circle and a small angle.

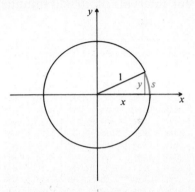

Since we are assuming *s* to be small, the length of arc *s* is very nearly the same as y. Also x is very nearly 1. Since y = sin *s*, we see that *s* ≈ sin *s*. Since tan *s* = y/x and x ≈ 1, we see that tan *s* ≈ *s*. These relations also hold if *s* is negative with small absolute value, and are useful in certain applications.

If |*s*| is a small number, *s* ≈ sin *s* ≈ tan *s* (a small angle—in radians—is approximately equal to its sine and its tangent).

DO EXERCISE 5.

[ii] Minutes Versus Tenths and Hundredths

It is usual to express the measure of an angle in degrees and minutes, but it is often useful to have the measure in degrees, tenths, hundredths, etc. Thus we will need to know how to convert between these.

Examples

3. Convert 16.35° to degrees and minutes.

$$16.35° = 16° + 0.35 \times 1°$$

Now

$$0.35 \times 1° = 0.35 \times 60' = 21'.$$

So

$$16.35° = 16°21'.$$

4. Convert $34°37'$ to degrees and decimal parts of degrees.

$$34°37' = 34° + \frac{37°}{60},$$

$$\frac{37}{60} = 0.617$$

so

$$34°37' = 34.617°.$$

DO EXERCISES 6 AND 7.

[iii] Interpolation

By using a procedure called *interpolation* we can find values between those listed in the table. Interpolation can be done in various ways, the simplest and most common being *linear* interpolation. We describe it now. What we say applies to a table for any continuous function.

Let us consider how a table of values for any function is made. We select members of the domain x_1, x_2, x_3, and so on. Then we compute or somehow determine the corresponding function values, $f(x_1)$, $f(x_2)$, $f(x_3)$, and so on. Then we tabulate the results. We might also graph the results.

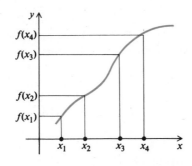

x	x_1	x_2	x_3	x_4	$\cdots$
$f(x)$	$f(x_1)$	$f(x_2)$	$f(x_3)$	$f(x_4)$	$\cdots$

Suppose we want to find the function value $f(x)$ for an x not in the table. If x is halfway between x_1 and x_2, then we can take the number halfway between $f(x_1)$ and $f(x_2)$ as an approximation to $f(x)$. If x is one-fifth of the way between x_2 and x_3, we take the number that is one-fifth of the way between $f(x_2)$ and $f(x_3)$ as an approximation to $f(x)$. What we do is divide the length from x_2 to x_3 in a certain ratio, and then divide the length from $f(x_2)$ to $f(x_3)$ in the same ratio. This is *linear interpolation*.

We can show this geometrically. The length from x_1 to x_2 is divided in a certain ratio by x. The length from $f(x_1)$ to $f(x_2)$ is divided in the

6. Convert $37.45°$ to degrees and minutes.

7. Convert $43°55'$ to degrees and decimal parts of degrees.

same ratio by y. The number y approximates $f(x)$ with the noted error.

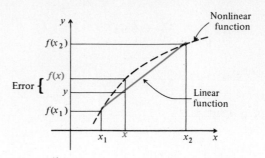

Note the slanted line in the figure. The approximation y comes from this line. This explains the use of the term *linear interpolation*.

Example 5 Find tan 27°43′.

$$
10′
\begin{cases}
3′\underline{}
\begin{cases}
\tan 27°40′ = 0.5243 \\
\tan 27°43′ = 0.52?? \\
\end{cases} \\
\tan 27°50′ = 0.5280
\end{cases}
$$

The tabular difference is 0.0037.

The tabular difference is 0.0037, and 43′ is $\frac{3}{10}$ of the way from 40′ to 50′, so we take $\frac{3}{10}$ of 0.0037. This is 0.0011. Thus

$$\tan 27°43′ = 0.5243 + 0.0011, \quad \text{or} \quad 0.5254.$$

Interpolation for decreasing functions is about the same as for increasing functions, but there is a slight difference as the next example shows.

Example 6 Find cot 29°44′.

$$
10′
\begin{cases}
4′\underline{}
\begin{cases}
\cot 29°40′ = 1.756 \\
\cot 29°44′ = 1.7?? \\
\end{cases} \\
\cot 29°50′ = 1.744
\end{cases}
$$

The tabular difference is 0.012.

We take 0.4 of the tabular difference, 0.012, and obtain 0.0048, or 0.005. Because the cotangent function is *decreasing* in the interval $(0°, 90°]$, we *subtract* 0.005 from 1.756. Thus cot 29°44′ = 1.751.

Once the process of interpolation is understood, it will not be necessary to write as much as we did in the above examples. After a bit of practice, you will find that interpolation is rather easy, and you can accomplish some of the steps without writing.

DO EXERCISES 8 AND 9.

[iv] Using the Table in Reverse

Let us look at an example of using the tables in reverse, that is, given a function value, to find the measure of an angle.

Example 7 Given $\tan B = 0.3727$, find B (between $0°$ and $90°$).

$$
\begin{array}{r}
10' \left[\begin{array}{l}
\tan 20°20' = 0.3706 \\
\tan B \quad\;\; = 0.3727 \\
\tan 20°30' = 0.3739
\end{array} \right.
\end{array}
\left.\begin{array}{l}
\Big]\, 0.0021 \\
\\
\end{array}\right\}
\begin{array}{l}
\text{The tabular difference} \\
\text{is } 0.0033.
\end{array}
$$

We find that B is $\frac{21}{33}$ or $\frac{7}{11}$ of the way from $20°20'$ to $20°30'$. The tabular difference is $10'$, so we take $\frac{7}{11}$ of $10'$ and obtain $6'$. Thus $B = 20°26'$, or 0.3566 radian.

DO EXERCISES 10 AND 11.

[v] Function Values for Angles of Any Size

Table 2 gives function values for angles from $0°$ to $90°$. For any other angle we can still use the table to find the function values. To do this, we first find the reference angle, which is the angle that the terminal side makes with the x-axis. We then look up the function values for the reference angle in the table and use the appropriate sign, depending on the quadrant in which the terminal side lies.

Example 8 Find $\sin 285°40'$.

We first find the reference angle. To do this we find the multiple of $180°$ that is nearest $285°40'$. The multiples of $180°$ are $180°$, $360°$, and so on. Now $285°40'$ is nearest $360°$. We subtract to find the reference angle.

$$
\begin{array}{r}
359°60' \\
-285°40' \\
\hline
74°20'
\end{array}
$$

Now we find that $\sin 74°20' = 0.9628$. The terminal side is in quadrant IV, where the sine is negative. Hence

$$\sin 285°41' = -0.9628.$$

Use Table 2 for the following.

8. Find $\sin 38°47'$.

9. Find $\cot 27°45'$.

10. Find the acute angle for which $\tan \theta = 2.394$, in both degrees and radians.

11. Find the acute angle for which $\cot \theta = 1.819$, in both degrees and radians.

Use Table 2 for the following.

12. Find cos 563°20′.

It is helpful to make a diagram, like the following, showing the reference angle.

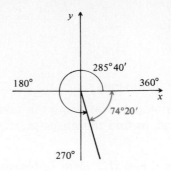

DO EXERCISES 12 AND 13.

Example 9 Given that tan $A = -0.3727$, find A between 270° and 360°.

We first find the reference angle, exactly as in Example 7, ignoring the fact that tan A is negative. The reference angle is thus 20°26′.

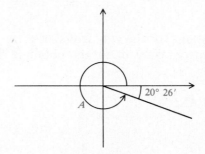

13. Find cos (−729°50′).

The angle A itself is found by subtracting 20°26′ from 360°.

$$\begin{array}{r} 359°60′ \\ -20°26′ \\ \hline 339°34′ \end{array} \quad \text{Subtracting}$$

14. Given that sin $A = -0.7304$, find A between 180° and 270°.

DO EXERCISE 14.

▦ The Use of Calculators

Scientific calculators contain tables of trigonometric functions—usually the sine, cosine, and tangent only. You can enter an angle and obtain any of these function values, and by taking their reciprocals, find the other three trigonometric function values.

Important: Some calculators require angles to be entered in degrees. On others either degrees or radians can be used. Be sure to use the unit appropriate to the calculator you are using.

To use the tables in reverse, e.g., to find an angle when its sine is given, first enter the sine value. Then press the key marked $\boxed{\sin^{-1}}$ (on some calculators). On other calculators the key is marked $\boxed{\text{arcsin}}$. On still others, two keys must be pressed in sequence: $\boxed{\text{arc}}$ and then $\boxed{\sin}$. Be sure to read the instructions for the calculator you are using.

EXERCISE SET 3.5

[i] Use Table 2 to find the following.

1. sin 13°20′   **2.** sin 41°40′   **3.** cos 56°30′   **4.** cos 71°50′

5. tan 0.3956   **6.** tan 0.7010   **7.** cot 0.9134   **8.** cot 1.0443

[ii] Convert to degrees and minutes.

9. 46.38°   **10.** 85.21°   **11.** 67.84°   **12.** 38.48°

Convert to degrees and decimal parts of degrees.

13. 45°45′   **14.** 36°17′   **15.** 76°53′   **16.** 12°23′

[iii] Use Table 2 to find the following.

17. sin 28°13′   **18.** sin 36°42′   **19.** cos 53°53′   **20.** cos 80°33′

21. cos 75.15°   **22.** cos 81.91°   **23.** sin 17.43°   **24.** sin 38.72°

25. $\sin \dfrac{5\pi}{7}$ **26.** $\cos \dfrac{4\pi}{5}$

[iv] For each of the following, find θ in degrees and minutes between 0° and 90°.

27. sin θ = 0.2363   **28.** sin θ = 0.3854   **29.** cos θ = 0.3719

30. cos θ = 0.6361   **31.** cos θ = 0.3538   **32.** cos θ = 0.9678

For each of the following, find θ in radians between 0 and $\pi/2$.

33. tan θ = 0.4699   **34.** cot θ = 1.621

35. sec θ = 1.457   **36.** csc θ = 1.173

[v] Use Table 2 to find the following.

37. sin 292°40′   **38.** cos 472°40′   **39.** cos 514°30′   **40.** tan 349°28′

Find θ in degrees and minutes in the interval indicated.

41. sin θ = 0.2363,   (90°, 180°)   **42.** sin θ = −0.3854,   (270°, 360°)

43. cos θ = −0.3719,   (180°, 270°)   **44.** cos θ = 0.6361,   (360°, 450°)

Find θ in radians in the interval indicated.

45. cos θ = −0.3538, $\left(\dfrac{\pi}{2}, \pi\right)$ **46.** cos θ = 0.9678, $\left(-\dfrac{\pi}{2}, 0\right)$

47. sec θ = 1.457, $\left(\dfrac{3\pi}{2}, 2\pi\right)$ **48.** csc θ = −1.173, $\left(\pi, \dfrac{3\pi}{2}\right)$

☆ ───

49. How big must an angle be in order that it first differs from its tangent in the fourth decimal place?

50. How big must an angle be in order that it first differs from its tangent in the third decimal place?

51. ▦ Angles are measured in degrees, minutes (60′ = 1 degree), and seconds (60 seconds = 1′). Seconds are denoted ″. Convert radian measure of $\pi/7$ to degrees, minutes, and seconds.

52. ▦ Convert 61°38′22″ to degrees and decimal parts of degrees.

53. ▦ The formula

$$\sin x = x - \frac{x^3}{6} + \frac{x^5}{120}$$

gives an approximation for sine values when x is in radians. Calculate sin 0.5 and compare your answer with Table 2.

54. Using values from the table, graph the tangent function between −90° and 90°. What is the domain of the tangent function?

55. Use the fact that $\sin \theta \approx \theta$ when θ is small to calculate the diameter of the sun. The sun is 93 million miles from the earth and the angle it subtends at the earth's surface is about 31′59″.

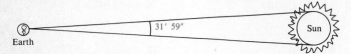

56. When light passes from one substance to another, rays are bent, depending on the speed of light in those substances. For example, light traveling in air is bent as it enters water. Angle *i* is called the angle of *incidence* and angle *r* is called the angle of *refraction*. Snell's law states that $\sin i / \sin r$ is constant. The constant is called the *index of refraction*. The index of refraction of a certain crystal is 1.52. Light strikes it, making an angle of incidence of 27°. What is the angle of refraction?

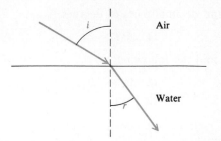

57. ▦ The function $\sin x / x$ (with x in radians) approaches a limit as x approaches 0. What is it?

OBJECTIVES

You should be able to:

[i] Given a function value and a quadrant, find the other trigonometric function values.
[ii] Determine whether trigonometric functions are even or odd.
[iii] Construct graphs of the trigonometric functions and determine the domain, range, and period.
[iv] Derive cofunction identities.

3.6 FURTHER RELATIONS AMONG THE FUNCTIONS (OPTIONAL)*

[i] Some Important Identities

Consider the unit circle, whose equation is $x^2 + y^2 = 1$. For any point (x, y) on the circle, the second coordinate is $\sin \theta$ and the first coordinate is $\cos \theta$, since the length of the hypotenuse is 1. Thus we have the following.

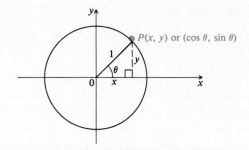

For any point *P* on the unit circle, the coordinates are $(\cos \theta, \sin \theta)$, where θ is any angle having $\overrightarrow{OP}$ for its terminal side.

If we substitute $\cos \theta$ and $\sin \theta$ for x and y, respectively, in the equation of the unit circle, we obtain the following important identity.

$$\sin^2 \theta + \cos^2 \theta \equiv 1$$

*This section should be omitted, except in the "minimal course" (see the preface).

For any point P on the unit circle, $\tan \theta = y/x$ and $\cot \theta = x/y$, provided denominators do not become 0. Since $x = \cos \theta$ and $y = \sin \theta$, we have the following:

$$\tan \theta = \frac{y}{x} = \frac{\sin \theta}{\cos \theta} \quad \text{and} \quad \cot \theta = \frac{x}{y} = \frac{\cos \theta}{\sin \theta}.$$

Since the secant and cosine values are reciprocals and the cosecant and sine values are reciprocals, we have the following:

$$\sec \theta = \frac{1}{\cos \theta} \quad \text{and} \quad \csc \theta = \frac{1}{\sin \theta}.$$

We now list four useful identities.

$$\tan \theta \equiv \frac{\sin \theta}{\cos \theta} \qquad \cot \theta \equiv \frac{\cos \theta}{\sin \theta}$$

$$\sec \theta \equiv \frac{1}{\cos \theta} \qquad \csc \theta \equiv \frac{1}{\sin \theta}$$

Example 1 Given that $\cos \theta = -0.61$ and that θ is in the second quadrant, use identities to find the other function values.

We begin with $\sin^2 \theta + \cos^2 \theta \equiv 1$, solving for $\sin \theta$:

$\sin \theta \equiv \pm \sqrt{1 - \cos^2 \theta}$ The sign is to be chosen according to the quadrant—in this case, $+$.

$\sin \theta = \sqrt{1 - (-0.61)^2} = 0.7924.$

We proceed to calculate, using the appropriate identities:

$$\tan \theta = \frac{\sin \theta}{\cos \theta} = \frac{0.7924}{-0.61} = -1.2990,$$

$$\cot \theta = \frac{\cos \theta}{\sin \theta} = \frac{-0.61}{0.7924} = -0.7698,$$

$$\sec \theta = \frac{1}{\cos \theta} = \frac{1}{-0.61} = -1.6393,$$

$$\csc \theta = \frac{1}{\sin \theta} = \frac{1}{0.7924} = 1.2620.$$

DO EXERCISE 1.

Some More Identities

The identity $\sin^2 \theta + \cos^2 \theta \equiv 1$ is one of the so-called *Pythagorean* identities, because it can be derived from the Pythagorean property of right triangles. There are two other Pythagorean identities, obtained from this one by dividing. In one case we divide by $\sin^2 \theta$ and in the other we divide by $\cos^2 \theta$.

$$\frac{\sin^2 \theta}{\sin^2 \theta} + \frac{\cos^2 \theta}{\sin^2 \theta} \equiv \frac{1}{\sin^2 \theta} \qquad \text{Dividing by } \sin^2 \theta$$

1. Given that $\sin \theta = -0.47$ and that θ is in quadrant IV, use identities to find the other function values.

2. Given that $\tan \theta = 1.43$ and that θ is in quadrant III, find the other function values.

3. Make a diagram to show that $\sin(-\theta) = -\sin \theta$ when θ is between 0° and 90°.

Simplifying, we obtain $1 + \cot^2 \theta \equiv \csc^2 \theta$. We list the Pythagorean identities thus derived:

$$1 + \tan^2 \theta \equiv \sec^2 \theta, \qquad 1 + \cot^2 \theta \equiv \csc^2 \theta.$$

DO EXERCISE 2.

[ii] Even and Odd Functions

Some of the trigonometric functions are even and some of them are odd. This fact gives us some other useful identities. We determine evenness or oddness for the sine and cosine functions by considering an angle θ and its inverse $-\theta$.

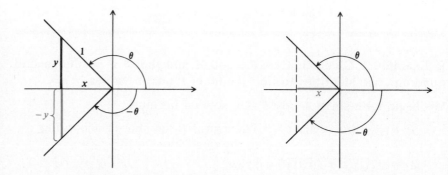

The diagrams show only special cases, but it is true in general that $\sin(-\theta) = -\sin \theta$ and $\cos(-\theta) = \cos \theta$. Thus the cosine function is even and the sine function is odd.

$$\sin(-\theta) \equiv -\sin \theta, \qquad \cos(-\theta) \equiv \cos \theta$$

DO EXERCISE 3.

Example 2 Determine whether the tangent function is even or odd.

We know that $\tan \theta \equiv \sin \theta / \cos \theta$. We substitute from the identities just derived:

$$\tan(-\theta) \equiv \frac{\sin(-\theta)}{\cos(-\theta)}$$
$$\equiv \frac{-\sin \theta}{\cos \theta}$$
$$\equiv -\frac{\sin \theta}{\cos \theta}$$
$$\equiv -\tan \theta.$$

The tangent function is therefore odd.

A similar derivation shows that the cotangent function is also odd. We thus have the following:

$$\tan(-\theta) \equiv -\tan\theta,$$
$$\cot(-\theta) \equiv -\cot\theta.$$

[iii] Graphs of the Functions

Graphs of the trigonometric functions can be constructed by plotting and connecting points, using values obtained from the tables and keeping in mind the signs of the functions in the various quadrants. Following are graphs of four of the functions. The angles are in radians, and we have chosen to use x instead of θ, as is customary when graphing functions.

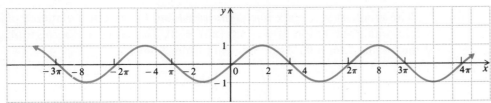

The sine function

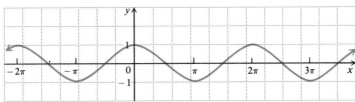

The cosine function

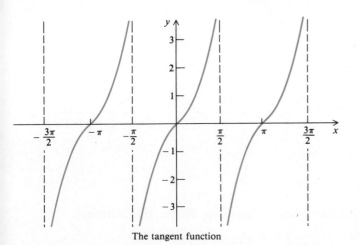

The tangent function

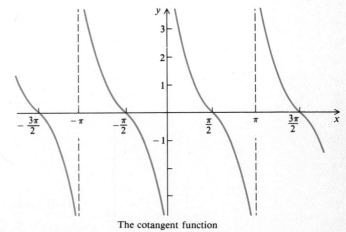

The cotangent function

4. Using values from Table 2, plot points and draw a graph of $y = \tan \theta$ between $-\pi/4$ and $\pi/4$, θ in radians.

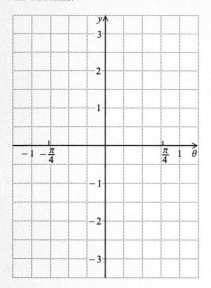

DO EXERCISE 4.

[iv] **The Cofunction Identities**

Another class of identities gives functions in terms of their cofunctions at numbers differing by $\pi/2$. We consider first the sine and cosine. Consider this graph.

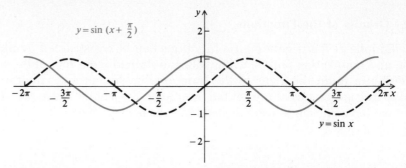

The graph of $y = \sin x$ has been translated to the left a distance of $\pi/2$, to obtain the graph of $y = \sin (x + \pi/2)$. The latter is also a graph of the cosine function. Thus we obtain the identity

$$\sin\left(x + \frac{\pi}{2}\right) \equiv \cos x.$$

By means similar to that above, we obtain the identity

$$\cos\left(x - \frac{\pi}{2}\right) \equiv \sin x.$$

Consider the following graph. The graph of $y = \sin x$ has been translated to the right a distance of $\pi/2$, to obtain the graph of $y = \sin (x - \pi/2)$. The latter is a reflection of the cosine function across the x-axis. In other words, it is a graph of $y = -\cos x$. We thus obtain the following identity:

$$\sin\left(x - \frac{\pi}{2}\right) \equiv -\cos x.$$

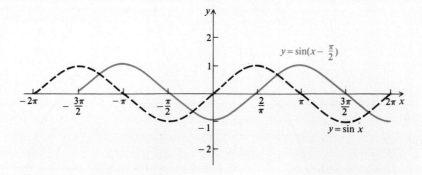

By a means similar to that above, we obtain the identity

$$\cos\left(x + \frac{\pi}{2}\right) \equiv -\sin x.$$

We have now established the following cofunction identities:*

To Be Memorized

$$\sin\left(x \pm \frac{\pi}{2}\right) \equiv \pm\cos x,$$

$$\cos\left(x \pm \frac{\pi}{2}\right) \equiv \mp\sin x.$$

These should be learned. The other cofunction identities can be obtained easily from them.

Example 3 Derive an identity relating $\sin(\pi/2 - x)$ to $\cos x$.

Since the sine function is odd, we know that

$$\sin\left(\frac{\pi}{2} - x\right) \equiv \sin-\left(x - \frac{\pi}{2}\right) \equiv -\sin\left(x - \frac{\pi}{2}\right).$$

Now consider the identity already established,

$$\sin\left(x - \frac{\pi}{2}\right) \equiv -\cos x.$$

This is equivalent to

$$-\sin\left(x - \frac{\pi}{2}\right) \equiv \cos x,$$

and we now have

$$\sin\left(\frac{\pi}{2} - x\right) \equiv \cos x.$$

DO EXERCISE 5.

5. Find an identity for $\cot(\pi/2 - x)$.

EXERCISE SET 3.6

[i]

1. ▦ Given that $\sin x = 0.1425$ in quadrant II, find the other circular function values, using identities.

3. ▦ Given that $\cot x = 0.7534$ and $\sin x > 0$, find the other circular function values, using identities.

2. ▦ Given that $\tan x = 0.8012$ in quadrant III, find the other circular function values, using identities.

4. ▦ Given that $\sin\frac{\pi}{8} = 0.38268$, use identities to find:

   a) $\cos\frac{\pi}{8}$;   b) $\cos\frac{5\pi}{8}$;   c) $\sin\frac{5\pi}{8}$;

   d) $\sin\frac{-3\pi}{8}$;   e) $\cos\frac{-3\pi}{8}$.

[ii]

5. Show that the cotangent function is odd.

7. Make a diagram to show that $\sin(-\theta) = -\sin\theta$ when θ is between $180°$ and $270°$.

6. Determine whether the secant and cosecant functions are even or odd.

8. Make a diagram to show that $\cos(-\theta) = \cos\theta$ when θ is between $-360°$ and $-540°$.

---

*There are four identities in this list. Two of them are obtained by taking the top signs, the other two by taking the bottom signs.

[iii]

9. Using function values from a table, construct a graph of the sine function between -2π and 2π.

10. Using function values from a table, construct a graph of the cosine function between -2π and 2π.

11. For the sine function, determine the domain, range, and period.

12. For the cosine function, determine the domain, range, and period.

13. For the tangent function, determine the domain, range, and period.

14. For the cotangent function, determine the domain, range, and period.

[iv]

15. Find an identity for $\cos\left(\dfrac{\pi}{2} - x\right)$.

16. Find an identity for $\tan\left(\dfrac{\pi}{2} - x\right)$.

17. Find an identity for $\tan\left(x - \dfrac{\pi}{2}\right)$.

18. Find an identity for $\tan\left(x + \dfrac{\pi}{2}\right)$.

19. Find an identity for $\sec\left(x - \dfrac{\pi}{2}\right)$.

20. Find an identity for $\csc\left(\dfrac{\pi}{2} - x\right)$.

☆ ───

21. Graph the secant function.

22. Graph the cosecant function.

23. Graph $y = \tan\dfrac{x}{2}$.

24. Graph $y = \cot 2x$.

25. The coordinates of any point on the unit circle are $(\cos s, \sin s)$, for some suitable number s. Show that for a circle with center at the origin of radius r, the coordinates of any point on the circle are $(r\cos s, r\sin s)$.

CHAPTER 3 REVIEW

[3.1, i] For angles of the following measures, state in which quadrant the terminal side lies, convert to radian measure in terms of π, and convert to radian measure not in terms of π. Use 3.14 for π.

[3.1, ii] **1.** 87° **2.** 145° **3.** 30° **4.** −30°

Convert to degree measure.

[3.1, ii] **5.** $\dfrac{3\pi}{2}$ **6.** 4π

[3.1, iii] **7.** Find the length of an arc of a circle, given a central angle of $\pi/4$ and a radius of 7 cm.

[3.1, iii] **8.** An arc 18 m long on a circle of radius 8 m subtends an angle of how many radians? how many degrees, to the nearest degree?

[3.2, i] **9.** A phonograph record revolves at 45 rpm. What is the linear velocity in cm/min of a point 4 cm from the center?

[3.2, i] **10.** An automobile wheel has a diameter of 26 in. If the car travels 30 mph, what is the angular velocity in radians/hr of a point on the edge of the wheel?

[3.3, i] **11.** Find the six trigonometric function values for the angle θ as shown.

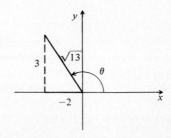

[3.3, ii]   **12.** Complete the following table.

θ	0°	30°	45°	60°	90°	270
$\sin \theta$						
$\cos \theta$						
$\tan \theta$						
$\cot \theta$						
$\sec \theta$						
$\csc \theta$						

[3.3, ii]   Find the following exactly. Do not use the table.

13. sin 495°   **14.** tan −315°   **15.** cot 210°

16. Graph the sine function and state its domain, **range**, and period.

[3.4, i]   Given that $\tan \theta = 2/\sqrt{5}$ and that the terminal side is in quadrant III, find the following.

17. sin θ **18.** cos θ **19.** cot θ **20.** sec θ **21.** csc θ

[3.4, ii]   Given that $\sin \theta = 0.6820$, $\cos \theta = 0.7314$, $\tan \theta = 0.9325$, $\cot \theta = 1.0724$, $\sec \theta = 1.3673$, and $\csc \theta = 1.4663$, find the following.

22. sin (90° − θ)   **23.** cos (90° − θ)   **24.** tan (90° − θ)

25. cot (90° − θ)   **26.** sec (90° − θ)   **27.** csc (90° − θ)

[3.5, ii]   **28.** Convert to degrees and minutes: 22.20°.

29. Convert to degrees and decimal parts of degrees: 47°33'.

Use Table 2 to find the following.

[3.5, i]   **30.** cos 8°20'   **31.** tan 27°14'   **32.** sec 44°30'

[3.5, iii]   **33.** $\cot \theta = 3.450$ and $0° < \theta < 90°$. Find θ.

34. $\sin \theta = 0.6293$ and $0° < \theta < 90°$. Find θ.

[3.6, i]   **35.** Given that $\cos x = -0.1425$ in quadrant II, find the other circular function values, using identities.

[3.6, ii]   **36.** a) List the circular functions that are even.

b) List the circular functions that are odd.

[3.6, iii] **37.** Graph the sine and cosine functions using the same axes.

[3.6, iv]   **38.** Find an identity for $\csc \left(x - \dfrac{\pi}{2} \right)$.

39. Graph $y = 3 \sin \dfrac{x}{2}$ and determine the domain, range, and period.

4

TRIGONOMETRIC IDENTITIES, INVERSE FUNCTIONS, AND EQUATIONS

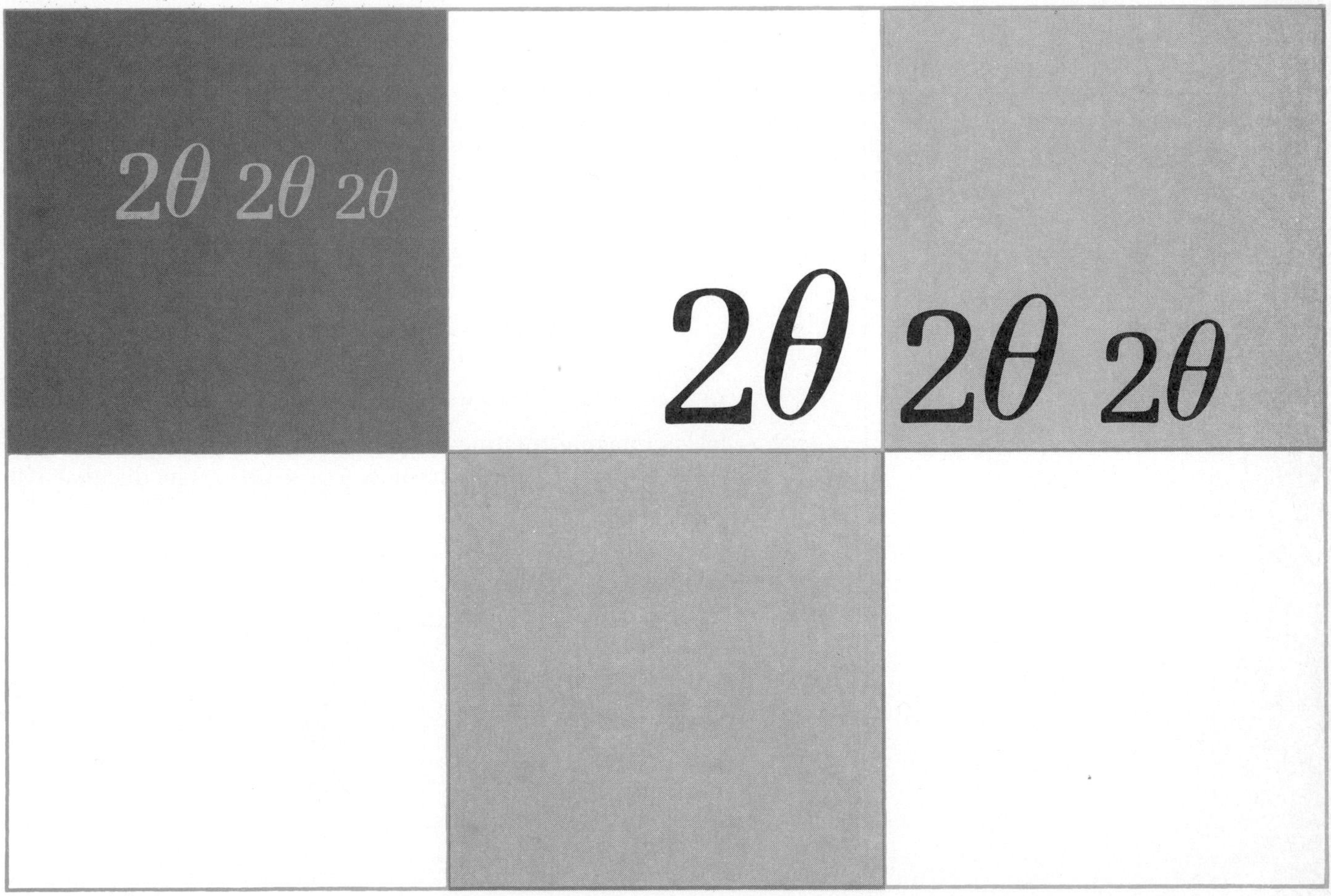

2θ 2θ 2θ

2θ 2θ 2θ

OBJECTIVES

You should be able to:

[i] Use the sum and difference identities to find function values.
[ii] Simplify expressions using the sum and difference identities.
[iii] Find angles between lines, knowing their slopes.

4.1 SUM AND DIFFERENCE FORMULAS

[i] Difference Identities

We now develop some important identities involving sums or differences of two numbers (or angles), first an identity for the cosine of the difference of two numbers. We shall use the Greek letters α (alpha) and β (beta) for these numbers. Let us consider a real number α in the interval $[\pi/2, \pi]$ and a real number β in the interval $[0, \pi/2]$. These determine points A and B on the unit circle as shown. The arc length s is $\alpha - \beta$, and we know that $0 \leq s \leq \pi$. Recall that the coordinates of A are $(\cos \alpha, \sin \alpha)$, and the coordinates of B are $(\cos \beta, \sin \beta)$.

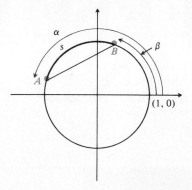

Using the distance formula, we can write an expression for the square of the distance AB:

$$AB^2 = (\cos \alpha - \cos \beta)^2 + (\sin \alpha - \sin \beta)^2.$$

This can be simplified as follows:

$$
\begin{aligned}
AB^2 &= \cos^2 \alpha - 2 \cos \alpha \cos \beta + \cos^2 \beta \\
&\quad + \sin^2 \alpha - 2 \sin \alpha \sin \beta + \sin^2 \beta \\
&= (\sin^2 \alpha + \cos^2 \alpha) + (\sin^2 \beta + \cos^2 \beta) \\
&\quad - 2(\cos \alpha \cos \beta + \sin \alpha \sin \beta) \\
&= 2 - 2(\cos \alpha \cos \beta + \sin \alpha \sin \beta).
\end{aligned}
$$

Now let us imagine rotating the circle above so that point B is at $(1, 0)$. The coordinates of point A are now $(\cos s, \sin s)$. The distance AB has not changed.

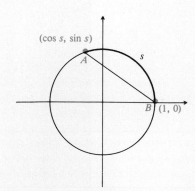

Again we use the distance formula to write an expression for the square of AB:

$$AB^2 = (\cos s - 1)^2 + (\sin s - 0)^2.$$

This simplifies as follows:

$$\begin{aligned} AB^2 &= \cos^2 s - 2\cos s + 1 + \sin^2 s \\ &= (\sin^2 s + \cos^2 s) + 1 - 2\cos s \\ &= 2 - 2\cos s. \end{aligned}$$

Equating our two expressions for AB^2 and simplifying, we obtain

$$\cos s = \cos \alpha \cos \beta + \sin \alpha \sin \beta. \qquad (1)$$

But $s = \alpha - \beta$, so we have the equation

$$\cos(\alpha - \beta) = \cos \alpha \cos \beta + \sin \alpha \sin \beta. \qquad (2)$$

This formula holds for any numbers α and β for which $\alpha - \beta$ is the length of the shortest arc from A to B; in other words, when $0 \le \alpha - \beta \le \pi$.

1. Find $\cos\left(\dfrac{\pi}{2} - \dfrac{\pi}{6}\right)$.

Example 1 Find $\cos\left(\dfrac{3\pi}{4} - \dfrac{\pi}{3}\right)$.

$$\begin{aligned} \cos\left(\frac{3\pi}{4} - \frac{\pi}{3}\right) &= \cos\frac{3\pi}{4}\cos\frac{\pi}{3} + \sin\frac{3\pi}{4}\sin\frac{\pi}{3} \\[2mm] &= -\frac{\sqrt{2}}{2}\cdot\frac{1}{2} + \frac{\sqrt{2}}{2}\cdot\frac{\sqrt{3}}{2} \\[2mm] &= \frac{\sqrt{2}}{4}(\sqrt{3} - 1) \quad \text{or} \quad \frac{\sqrt{6} - \sqrt{2}}{4} \end{aligned}$$

2. Find $\cos 105°$ as $\cos(150° - 45°)$.

Example 2 Find $\cos 15°$.

$$\begin{aligned} \cos 15° &= \cos(45° - 30°) \\ &= \cos 45° \cos 30° + \sin 45° \sin 30° \\[2mm] &= \frac{\sqrt{2}}{2}\cdot\frac{\sqrt{3}}{2} + \frac{\sqrt{2}}{2}\cdot\frac{1}{2} \\[2mm] &= \frac{\sqrt{2}}{4}(\sqrt{3} + 1) \quad \text{or} \quad \frac{\sqrt{6} + \sqrt{2}}{4} \end{aligned}$$

DO EXERCISES 1 AND 2.

Formula (1) above holds when s is the length of the shortest arc from A to B. Given any real numbers α and β, the length of the shortest arc from A to B is not always $\alpha - \beta$. However, $\cos s$ is always equal to $\cos(\alpha - \beta)$. We will not prove this in detail, but the following diagrams help to show how a proof would be done. The length of the shortest arc from A to B always turns out to be $\alpha - \beta$ or $\beta - \alpha$, plus some multiple of 2π.

Simplify.

3. $\sin \dfrac{\pi}{3} \sin \left(-\dfrac{\pi}{4}\right) + \cos \left(-\dfrac{\pi}{4}\right) \cos \dfrac{\pi}{3}$

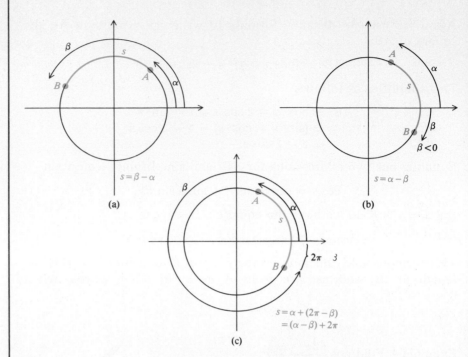

(a)

(b)

(c)

4. $\cos 37° \cos 12° + \sin 12° \sin 37°$

Since $s = |\alpha - \beta| + 2k\pi$, where k is an integer, we have

$$\cos s = \cos (|\alpha - \beta| + 2k\pi)$$
$$= \cos |\alpha - \beta|,$$

since the cosine function has a period of 2π. Now let us recall the identity $\cos (-s) \equiv \cos s$. From this identity we see that we can drop the absolute value signs and then we know that $\cos s = \cos (\alpha - \beta)$ for any real numbers α and β. Thus formula (2) holds for all real numbers α and β. That formula is thus the identity we sought.

$$\cos (\alpha - \beta) \equiv \cos \alpha \cos \beta + \sin \alpha \sin \beta$$

5. $\cos \alpha \cos (-\beta) + \sin \alpha \sin (-\beta)$

[ii] **Simplification**

Example 3 Simplify

$$\sin \left(-\dfrac{5\pi}{2}\right) \sin \dfrac{\pi}{2} + \cos \dfrac{\pi}{2} \cos \left(-\dfrac{5\pi}{2}\right).$$

This is equal to

$$\cos \dfrac{\pi}{2} \cos \left(-\dfrac{5\pi}{2}\right) + \sin \dfrac{\pi}{2} \sin \left(-\dfrac{5\pi}{2}\right).$$

By the identity above we can simplify to

$$\cos \left[\dfrac{\pi}{2} - \left(-\dfrac{5\pi}{2}\right)\right], \quad \text{or} \quad \cos \dfrac{6\pi}{2}, \quad \text{or} \quad \cos 3\pi,$$

which is -1.

DO EXERCISES 3–5.

[i], [ii] **Other Formulas**

The other sum and difference formulas we seek follow easily from the one we have just derived. Let us consider cos $(\alpha + \beta)$. This is equal to cos $[\alpha - (-\beta)]$, and by the identity on the preceding page, we have

$$\cos (\alpha + \beta) \equiv \cos \alpha \cos (-\beta) + \sin \alpha \sin (-\beta).$$

But cos $(-\beta) \equiv \cos \beta$ and sin $(-\beta) \equiv -\sin \beta$, so the identity we seek is the following.

$$\cos (\alpha + \beta) \equiv \cos \alpha \cos \beta - \sin \alpha \sin \beta.$$

To develop an identity for the sine of a sum, we recall the cofunction identity sin $\theta \equiv \cos (\pi/2 - \theta)$. In this identity we shall substitute $\alpha + \beta$ for θ, obtaining sin $(\alpha + \beta) \equiv \cos [\pi/2 - (\alpha + \beta)]$. We can now use the identity for the cosine of a difference. We get

$$\sin (\alpha + \beta) \equiv \cos \left[\frac{\pi}{2} - (\alpha + \beta) \right]$$

$$\equiv \cos \left[\left(\frac{\pi}{2} - \alpha \right) - \beta \right]$$

$$\equiv \cos \left(\frac{\pi}{2} - \alpha \right) \cos \beta + \sin \left(\frac{\pi}{2} - \alpha \right) \sin \beta$$

$$\equiv \sin \alpha \cos \beta + \cos \alpha \sin \beta.$$

Thus the identity we seek is

$$\sin (\alpha + \beta) \equiv \sin \alpha \cos \beta + \cos \alpha \sin \beta.$$

DO EXERCISES 6 AND 7.

To find a formula for the sine of a difference, we can use the identity just derived, substituting $-\beta$ for β. We obtain

$$\sin (\alpha - \beta) \equiv \sin \alpha \cos \beta - \cos \alpha \sin \beta.$$

A formula for the tangent of a sum can be derived as follows, using identities already established:

$$\tan (\alpha + \beta) \equiv \frac{\sin (\alpha + \beta)}{\cos (\alpha + \beta)}$$

$$\equiv \frac{\sin \alpha \cos \beta + \cos \alpha \sin \beta}{\cos \alpha \cos \beta - \sin \alpha \sin \beta} \cdot \frac{\dfrac{1}{\cos \alpha \cos \beta}}{\dfrac{1}{\cos \alpha \cos \beta}}$$

$$\equiv \frac{\dfrac{\sin \alpha \cos \beta}{\cos \alpha \cos \beta} + \dfrac{\cos \alpha \sin \beta}{\cos \alpha \cos \beta}}{\dfrac{\cos \alpha \cos \beta}{\cos \alpha \cos \beta} - \dfrac{\sin \alpha \sin \beta}{\cos \alpha \cos \beta}}$$

$$\equiv \frac{\dfrac{\sin \alpha}{\cos \alpha} + \dfrac{\sin \beta}{\cos \beta}}{1 - \dfrac{\sin \alpha \sin \beta}{\cos \alpha \cos \beta}} \equiv \frac{\tan \alpha + \tan \beta}{1 - \tan \alpha \tan \beta}.$$

6. Find $\sin \left(\dfrac{\pi}{4} + \dfrac{\pi}{3} \right)$.

7. Simplify.

$$\sin \alpha \cos (-\beta) + \cos \alpha \sin (-\beta)$$

8. Derive the formula for $\tan (\alpha - \beta)$.

Similarly, a formula for the tangent of a difference can be established. Following is a list of the sum and difference formulas. These should be memorized.

$$\cos (\alpha \mp \beta) \equiv \cos \alpha \cos \beta \pm \sin \alpha \sin \beta^*$$
$$\sin (\alpha \mp \beta) \equiv \sin \alpha \cos \beta \mp \cos \alpha \sin \beta$$
$$\tan (\alpha \mp \beta) \equiv \frac{\tan \alpha \mp \tan \beta}{1 \pm \tan \alpha \tan \beta}$$

DO EXERCISE 8.

The identities involving sines and tangents can be used in the same way as those involving cosines in the earlier examples.

9. Find $\tan 75°$ as $\tan (45° + 30°)$.

Example 4 Find $\tan 15°$.

$$\tan 15° = \tan (45° - 30°) = \frac{\tan 45° - \tan 30°}{1 + \tan 45° \tan 30°}$$
$$= \frac{1 - \sqrt{3}/3}{1 + \sqrt{3}/3} = \frac{3 - \sqrt{3}}{3 + \sqrt{3}}$$

Example 5 Simplify $\sin \frac{\pi}{3} \cos \pi + \sin \pi \cos \frac{\pi}{3}$.

This is equal to

$$\sin \frac{\pi}{3} \cos \pi + \cos \frac{\pi}{3} \sin \pi.$$

Thus by the fourth identity in the list above, we can simplify to

$$\sin \left(\frac{\pi}{3} + \pi \right), \quad \text{or} \quad \sin \frac{4\pi}{3}, \quad \text{or} \quad -\frac{\sqrt{3}}{2}.$$

10. Simplify $\sin \frac{\pi}{2} \cos \frac{\pi}{3} - \sin \frac{\pi}{3} \cos \frac{\pi}{2}$.

DO EXERCISES 9 AND 10.

[iii] Angles Between Lines

Recall that a nonvertical line has an equation $y = mx + b$, where m is the slope. Note that such a line makes an angle with the positive half of the x-axis whose tangent is the slope of the line, as in $\tan \theta = m$.

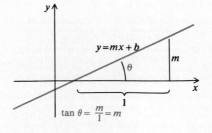

*There are six identities here, half of them obtained by using the colored signs.

One of the identities just developed gives us an easy way to find an angle formed by two lines. We shall consider the case in which neither line is vertical. Thus we shall consider two lines with equations:

$$l_1: y = m_1 x + b_1 \quad \text{and} \quad l_2: y = m_2 x + b_2.$$

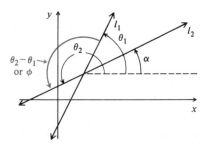

The slopes m_1 and m_2 are the tangents of the angles θ_1 and θ_2 that the lines form with the positive half of the x-axis. Thus we have $m_1 = \tan \theta_1$ and $m_2 = \tan \theta_2$. We wish to find the measure of the smallest angle through which l_1 can be rotated in the positive direction to get to l_2. In this case that angle is $\theta_2 - \theta_1$, or ϕ. We proceed as follows:

$$\tan \phi = \tan (\theta_2 - \theta_1) = \frac{\tan \theta_2 - \tan \theta_1}{1 + \tan \theta_2 \tan \theta_1}$$

$$= \frac{m_2 - m_1}{1 + m_2 m_1}.$$

Suppose we had taken the lines in the reverse order. Then the smallest *positive* angle from l_1 to l_2 would be ϕ, as shown here. Note that $\tan \theta_1 = m_1$.

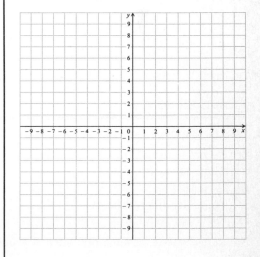

Since $\theta_2 = \alpha + 180°$, $\tan \theta_2 = \tan \alpha$ and $\tan \theta_2 = m_2$. Thus the same formula for $\tan \phi$ derived above holds. In the first case ϕ is an acute angle, so $\tan \phi$ will be positive. In the second case ϕ is obtuse, so $\tan \phi$ will be negative. Thus we have a general formula for finding the angle from one line to another.

THEOREM 1

The smallest positive angle ϕ from a line l_1 to a line l_2 (where neither line is vertical) can be obtained from

$$\tan \phi = \frac{m_2 - m_1}{1 + m_2 m_1},$$

where m_1 and m_2 are the slopes of l_1 and l_2, respectively. If $\tan \phi$ is positive, ϕ is acute. If $\tan \phi$ is negative, then ϕ is obtuse.

11. Find the smallest positive angle from l_1 to l_2. Sketch a graph and show this angle.

$$l_1: 3y = \sqrt{3}x + 6,$$
$$l_2: y + 5 = \sqrt{3}x.$$

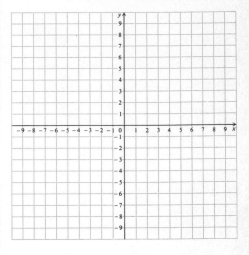

12. Find the smallest positive angle from l_1 to l_2. Sketch a graph and show this angle.

$$l_1: y = \sqrt{3}x - 4,$$
$$l_2: 3y = \sqrt{3}x + 3.$$

13. A loading ramp rises 4 ft with a run of 10 ft. A wire from the top of the building is attached to the ground 8 ft from the building. Find the angle θ that the wire makes with the ramp.

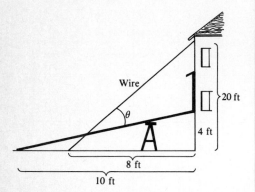

Example 6 Two lines have slopes as follows: l_1 has slope $\sqrt{3}/3$, l_2 has slope $\sqrt{3}$. How large is the smallest positive angle from l_1 to l_2?

We have $m_1 = \sqrt{3}/3$ and $m_2 = \sqrt{3}$. Then

$$\tan \phi = \frac{\sqrt{3} - \dfrac{\sqrt{3}}{3}}{1 + \sqrt{3} \cdot \dfrac{\sqrt{3}}{3}} = \frac{\dfrac{3\sqrt{3} - \sqrt{3}}{3}}{\dfrac{3 + 3}{3}} = \frac{2\sqrt{3}}{6} = \frac{\sqrt{3}}{3}.$$

Since $\tan \phi$ is positive, ϕ is acute. Hence

$$\phi = \frac{\pi}{6} \quad \text{or} \quad 30°.$$

Example 7 Find the smallest positive angle from l_1 to l_2:

$$l_1 : 2y = x + 3, \qquad l_2 : 3y + 4x + 6 = 0.$$

We first solve for y to obtain the slopes:

$$l_1 : y = \frac{1}{2}x + \frac{3}{2}, \quad \text{so} \quad m_1 = \frac{1}{2};$$

$$l_2 : y = -\frac{4}{3}x - 2, \quad \text{so} \quad m_2 = -\frac{4}{3};$$

$$\tan \phi = \frac{-\dfrac{4}{3} - \dfrac{1}{2}}{1 + \left(-\dfrac{4}{3}\right)\left(\dfrac{1}{2}\right)} = \frac{\dfrac{-8 - 3}{6}}{\dfrac{6 - 4}{6}} = -\frac{11}{2} = -5.5.$$

Since $\tan \phi$ is negative, we know that ϕ is obtuse. From the table we find an angle whose tangent is 5.5 to be about $79°40'$. This is a reference angle. Subtracting from $180°$, we obtain

$$\phi = 100°20', \qquad \text{approximately.}$$

DO EXERCISES 11–13. (NOTE THAT EXERCISES 11 AND 12 ARE ON THE PRECEDING PAGE.)

EXERCISE SET 4.1

[i] In Exercises 1–8, use the sum and difference identities to evaluate.

1. $\sin 75°$
(*Hint:* $75° = 45° + 30°$.)

2. $\cos 75°$

3. $\sin 15°$
(*Hint:* $15° = 45° - 30°$.)

4. $\cos 15°$

5. $\sin 105°$
(*Hint:* Use the results of Exercises 1 and 2.)

6. $\cos 105°$

7. $\tan 75°$

8. $\cot 15°$

In Exercises 9–14, assume that $\sin u = \frac{3}{5}$ and $\sin v = \frac{4}{5}$ and that u and v are between 0 and $\pi/2$. Then evaluate.

9. $\sin (u + v)$

10. $\sin (u - v)$

11. $\cos (u + v)$

12. $\cos (u - v)$

13. $\tan (u + v)$

14. $\tan (u - v)$

In Exercises 15 and 16, assume that $\sin \theta = 0.6249$ and $\cos \phi = 0.1102$, and that θ and ϕ are both first-quadrant angles. Evaluate.

15. ▦ $\sin (\theta + \phi)$

16. ▦ $\cos (\theta + \phi)$

[ii] In Exercises 17–26, simplify or evaluate.

17. $\sin 37° \cos 22° + \cos 37° \sin 22°$

18. $\cos 37° \cos 22° - \sin 22° \sin 37°$

19. $\dfrac{\tan 20° + \tan 32°}{1 - \tan 20° \tan 32°}$

20. $\dfrac{\tan 35° - \tan 12°}{1 + \tan 35° \tan 12°}$

21. $\sin (\alpha + \beta) + \sin (\alpha - \beta)$

22. $\sin (\alpha + \beta) - \sin (\alpha - \beta)$

23. $\cos (\alpha + \beta) + \cos (\alpha - \beta)$

24. $\cos (\alpha + \beta) - \cos (\alpha - \beta)$

25. $\cos (u + v) \cos v + \sin (u + v) \sin v$

26. $\sin (u - v) \cos v + \cos (u - v) \sin v$

[iii] In Exercises 27–34, find the angle from l_1 to l_2.

27. $l_1 : 3y = \sqrt{3}x + 2,$
$l_2 : 3y + \sqrt{3}x = -3$

28. $l_1 : 3y = \sqrt{3}x + 3,$
$l_2 : y = \sqrt{3}x + 2$

29. $l_1 : 2x = 3 - 2y,$
$l_2 : x + y = 5$

30. $l_1 : 2x = 3 + 2y,$
$l_2 : x - y = 5$

31. $l_1 : 2x - 5y + 1 = 0,$
$l_2 : 3x + y - 7 = 0$

32. $l_1 : 2x + y - 4 = 0,$
$l_2 : y - 2x + 5 = 0$

33. $l_1 : y = 3,$
$l_2 : x + y = 5$

34. $l_1 : y = 5,$
$l_2 : x - y = 2$

☆ ───────────────────────────────────

35. Find an identity for $\sin 2\theta$.
(Hint: $2\theta = \theta + \theta$.)

36. Find an identity for $\cos 2\theta$.
(Hint: $2\theta = \theta + \theta$.)

37. Derive an identity for $\cot (\alpha + \beta)$ in terms of $\cot \alpha$ and $\cot \beta$.

38. Derive an identity for $\cot (\alpha - \beta)$ in terms of $\cot \alpha$ and $\cot \beta$.

The cofunction identities can be derived from the sum and difference formulas. Derive identities for the following.

39. $\sin \left(\dfrac{\pi}{2} - x \right)$

40. $\cos \left(\dfrac{\pi}{2} - x \right)$

Find the slope of line l_1, where m_2 is the slope of line l_2 and ϕ is the smallest positive angle from l_1 to l_2.

41. $m_2 = \dfrac{4}{3}, \quad \phi = 45°$

42. $m_2 = \dfrac{2}{3}, \quad \phi$ is the smallest angle having slope $\dfrac{5}{2}$

43. Line l_1 contains the points $(-2, 4)$ and $(5, -1)$. Find the slope of line l_2 such that the angle from l_1 to l_2 is $45°$.

44. Line l_1 contains $(3, -1)$ and $(-4, 2)$. Find the slope of line l_2 such that the angle from l_1 to l_2 is $-45°$.

45. ▦ Line l_1 contains the points $(-2.123, 3.899)$ and $(-4.892, -0.9012)$ and line l_2 contains $(0, -3.814)$ and $(5.925, 4.013)$. Find the smallest positive angle from l_1 to l_2.

46. Line l_1 contains the points $(-3, 7)$ and $(-3, -2)$. Line l_2 contains $(0, -4)$ and $(2, 6)$. Find the smallest positive angle from l_1 to l_2.

47. In a circus a guy wire A is attached to the top of a 30-ft pole. Wire B is used for performers to walk up to the tight wire, 10 ft above the ground. Find the angle ϕ between the wires.

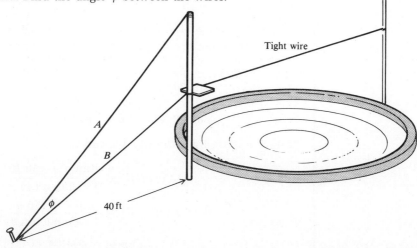

Tight wire

A

B

ϕ 40 ft

★ ───────────────────────────────────

48. Find an identity for $\cos (\alpha + \beta)$ involving only cosines.

49. Find an identity for $\sin (\alpha + \beta + \gamma)$.

OBJECTIVES

You should be able to:

[i] Use the double-angle identities to find function values of twice an angle when one function value is known for that angle.

[ii] Use the half-angle identities to find the function values of half an angle when one function value is known for that angle.

[iii] Simplify certain trigonometric expressions using the half-angle and double-angle formulas.

1. Given $\sin \theta = \frac{3}{5}$ and that θ is in the first quadrant, what is $\sin 2\theta$?

4.2 SOME IMPORTANT IDENTITIES

Two important classes of trigonometric identities are known as the half-angle identities and the double-angle identities.

[i] Double-Angle Identities

To develop these identities we shall use the sum formulas from the preceding section. We first develop a formula for $\sin 2\theta$. Recall that

$$\sin (\alpha + \beta) \equiv \sin \alpha \cos \beta + \cos \alpha \sin \beta.$$

We shall consider a number θ and substitute it for both α and β in this identity. We obtain

$$\sin (\theta + \theta) \equiv \sin 2\theta \equiv \sin \theta \cos \theta + \cos \theta \sin \theta$$
$$\equiv 2 \sin \theta \cos \theta.$$

The identity we seek is

$$\sin 2\theta \equiv 2 \sin \theta \cos \theta.$$

Example 1 If $\sin \theta = \frac{3}{8}$ and θ is in the first quadrant, what is $\sin 2\theta$?

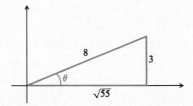

From the figure we see that $\cos \theta = \sqrt{55}/8$. Thus

$$\sin 2\theta = 2 \sin \theta \cos \theta = 2 \cdot \frac{3}{8} \cdot \frac{\sqrt{55}}{8} = \frac{3\sqrt{55}}{32}.$$

DO EXERCISE 1.

Double-angle identities for the cosine and tangent functions can be derived in much the same way as the above identity:

$$\cos (\alpha + \beta) \equiv \cos \alpha \cos \beta - \sin \alpha \sin \beta$$
$$\cos 2\theta \equiv \cos (\theta + \theta) \equiv \cos \theta \cos \theta - \sin \theta \sin \theta$$
$$\equiv \cos^2 \theta - \sin^2 \theta$$
$$\cos 2\theta \equiv \cos^2 \theta - \sin^2 \theta.$$

Now we derive an identity for $\tan 2\theta$:

$$\tan (\alpha + \beta) \equiv \frac{\tan \alpha + \tan \beta}{1 - \tan \alpha \tan \beta}$$

$$\tan 2\theta \equiv \tan (\theta + \theta) \equiv \frac{\tan \theta + \tan \theta}{1 - \tan \theta \tan \theta} \equiv \frac{2 \tan \theta}{1 - \tan^2 \theta}$$

$$\tan 2\theta \equiv \frac{2 \tan \theta}{1 - \tan^2 \theta}.$$

Example 2 Given that $\tan \theta = -\frac{3}{4}$ and θ is in the second quadrant, find $\sin 2\theta$, $\cos 2\theta$, $\tan 2\theta$, and the quadrant in which 2θ lies.

By drawing a diagram as shown, we find that $\sin \theta = \frac{3}{5}$ and $\cos \theta = -\frac{4}{5}$.

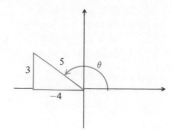

Now,

$$\sin 2\theta = 2 \sin \theta \cos \theta$$

$$= 2 \cdot \frac{3}{5} \cdot \left(-\frac{4}{5}\right) = -\frac{24}{25};$$

$$\cos 2\theta = \cos^2 \theta - \sin^2 \theta = \left(-\frac{4}{5}\right)^2 - \left(\frac{3}{5}\right)^2$$

$$= \frac{16}{25} - \frac{9}{25} = \frac{7}{25};$$

$$\tan 2\theta = \frac{2 \tan \theta}{1 - \tan^2 \theta} = \frac{2 \cdot \left(-\frac{3}{4}\right)}{1 - \left(-\frac{3}{4}\right)^2}$$

$$= -\frac{\frac{3}{2}}{1 - \frac{9}{16}} = -\frac{24}{7}.$$

Since $\sin 2\theta$ is negative and $\cos 2\theta$ is positive, we know that 2θ is in quadrant IV. Note that $\tan 2\theta$ could have been found more easily in this case by dividing the values of $\sin 2\theta$ and $\cos 2\theta$.

DO EXERCISE 2.

Two other useful identities for $\cos 2\theta$ can be derived easily, as follows:

$$\cos 2\theta \equiv \cos^2 \theta - \sin^2 \theta$$
$$\equiv (1 - \sin^2 \theta) - \sin^2 \theta \qquad \text{Using } \sin^2 \theta + \cos^2 \theta \equiv 1$$
$$\equiv 1 - 2 \sin^2 \theta.$$

Similarly,

$$\cos 2\theta \equiv \cos^2 \theta - \sin^2 \theta$$
$$\equiv \cos^2 \theta - (1 - \cos^2 \theta)$$
$$\equiv 2 \cos^2 \theta - 1.$$

2. Given $\cos \theta = -\frac{5}{13}$ and that θ is in the third quadrant, find $\sin 2\theta$, $\cos 2\theta$, and $\tan 2\theta$. Also determine the quadrant in which 2θ lies.

3. Find a formula for $\cos 3\theta$ in terms of function values of θ.

Solving these two identities for $\sin^2 \theta$ and $\cos^2 \theta$, respectively, we obtain two more identities that are often useful. Following is a list of double-angle identities. It should be memorized.

$$\sin 2\theta \equiv 2 \sin \theta \cos \theta$$
$$\cos 2\theta \equiv \cos^2 \theta - \sin^2 \theta$$
$$\equiv 1 - 2 \sin^2 \theta$$
$$\equiv 2 \cos^2 \theta - 1$$
$$\tan 2\theta \equiv \frac{2 \tan \theta}{1 - \tan^2 \theta}$$
$$\sin^2 \theta \equiv \frac{1 - \cos 2\theta}{2}$$
$$\cos^2 \theta \equiv \frac{1 + \cos 2\theta}{2}$$

By division and the last two identities, it is always easy to deduce the following identity, which is also often useful:

$$\tan^2 \theta \equiv \frac{1 - \cos 2\theta}{1 + \cos 2\theta}.$$

From the basic identities (in the lists to be memorized), others can be obtained.

Example 3 Find a formula for $\sin 3\theta$ in terms of function values of θ.

$$\sin 3\theta \equiv \sin (2\theta + \theta)$$
$$\equiv \sin 2\theta \cos \theta + \cos 2\theta \sin \theta$$
$$\equiv (2 \sin \theta \cos \theta) \cos \theta + (2 \cos^2 \theta - 1) \sin \theta$$
$$\equiv 2 \sin \theta \cos^2 \theta + 2 \sin \theta \cos^2 \theta - \sin \theta$$
$$\sin 3\theta \equiv 4 \sin \theta \cos^2 \theta - \sin \theta$$

4. Find a formula for $\sin^3 x$ in terms of function values of x or 2x, raised only to the first power.

Example 4 Find a formula for $\cos^3 x$ in terms of function values of x or 2x, raised only to the first power.

$$\cos^3 x \equiv \cos^2 x \cos x \equiv \frac{1 + \cos 2x}{2} \cos x$$

DO EXERCISES 3 AND 4.

[ii] Half-Angle Identities

To develop these identities, we use three of the previously developed ones. As shown below, we take square roots and replace θ by $\phi/2$.

$$\sin^2 \theta \equiv \frac{1 - \cos 2\theta}{2} \longrightarrow \left| \sin \frac{\phi}{2} \right| \equiv \sqrt{\frac{1 - \cos \phi}{2}}$$

$$\cos^2 \theta \equiv \frac{1 + \cos 2\theta}{2} \longrightarrow \left| \cos \frac{\phi}{2} \right| \equiv \sqrt{\frac{1 + \cos \phi}{2}}$$

$$\tan^2 \theta \equiv \frac{1 - \cos 2\theta}{1 + \cos 2\theta} \longrightarrow \left| \tan \frac{\phi}{2} \right| \equiv \sqrt{\frac{1 - \cos \phi}{1 + \cos \phi}}$$

The half-angle formulas are those on the right above. We can eliminate the absolute value signs by introducing $\pm$ signs, with the understanding that we use $+$ or $-$ depending on the quadrant in which the angle lies. We thus obtain these formulas in the following form.

$$\sin\frac{\phi}{2} \equiv \pm\sqrt{\frac{1-\cos\phi}{2}}$$

$$\cos\frac{\phi}{2} \equiv \pm\sqrt{\frac{1+\cos\phi}{2}}$$

$$\tan\frac{\phi}{2} \equiv \pm\sqrt{\frac{1-\cos\phi}{1+\cos\phi}}$$

These formulas should be memorized.

There are two other formulas for $\tan(\phi/2)$ that are often useful. They can be obtained as follows:

$$\left|\tan\frac{\phi}{2}\right| \equiv \sqrt{\frac{1-\cos\phi}{1+\cos\phi}} \equiv \sqrt{\frac{1-\cos\phi}{1+\cos\phi}\cdot\frac{1+\cos\phi}{1+\cos\phi}}$$

$$\equiv \sqrt{\frac{1-\cos^2\phi}{(1+\cos\phi)^2}}$$

$$\equiv \sqrt{\frac{\sin^2\phi}{(1+\cos\phi)^2}} \equiv \frac{|\sin\phi|}{|1+\cos\phi|}.$$

Now $1+\cos\phi$ cannot be negative because $\cos\phi$ is never less than -1. Hence the absolute value signs are not necessary in the denominator. As the following graph shows, $\tan(\phi/2)$ and $\sin\phi$ have the same sign for all ϕ for which $\tan(\phi/2)$ is defined.

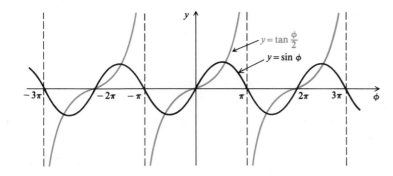

Thus we can dispense with the other absolute value signs, and obtain the formula we seek. A second formula can be obtained similarly.

$$\tan\frac{\phi}{2} \equiv \frac{\sin\phi}{1+\cos\phi}, \qquad \tan\frac{\phi}{2} \equiv \frac{1-\cos\phi}{\sin\phi}$$

These formulas have the advantage that they give the sign of $\tan(\phi/2)$ directly.

5. Find $\cos 15°$.

6. Find $\tan \dfrac{\pi}{12}$.

Example 5 Find $\sin 15°$.

$$\sin 15° = \sin \frac{30°}{2} = \pm\sqrt{\frac{1 - \cos 30°}{2}}$$

$$= \pm\sqrt{\frac{1 - (\sqrt{3}/2)}{2}}$$

$$= \pm\sqrt{\frac{2 - \sqrt{3}}{4}} = \frac{\sqrt{2 - \sqrt{3}}}{2}$$

(The expression is positive, because 15° is in the first quadrant.)

Example 6 Find $\tan \dfrac{\pi}{8}$.

$$\tan \frac{\pi}{8} = \tan \frac{\frac{\pi}{4}}{2} = \frac{\sin \frac{\pi}{4}}{1 + \cos \frac{\pi}{4}} = \frac{\frac{\sqrt{2}}{2}}{1 + \frac{\sqrt{2}}{2}}$$

$$= \frac{\sqrt{2}}{2 + \sqrt{2}}$$

$$= \sqrt{2} - 1 \qquad \text{Rationalizing the denominator}$$

DO EXERCISES 5 AND 6.

[iii] Simplification

Many simplifications of trigonometric expressions are possible through the use of the identities we have developed.

Example 7 Simplify

$$\frac{1 - \cos 2x}{4 \sin x \cos x}.$$

We search the list of identities, attempting to find some substitution that will simplify the expression. In this case, one might note that the denominator is $2(2 \sin x \cos x)$ and thus find a simplification, since $2 \sin x \cos x \equiv \sin 2x$. We now have

$$\frac{1 - \cos 2x}{2 \sin 2x}.$$

Now

$$\frac{1 - \cos \phi}{\sin \phi} \equiv \tan \frac{\phi}{2}.$$

Using this we obtain

$$\frac{1}{2} \cdot \frac{1 - \cos 2x}{\sin 2x} \equiv \frac{1}{2} \tan x.$$

Sum–Product Identities

On occasion, it is convenient to convert a product of trigonometric expressions to a sum, or the reverse. The following identities are useful in this connection. Proofs are left as exercises.

$$\sin u \cdot \cos v \equiv \frac{1}{2}[\sin (u + v) + \sin (u - v)]$$

$$\cos u \cdot \sin v \equiv \frac{1}{2}[\sin (u + v) - \sin (u - v)]$$

$$\cos u \cdot \cos v \equiv \frac{1}{2}[\cos (u - v) + \cos (u + v)]$$

$$\sin u \cdot \sin v \equiv \frac{1}{2}[\cos (u - v) - \cos (u + v)]$$

$$\sin x + \sin y \equiv 2 \sin \frac{x + y}{2} \cos \frac{x - y}{2}$$

$$\sin x - \sin y \equiv 2 \cos \frac{x + y}{2} \sin \frac{x - y}{2}$$

$$\cos y + \cos x \equiv 2 \cos \frac{x + y}{2} \cos \frac{x - y}{2}$$

$$\cos y - \cos x \equiv 2 \sin \frac{x + y}{2} \sin \frac{x - y}{2}$$

EXERCISE SET 4.2

[i] In Exercises 1–6, find sin 2θ, cos 2θ, tan 2θ, and the quadrant in which 2θ lies.

1. $\sin \theta = \frac{4}{5}$ (θ in quadrant I)

2. $\sin \theta = \frac{5}{13}$ (θ in quadrant I)

3. $\cos \theta = -\frac{4}{5}$ (θ in quadrant III)

4. $\cos \theta = -\frac{3}{5}$ (θ in quadrant III)

5. $\tan \theta = \frac{4}{3}$ (θ in quadrant III)

6. $\tan \theta = \frac{3}{4}$ (θ in quadrant III)

7. Find a formula for sin 4θ in terms of function values of θ.

8. Find a formula for cos 4θ in terms of function values of θ.

9. Find a formula for $\sin^4 \theta$ in terms of function values of θ or 2θ or 4θ, raised only to the first power.

10. Find a formula for $\cos^4 \theta$ in terms of function values of θ or 2θ or 4θ, raised only to the first power.

[ii]

11. Find sin 75° without using tables. (*Hint:* 75 = 150/2.)

12. Find cos 75° without using tables.

13. Find tan 75° without using tables.

14. Find tan 67.5° without using tables. (*Hint:* 67.5 = 135/2.)

15. Find $\sin \frac{5\pi}{8}$ without using tables.

16. Find $\cos \frac{5\pi}{8}$ without using tables.

[i], [ii] In Exercises 17–22, given that sin θ = 0.3416 and that θ is in the first quadrant, find:

17. ▦ sin 2θ.

18. ▦ cos 2θ.

19. ▦ sin 4θ.

20. ▦ cos 4θ.

21. ▦ $\sin \frac{\theta}{2}$.

22. ▦ $\cos \frac{\theta}{2}$.

[iii] Simplify.

23. $\dfrac{\sin 2x}{2 \sin x}$

24. $\dfrac{\sin 2x}{2 \cos x}$

25. $1 - 2 \sin^2 \dfrac{x}{2}$

26. $2 \cos^2 \dfrac{x}{2} - 1$

27. $2 \sin \dfrac{x}{2} \cos \dfrac{x}{2}$

28. $2 \sin 2x \cos 2x$

29. $\cos^2 \dfrac{x}{2} - \sin^2 \dfrac{x}{2}$

30. $\cos^4 x - \sin^4 x$

31. $(\sin x + \cos x)^2 - \sin 2x$

32. $(\sin x - \cos x)^2 + \sin 2x$

33. $2 \sin^2 \dfrac{x}{2} + \cos x$

34. $2 \cos^2 \dfrac{x}{2} - \cos x$

35. $(-4 \cos x \sin x + 2 \cos 2x)^2 + (2 \cos 2x + 4 \sin x \cos x)^2$

36. $(-4 \cos 2x + 8 \cos x \sin x)^2 + (8 \sin x \cos x + 4 \cos 2x)^2$

37. $2 \sin x \cos^3 x + 2 \sin^3 x \cos x$

38. $2 \sin x \cos^3 x - 2 \sin^3 x \cos x$

☆ ───

39. Prove the first four of the sum–product identities. (*Hint:* They follow from the sum and difference formulas given in Section 4.1.)

40. Prove the last four of the sum–product identities. (*Hint:* Use the results of Exercise 39.)

★ ───

41. Graph $f(x) = \cos^2 x - \sin^2 x$.

42. Graph $y = |\sin x \cos x|$.

OBJECTIVES

You should be able to:

[i] Use the basic identities to prove other identities.

[ii] Given an expression $c \sin ax + d \cos ax$, find an equivalent expression $A \sin (ax + b)$.

[iii] Graph an equation

$$y = c \sin ax + d \cos ax$$

by first expressing the right-hand side in terms of the sine function only.

4.3 PROVING IDENTITIES

Basic Identities

One should remember certain trigonometric identities. Then as the occasion arises, other identities can be proved using those memorized. Following is a minimal list of identities that should be learned. Note that most formulas involving cotangents, secants, and cosecants are not in the list. That is because these functions are reciprocals of the sine, cosine, and tangent, and thus formulas for the former can easily be derived from those for the latter.

$$\sin (-x) \equiv -\sin x,$$
$$\cos (-x) \equiv \cos x,$$
$$\tan (-x) \equiv -\tan x$$

Pythagorean identities

$$\sin^2 x + \cos^2 x \equiv 1,$$
$$1 + \tan^2 x \equiv \sec^2 x,$$
$$1 + \cot^2 x \equiv \csc^2 x$$

Cofunction identities

$$\sin \left(x \pm \frac{\pi}{2} \right) \equiv \pm \cos x,$$
$$\cos \left(x \pm \frac{\pi}{2} \right) \equiv \mp \sin x$$

Sum and difference identities

$$\sin(\alpha \pm \beta) \equiv \sin\alpha\cos\beta \pm \cos\alpha\sin\beta,$$
$$\cos(\alpha \pm \beta) \equiv \cos\alpha\cos\beta \mp \sin\alpha\sin\beta,$$
$$\tan(\alpha \pm \beta) \equiv \frac{\tan\alpha \pm \tan\beta}{1 \mp \tan\alpha\tan\beta}$$

Double-angle identities

$$\sin 2x \equiv 2\sin x\cos x,$$
$$\cos 2x \equiv \cos^2 x - \sin^2 x \equiv 1 - 2\sin^2 x$$
$$\equiv 2\cos^2 x - 1,$$
$$\tan 2x \equiv \frac{2\tan x}{1 - \tan^2 x},$$
$$\sin^2 x \equiv \frac{1 - \cos 2x}{2},$$
$$\cos^2 x \equiv \frac{1 + \cos 2x}{2}$$

Half-angle identities

$$\sin\frac{x}{2} \equiv \pm\sqrt{\frac{1 - \cos x}{2}},$$
$$\cos\frac{x}{2} \equiv \pm\sqrt{\frac{1 + \cos x}{2}},$$
$$\tan\frac{x}{2} \equiv \pm\sqrt{\frac{1 - \cos x}{1 + \cos x}} \equiv \frac{\sin x}{1 + \cos x} \equiv \frac{1 - \cos x}{\sin x}$$

[i] **Proving Identities**

In proving other identities, using these, it is wise to consult the list to gain an idea of how to proceed. It is usually helpful to express all function values in terms of sines and cosines. *It is most important* in proving an identity to work with one side at a time. The idea is to simplify each side separately until the same expression is obtained in the two cases.

Example 1 Prove the following identity.

$$\tan^2 x - \sin^2 x \equiv \sin^2 x \tan^2 x$$

$\dfrac{\sin^2 x}{\cos^2 x} - \sin^2 x$	$\sin^2 x\,\dfrac{\sin^2 x}{\cos^2 x}$
$\dfrac{\sin^2 x - \sin^2 x\cos^2 x}{\cos^2 x}$	$\dfrac{\sin^4 x}{\cos^2 x}$
$\dfrac{\sin^2 x(1 - \cos^2 x)}{\cos^2 x}$	
$\dfrac{\sin^2 x\sin^2 x}{\cos^2 x}$	
$\dfrac{\sin^4 x}{\cos^2 x}$	

Prove the following identities.

1. $\cot^2 x - \cos^2 x \equiv \cos^2 x \cot^2 x$

2. $\dfrac{\sin 2\theta + \sin \theta}{\cos 2\theta + \cos \theta + 1} \equiv \tan \theta$

In the above proof, a vertical line has been drawn as a reminder to work separately with the two sides. Everything has been expressed in terms of sines and cosines and then the two sides are simplified. We work with each side separately because "doing the same thing to both sides of an equation" does not always yield equivalent equations.

Example 2 Prove the following identity.

$$\frac{\sin 2\theta}{\sin \theta} - \frac{\cos 2\theta}{\cos \theta} \equiv \sec \theta$$

$$\frac{2 \sin \theta \cos \theta}{\sin \theta} - \frac{\cos^2 \theta - \sin^2 \theta}{\cos \theta} \qquad \left| \quad \frac{1}{\cos \theta} \right.$$

$$\frac{2 \cos^2 \theta - \cos^2 \theta + \sin^2 \theta}{\cos \theta}$$

$$\frac{\cos^2 \theta + \sin^2 \theta}{\cos \theta}$$

$$\frac{1}{\cos \theta}$$

DO EXERCISES 1 AND 2.

> **HINTS FOR PROVING IDENTITIES**
>
> 1. **Work with one side at a time.**
> 2. **Work with the more complex side first.**
> 3. **Do the algebraic manipulations, such as adding and subtracting.**
> 4. **Converting all expressions to sines and cosines is often helpful.**
> 5. **Try something! Identities are never proved by staring.**

[ii] An Identity for $c \sin ax + d \cos ax$

An expression $c \sin ax + d \cos ax$, where a, c, and d are constants, is equivalent to an expression $A \sin (ax + b)$, where A and b are also constants. To prove this we shall look for the numbers A and b. First, let us consider the numbers $c/\sqrt{c^2 + d^2}$ and $d/\sqrt{c^2 + d^2}$. These numbers are the coordinates of a point on the unit circle because

$$\left(\frac{c}{\sqrt{c^2 + d^2}}\right)^2 + \left(\frac{d}{\sqrt{c^2 + d^2}}\right)^2 = 1.$$

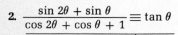

Thus there is a number b for which these numbers are the cosine and sine, respectively. In other words,

$$\frac{c}{\sqrt{c^2 + d^2}} = \cos b,$$

and

$$\frac{d}{\sqrt{c^2 + d^2}} = \sin b,$$

for some number b. Now we will consider the expression $c \sin ax + d \cos ax$, and multiply it by 1, as follows:

$$\frac{\sqrt{c^2 + d^2}}{\sqrt{c^2 + d^2}} [c \sin ax + d \cos ax]$$

$$\equiv \sqrt{c^2 + d^2} \left[(\sin ax) \cdot \frac{c}{\sqrt{c^2 + d^2}} + (\cos ax) \frac{d}{\sqrt{c^2 + d^2}} \right]$$

$$\equiv \sqrt{c^2 + d^2} \, [\sin ax \cos b + \cos ax \sin b].$$

Now, by the identity for the sine of a sum, we obtain

$$\sqrt{c^2 + d^2} \sin (ax + b).$$

Thus we have an identity as follows.

THEOREM 2

For any numbers a, c, and d,

$$c \sin ax + d \cos ax \equiv A \sin (ax + b),$$

where $A = \sqrt{c^2 + d^2}$ and b is a number whose cosine is c/A and whose sine is d/A.

Example 3 Find an expression equivalent to $\sqrt{3} \sin x + \cos x$ that involves only the sine function.

In this expression, referring to the theorem above, we see that $c = \sqrt{3}$, $d = 1$, and $a = 1$. From this we find A and b:

$$A = \sqrt{c^2 + d^2} = \sqrt{3 + 1} = 2, \quad \cos b = \frac{\sqrt{3}}{2}, \quad \text{and} \quad \sin b = \frac{1}{2}.$$

Thus we can use 30° or $\pi/6$ for b. Thus

$$\sqrt{3} \sin x + \cos x \equiv 2 \sin \left(x + \frac{\pi}{6} \right).$$

Example 4 Find an expression equivalent to $\sqrt{3} \sin (\pi/4)t + \cos (\pi/4)t$ involving only the sine function.

In this case $c = \sqrt{3}$, $d = 1$, and $a = \pi/4$.

$$A = \sqrt{3 + 1} = 2, \quad \cos b = \frac{\sqrt{3}}{2}, \quad \text{and} \quad \sin b = \frac{1}{2}.$$

Thus we have

$$\sqrt{3} \sin \frac{\pi}{4} t + \cos \frac{\pi}{4} t \equiv -2 \sin \left(\frac{\pi}{4} t + \frac{\pi}{6} \right).$$

Find an expression involving only the sine function, equivalent to each of the following.

3. $\sin 2x + \cos 2x$

4. $12 \sin x - 5 \cos x$

5. Graph $y = \sqrt{3} \sin x + \cos x$.

Example 5 Find an expression equivalent to $3 \sin 2x - 4 \cos 2x$ involving only the sine function.

In this case $c = 3$, $d = -4$, and $a = 2$.

$$A = \sqrt{3^2 + (-4)^2} = 5, \quad \cos b = \frac{3}{5}, \quad \text{and} \quad \sin b = -\frac{4}{5}.$$

Thus we have $3 \sin 2x - 4 \cos 2x \equiv 5 \sin (2x + b)$, where b is a number whose cosine is $\frac{3}{5}$ and whose sine is $-\frac{4}{5}$.

DO EXERCISES 3 AND 4.

[iii] **Graphing**

Example 6 Graph $y = \sin x - \cos x$.

We first transform the right-hand side to an expression involving only the sine function. Since $c = 1$ and $d = -1$, we have

$$A = \sqrt{1^2 + (-1)^2} = \sqrt{2}, \quad \cos b = \frac{1}{\sqrt{2}}, \quad \text{and} \quad \sin b = -\frac{1}{\sqrt{2}}.$$

So $b = -\pi/4$ and we have $y = \sqrt{2} \sin (x - \pi/4)$. This can be graphed by translating the sine graph $\pi/4$ units to the right and stretching it vertically.

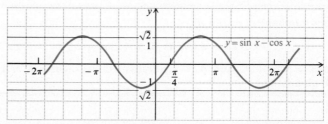

Skill in making graphs like this is important in such fields as engineering, electricity, and physics.

DO EXERCISE 5.

EXERCISE SET 4.3

[i] In Exercises 1–32, prove the identities.

1. $\csc x - \cos x \cot x \equiv \sin x$

2. $\sec x - \sin x \tan x \equiv \cos x$

3. $\dfrac{1 + \cos \theta}{\sin \theta} + \dfrac{\sin \theta}{\cos \theta} \equiv \dfrac{\cos \theta + 1}{\sin \theta \cos \theta}$

4. $\dfrac{1}{\sin \theta \cos \theta} - \dfrac{\cos \theta}{\sin \theta} \equiv \dfrac{\sin \theta \cos \theta}{1 - \sin^2 \theta}$

5. $\dfrac{1 - \sin x}{\cos x} \equiv \dfrac{\cos x}{1 + \sin x}$

6. $\dfrac{1 - \cos x}{\sin x} \equiv \dfrac{\sin x}{1 + \cos x}$

7. $\dfrac{1 + \tan \theta}{1 + \cot \theta} \equiv \dfrac{\sec \theta}{\csc \theta}$

8. $\dfrac{\cot \theta - 1}{1 - \tan \theta} \equiv \dfrac{\csc \theta}{\sec \theta}$

9. $\dfrac{\sin x + \cos x}{\sec x + \csc x} \equiv \dfrac{\sin x}{\sec x}$

10. $\dfrac{\sin x - \cos x}{\sec x - \csc x} \equiv \dfrac{\cos x}{\csc x}$

11. $\dfrac{1 + \tan\theta}{1 - \tan\theta} + \dfrac{1 + \cot\theta}{1 - \cot\theta} \equiv 0$

12. $\dfrac{\cos^2\theta + \cot\theta}{\cos^2\theta - \cot\theta} \equiv \dfrac{\cos^2\theta\tan\theta + 1}{\cos^2\theta\tan\theta - 1}$

13. $\dfrac{1 + \cos 2\theta}{\sin 2\theta} \equiv \cot\theta$

14. $\dfrac{2\tan\theta}{1 + \tan^2\theta} \equiv \sin 2\theta$

15. $\sec 2\theta \equiv \dfrac{\sec^2\theta}{2 - \sec^2\theta}$

16. $\cot 2\theta \equiv \dfrac{\cot^2\theta - 1}{2\cot\theta}$

17. $\dfrac{\sin(\alpha + \beta)}{\cos\alpha\cos\beta} \equiv \tan\alpha + \tan\beta$

18. $\dfrac{\cos(\alpha - \beta)}{\cos\alpha\sin\beta} \equiv \tan\alpha + \cot\beta$

19. $1 - \cos 5\theta\cos 3\theta - \sin 5\theta\sin 3\theta \equiv 2\sin^2\theta$

20. $2\sin\theta\cos^3\theta + 2\sin^3\theta\cos\theta \equiv \sin 2\theta$

21. $\dfrac{\tan\theta + \sin\theta}{2\tan\theta} \equiv \cos^2\dfrac{\theta}{2}$

22. $\dfrac{\tan\theta - \sin\theta}{2\tan\theta} \equiv \sin^2\dfrac{\theta}{2}$

23. $\cos^4 x - \sin^4 x \equiv \cos 2x$

24. $\dfrac{\cos^4 x - \sin^4 x}{1 - \tan^4 x} \equiv \cos^4 x$

25. $\dfrac{\tan 3\theta - \tan\theta}{1 + \tan 3\theta\tan\theta} \equiv \dfrac{2\tan\theta}{1 - \tan^2\theta}$

26. $\left(\dfrac{1 + \tan\theta}{1 - \tan\theta}\right)^2 \equiv \dfrac{1 + \sin 2\theta}{1 - \sin 2\theta}$

27. $\dfrac{\cos^3 x - \sin^3 x}{\cos x - \sin x} \equiv \dfrac{2 + \sin 2x}{2}$

28. $\dfrac{\sin^3 t + \cos^3 t}{\sin t + \cos t} \equiv \dfrac{2 - \sin 2t}{2}$

29. $\sin(\alpha + \beta)\sin(\alpha - \beta) \equiv \sin^2\alpha - \sin^2\beta$

30. $\cos(\alpha + \beta)\cos(\alpha - \beta) \equiv \cos^2\alpha - \sin^2\beta$

31. $\cos(\alpha + \beta) + \cos(\alpha - \beta) \equiv 2\cos\alpha\cos\beta$

32. $\sin(\alpha + \beta) + \sin(\alpha - \beta) \equiv 2\sin\alpha\cos\beta$

[ii] In Exercises 33–38, find an equivalent expression involving the sine function only.

33. $\sin 2x + \sqrt{3}\cos 2x$

34. $\sqrt{3}\sin 3x - \cos 3x$

35. $4\sin x + 3\cos x$

36. $4\sin 2x - 3\cos 2x$

37. $6.75\sin 0.374x + 4.08\cos 0.374x$

38. $97.81\sin 0.8081x - 4.89\cos 0.8081x$

[iii]

39. Graph $y = \sin 2x - \cos 2x$.

40. Graph $y = \sin x + \sqrt{3}\cos x$.

☆ ────────────────────────────────────

41. The following equation occurs in the study of mechanics:

$$\sin\theta = \frac{I_1\cos\phi}{\sqrt{(I_1\cos\phi)^2 + (I_2\sin\phi)^2}}.$$

It can happen that $I_1 = I_2$. Assuming that this happens, simplify the equation.

42. In the theory of alternating current, the following equation occurs:

$$R = \frac{1}{\omega C(\tan\theta + \tan\phi)}.$$

Show that this equation is equivalent to $R = \dfrac{\cos\theta\cos\phi}{\omega C\sin(\theta + \phi)}.$

★ ────────────────────────────────────

43. In electrical theory the following equations occur:

$$E_1 = \sqrt{2}\,E_t\cos\left(\theta + \frac{\pi}{P}\right), \quad E_2 = \sqrt{2}\,E_t\cos\left(\theta - \frac{\pi}{P}\right).$$

Assuming that these equations hold, show that

$$\frac{E_1 + E_2}{2} = \sqrt{2}\,E_t\cos\theta\cos\frac{\pi}{P} \quad \text{and} \quad \frac{E_1 - E_2}{2} = -\sqrt{2}\,E_t\sin\theta\sin\frac{\pi}{P}.$$

OBJECTIVES

You should be able to:

[i] Given a number a, find all values of arcsin a, arccos a, and arctan a, in degrees; and in radians, if a is a number for which use of the table is not required.

[ii] Find principal values of the inverses of the trigonometric functions.

1. Sketch a graph of $y = \cos^{-1} x$. Is this relation a function?

2. Sketch a graph of $y = \cot^{-1} x$. Is this relation a function?

4.4 INVERSES OF THE TRIGONOMETRIC FUNCTIONS

Recall that to obtain the inverse of any relation we interchange the first and second members of each ordered pair in the relation. If a relation is defined by an equation, say in x and y, interchanging x and y produces an equation of the inverse relation. The graphs of a relation and its inverse are reflections of each other across the line $y = x$. The x- and y-axes are interchanged in such a reflection. Let us consider the inverse of the sine function, $y = \sin x$. The inverse may be denoted several ways, as follows:

$$x = \sin y, \qquad y = \sin^{-1} x, \qquad y = \arcsin x.$$

Thus $\sin^{-1} x$ is a number whose sine is x.

The notation arcsin x arises because it is the length of an arc on the unit circle for which the sine is x. The notation $\sin^{-1} x$ is not exponential notation. It does *not* mean $1/\sin x$! Either of the latter two kinds of notation above can be read "the inverse sine of x" or "the arc sine of x" or "the number (or angle) whose sine is x." Notation is chosen similarly for the inverses of the other trigonometric functions: $\cos^{-1} x$ or arccos x, $\tan^{-1} x$ or arctan x, and so on.

Graphs of $y = \sin^{-1} x$ and $y = \tan^{-1} x$ are shown below. Note that these relations are not functions.

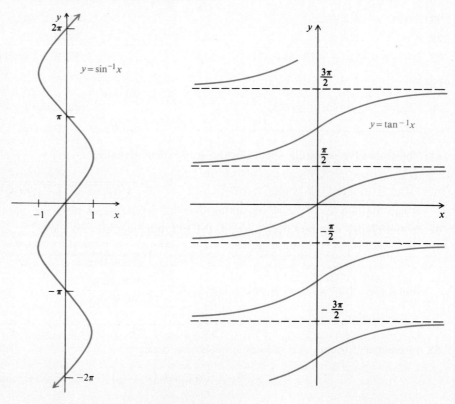

DO EXERCISES 1 AND 2.

[i] Finding Inverse Values

We illustrate finding inverse values using both a graph and the unit circle. In practice the unit circle is easier to use.

Example 1 Find all values of arcsin ½.

On the graph of $y = \arcsin x$, we draw a vertical line at $x = \frac{1}{2}$ as shown. It intersects the graph at points whose y-value is arcsin ½. Some of the numbers whose sine is ½ are seen to be $\pi/6$, $5\pi/6$, $-7\pi/6$, and so on. From the graph we can see that $\pi/6$ plus any multiple of 2π is such a number. Also $5\pi/6$ plus any multiple of 2π is such a number. The complete set of values is given by $\pi/6 + 2k\pi$, k an integer, and $5\pi/6 + 2k\pi$, k an integer.

Use a unit circle to find all values of the following.

3. $\arccos \dfrac{\sqrt{2}}{2}$

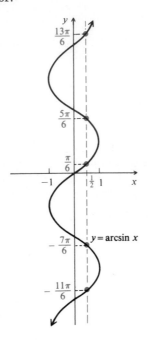

4. $\sin^{-1} \dfrac{\sqrt{3}}{2}$

Example 2 Find all values of arcsin ½.

On the unit circle there are two points at which the sine is ½. The arc length for the point in the first quadrant is $\pi/6$ plus any multiple of 2π. The arc length for the point in the second quadrant is $5\pi/6$ plus any multiple of 2π. Hence we obtain all values of arcsin ½ as follows:

$$\frac{\pi}{6} + 2k\pi \quad \text{or} \quad \frac{5\pi}{6} + 2k\pi, \qquad k \text{ an integer.}$$

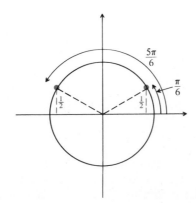

DO EXERCISES 3 AND 4.

5. Find, in degrees, all values of $\sin^{-1}(0.4226)$.

Example 3 Find all values of $\cos^{-1}(-0.9397)$ in degrees.

From Table 2 we find that the angle whose cosine is 0.9397 is 20°. This is the reference angle. We sketch this on a unit circle to find the two points where the cosine is −0.9397. The angles are 160° and 200°, plus any multiple of 360°. Thus the values of $\cos^{-1}(-0.9397)$ are

$$160° + k \cdot 360° \quad \text{or} \quad 200° + k \cdot 360°,$$

where k is any integer.

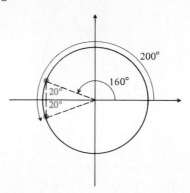

Example 4 Find all values of arctan 1 (see the figure below).

We find the two points on the unit circle at which the tangent is 1. These points are opposite ends of a diameter. Hence the arc lengths differ by π. Thus we have for all values of arctan 1,

$$\frac{\pi}{4} + k\pi, \qquad k \text{ an integer.}$$

6. Find all values of arctan (−1).

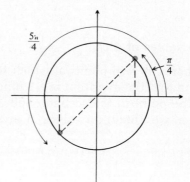

DO EXERCISES 5 AND 6.

[ii] Principal Values

The inverses of the trigonometric functions are not themselves functions. However, if we restrict the ranges of these relations, we can obtain functions. These graphs show how this restriction is made.

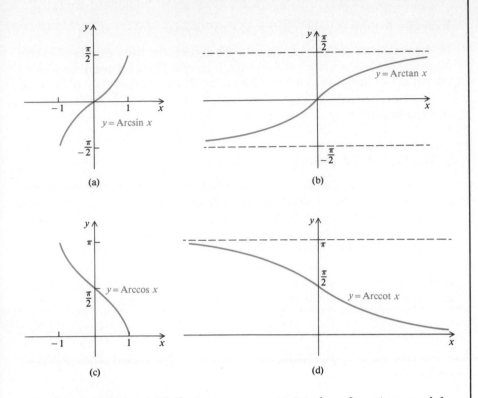

(a)

(b)

(c)

(d)

These relations with their ranges so restricted are functions, and the values in these restricted ranges are called *principal values*. To denote principal values we capitalize, as follows:

$$\text{Arcsin } x, \qquad \text{Sin}^{-1} x, \qquad \text{Arccos } x, \qquad \text{Cos}^{-1} x,$$

and so on. Thus whereas arcsin $(\frac{1}{2})$ represents an infinite set of numbers, Arcsin $(\frac{1}{2})$ represents the single number $\pi/6$.

Note that for the function $y = \text{Arcsin } x$ the range is the interval $[-\pi/2, \pi/2]$. For the function $y = \text{Arctan } x$ the range is $(-\pi/2, \pi/2)$. For the function $y = \text{Arccos } x$ the range is $[0, \pi]$, and for $y = \text{Arccot } x$ the range is $(0, \pi)$. These diagrams show where principal values are found on the unit circle. The restricted ranges should be memorized.

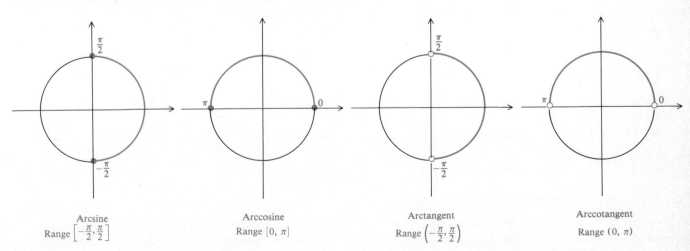

Arcsine
Range $\left[-\dfrac{\pi}{2}, \dfrac{\pi}{2}\right]$

Arccosine
Range $[0, \pi]$

Arctangent
Range $\left(-\dfrac{\pi}{2}, \dfrac{\pi}{2}\right)$

Arccotangent
Range $(0, \pi)$

Find the following.

7. Arcsin $\dfrac{\sqrt{3}}{2}$

8. $\text{Cos}^{-1}\left(-\dfrac{\sqrt{2}}{2}\right)$

9. Arccot (-1)

10. $\text{Tan}^{-1}(-1)$

Example 5 Find Arcsin $\sqrt{2}/2$ and $\text{Cos}^{-1}\left(-\frac{1}{2}\right)$.

In the restricted range as shown in the figure, the only number whose sine is $\sqrt{2}/2$ is $\pi/4$. Hence Arcsin $\sqrt{2}/2 = \pi/4$. The only number whose cosine is $-\frac{1}{2}$ in the restricted range is $2\pi/3$. Hence $\text{Cos}^{-1}\left(-\frac{1}{2}\right) = 2\pi/3$.

DO EXERCISES 7–10.

▦ The Use of Calculators

Scientific calculators can be used to find inverses of trigonometric function values. The values obtained from a calculator are *principal* values, and the typical calculator provides values for only the Arcsin, Arccos, and Arctan functions.

Some calculators give inverse function values in either radians or degrees. Some give values only in degrees. The key strokes involved in finding inverse function values vary with the calculator. Be sure to read the instructions for the calculator you are using.

EXERCISE SET 4.4

[i] Find all values of the following without using tables.

1. arcsin $\dfrac{\sqrt{2}}{2}$

2. arcsin $\dfrac{\sqrt{3}}{2}$

3. $\cos^{-1}\dfrac{\sqrt{2}}{2}$

4. $\cos^{-1}\dfrac{\sqrt{3}}{2}$

5. $\sin^{-1}\left(-\dfrac{\sqrt{2}}{2}\right)$

6. $\sin^{-1}\left(-\dfrac{\sqrt{3}}{2}\right)$

7. arccos $\left(-\dfrac{\sqrt{2}}{2}\right)$

8. arccos $\left(-\dfrac{\sqrt{3}}{2}\right)$

9. arctan $\sqrt{3}$

10. arctan $\dfrac{\sqrt{3}}{3}$

11. $\cot^{-1} 1$

12. $\cot^{-1} \sqrt{3}$

13. arctan $\left(-\dfrac{\sqrt{3}}{3}\right)$

14. arctan $(-\sqrt{3})$

15. arccot (-1)

16. arccot $(-\sqrt{3})$

17. arcsec 1

18. arcsec 2

19. $\csc^{-1} 1$

20. $\csc^{-1} 2$

Use tables to find, in degrees, all values of the following.

21. arcsin 0.3907

22. arcsin 0.9613

23. $\sin^{-1} 0.6293$

24. $\sin^{-1} 0.8746$

25. arccos 0.7990

26. arccos 0.9265

27. $\cos^{-1} 0.9310$

28. $\cos^{-1} 0.2735$

29. $\tan^{-1} 0.3673$

30. $\tan^{-1} 1.091$

31. $\cot^{-1} 1.265$

32. $\cot^{-1} 0.4770$

33. $\sec^{-1} 1.167$

34. $\sec^{-1} 1.440$

35. arccsc 6.277

36. arccsc 1.111

[ii] Find the following without using tables.

37. Arcsin $\dfrac{\sqrt{2}}{2}$

38. Arcsin $\dfrac{1}{2}$

39. $\text{Cos}^{-1}\dfrac{1}{2}$

40. $\text{Cos}^{-1}\dfrac{\sqrt{2}}{2}$

41. $\text{Sin}^{-1}\left(-\dfrac{\sqrt{3}}{2}\right)$

42. $\text{Sin}^{-1}\left(-\dfrac{1}{2}\right)$

43. Arccos $\left(-\dfrac{\sqrt{2}}{2}\right)$

44. Arccos $\left(-\dfrac{\sqrt{3}}{2}\right)$

45. $\text{Tan}^{-1}\left(-\dfrac{\sqrt{3}}{3}\right)$

46. $\text{Tan}^{-1}(-\sqrt{3})$

47. Arccot $\left(-\dfrac{\sqrt{3}}{3}\right)$

48. Arccot $(-\sqrt{3})$

Find the following, in degrees, using tables.

49. Arcsin 0.2334

50. Arcsin 0.4514

51. $\text{Sin}^{-1}(-0.6361)$

52. $\text{Sin}^{-1}(-0.8192)$

53. Arccos (-0.8897)

54. Arccos (-0.2924)

55. $\text{Tan}^{-1}(-0.4074)$

56. $\text{Tan}^{-1}(-0.2401)$

57. $\text{Cot}^{-1}(-5.396)$

58. $\text{Cot}^{-1}(-1.319)$

4.5 COMBINATIONS OF TRIGONOMETRIC FUNCTIONS AND THEIR INVERSES

[i] Immediate Simplification

In practice, various combinations of trigonometric functions and their inverses arise. We shall discuss evaluating and simplifying such expressions. Let us consider, for example,

sin Arcsin x (the sine of a number whose sine is x).

Recall, from Chapter 1, that if a function f has an inverse that is also a function, then $f(f^{-1}(x)) = x$ and also $f^{-1}(f(x)) = x$ for all x in the domains of the functions. Since the sine and arcsine relations are inverses of each other, we might suspect that

$$\text{sin Arcsin } x = x$$

for all x in the domain of the Arcsin relation. However, we cannot apply the result of Chapter 1 here because the inverse of the sine function is not a function. Let us study sin Arcsin x using a unit circle. Consider a number a in the domain of the Arcsine function (a number from −1 to 1). Then Arcsin a is the length of an arc s as shown. Its sine is a. In other words, s = Arcsin a.

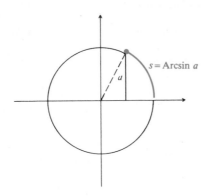

Recalling that s = Arcsin a, and taking the sine on both sides, we have

$$\text{sin } s = \text{sin Arcsin } a.$$

From the drawing we see that sin s = a. Substituting, we have

$$\text{sin Arcsin } a = a.$$

Thus it is true that sin Arcsin x $\equiv$ x. The sensible replacements are those numbers in the domain of the Arcsine function, that is, numbers from −1 to 1.

Similar identities hold for the other trigonometric functions:

sin Arcsin $x \equiv x$,	cot Arccot $x \equiv x$,
cos Arccos $x \equiv x$,	sec Arcsec $x \equiv x$,
tan Arctan $x \equiv x$,	csc Arccsc $x \equiv x$.

OBJECTIVES

You should be able to:

[i] Immediately simplify expressions such as sin Arcsin x to x, and simplify expressions like Arcsin sin x. If x is in the domain of the inverse function, immediately simplify to x.

[ii] Simplify expressions involving combinations such as sin Arccos ($\frac{1}{2}$), without using tables, and simplify expressions such as sin Arctan (a/b) by drawing a triangle and reading off appropriate ratios.

Find the following.

1. Arcsin sin $\dfrac{3\pi}{2}$

2. Sin^{-1} sin $\dfrac{\pi}{6}$

3. Arctan tan $\dfrac{3\pi}{4}$

4. Cos^{-1} cos $\dfrac{5\pi}{4}$

5. Arccos cos $\dfrac{5\pi}{6}$

6. sin Arctan 1

7. cos Arcsin $\dfrac{1}{2}$

8. Cos^{-1} sin $\dfrac{\pi}{6}$

9. Sin^{-1} tan $\dfrac{\pi}{4}$

Now let us consider Arcsin sin x. We might also suspect that this is equal to x for any x, but this is not true unless x is in the range of the Arcsine function.

Example 1 Find Arcsin sin $3\pi/4$.

We first find sin $3\pi/4$. It is $\sqrt{2}/2$. Next we find Arcsin $\sqrt{2}/2$. It is $\pi/4$. So Arcsin sin $3\pi/4 = \pi/4$.

DO EXERCISES 1–5.

For any x in the range of the Arcsine function, we do have Arcsin sin $x = x$. Similar conditions hold for the other functions.

THEOREM 3

> The following are true if x is in the range of the inverse function. Otherwise they are not.
>
> $$\text{Arcsin sin } x = x, \qquad \text{Arccot cot } x = x,$$
> $$\text{Arccos cos } x = x, \qquad \text{Arcsec sec } x = x,$$
> $$\text{Arctan tan } x = x, \qquad \text{Arccsc csc } x = x.$$

[ii] Simplifying Combinations

Now we consider some other kinds of combinations.

Example 2 Find sin Arctan (-1).

We first find Arctan (-1). It is $-\pi/4$. Now we find the sine of this, sin $(-\pi/4) = -\sqrt{2}/2$, so sin Arctan $(-1) = -\sqrt{2}/2$.

Example 3 Find Cos^{-1} sin $\pi/2$.

We first find sin $\pi/2$. It is 1. Now we find Cos^{-1} 1. It is 0, so Cos^{-1} sin $\pi/2 = 0$.

DO EXERCISES 6–9.

Now let us consider

$$\cos \text{Arcsin } \frac{3}{5}.$$

Without using tables, we cannot find Arcsin $\frac{3}{5}$. However, we can still evaluate the entire expression without using tables. In a case like this we sketch a triangle as shown. The angle θ in this triangle is an angle whose sine is $\frac{3}{5}$ (it is Arcsin $\frac{3}{5}$). We wish to find the cosine of this angle. Since the triangle is a right triangle, we can find the length of the base, b. It is 4. Thus we know that cos $\theta = b/5$ or $\frac{4}{5}$. Therefore

$$\cos \text{Arcsin } \frac{3}{5} = \frac{4}{5}.$$

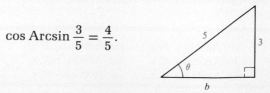

Example 4 Find sin Arccot x/2.

We draw a right triangle whose sides have lengths x and 2, so that cot θ = x/2. We find the length of the hypotenuse and then read off the sine ratio.

$$\sin \text{Arccot } \frac{x}{2} = \frac{2}{\sqrt{x^2 + 2^2}}$$

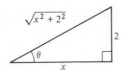

Example 5 Find cos Arctan p.

We draw a right triangle, two of whose sides have lengths p and 1, so that tan θ = p/1. We find the length of the other side and then read off the cosine ratio.

$$\cos \text{Arctan } p = \frac{1}{\sqrt{1 + p^2}}$$

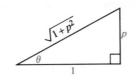

The idea in Examples 4 and 5 is to sketch a triangle, two of whose sides have the appropriate ratio. Then use the Pythagorean theorem to find the length of the third side and read off the desired ratio.

DO EXERCISES 10–12.

In some cases the use of certain identities is needed to evaluate expressions.

Example 6 Evaluate sin (Sin$^{-1}\frac{1}{2}$ + Cos$^{-1}\frac{4}{5}$).

To simplify the use of an identity we shall make substitutions. Let

$$u = \text{Sin}^{-1} \frac{1}{2} \quad \text{and} \quad v = \text{Cos}^{-1} \frac{4}{5}.$$

Then we have sin (u + v). By a sum formula, this is equivalent to

$$\sin u \cos v + \cos u \sin v.$$

Now we make the reverse substitutions and obtain

$$\sin \text{Sin}^{-1} \frac{1}{2} \cdot \cos \text{Cos}^{-1} \frac{4}{5} + \cos \text{Sin}^{-1} \frac{1}{2} \cdot \sin \text{Cos}^{-1} \frac{4}{5}.$$

This immediately simplifies to

$$\frac{1}{2} \cdot \frac{4}{5} + \cos \text{Sin}^{-1} \frac{1}{2} \cdot \sin \text{Cos}^{-1} \frac{4}{5}.$$

Find the following.

10. cos Arctan $\frac{b}{3}$

11. sin Arctan 2

$\left(Hint:\ 2 = \frac{2}{1}.\right)$

12. tan Arcsin t

Evaluate.

13. $\cos\left(\text{Sin}^{-1}\dfrac{\sqrt{3}}{2} - \text{Cos}^{-1}\dfrac{1}{2}\right)$

14. $\tan\left(\dfrac{1}{2}\text{Arcsin}\dfrac{3}{5}\right)$

(*Hint:* Let Arcsin $\frac{3}{5} = u$ and use a half-angle formula.)

Now $\cos \text{Sin}^{-1}\frac{1}{2}$ readily simplifies to $\sqrt{3}/2$. To find $\sin \text{Cos}^{-1}\frac{4}{5}$, we shall need a triangle.

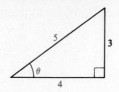

We find that $\sin \text{Cos}^{-1}\frac{4}{5} = \frac{3}{5}$. Our expression is now simplified to

$$\frac{1}{2}\cdot\frac{4}{5} + \frac{\sqrt{3}}{2}\cdot\frac{3}{5},$$

which simplifies to

$$\frac{4 + 3\sqrt{3}}{10}.$$

DO EXERCISES 13 AND 14.

EXERCISE SET 4.5

[i] Evaluate or simplify.

1. $\sin \text{Arcsin } 0.3$

2. $\cos \text{Arccos } 0.2$

3. $\tan \text{Tan}^{-1}(-4.2)$

4. $\cot \text{Cot}^{-1}(-1.5)$

5. $\text{Arcsin} \sin \dfrac{2\pi}{3}$

6. $\text{Arccos} \cos \dfrac{3\pi}{2}$

7. $\text{Sin}^{-1} \sin\left(-\dfrac{3\pi}{4}\right)$

8. $\text{Cos}^{-1} \cos\left(-\dfrac{\pi}{4}\right)$

9. $\text{Sin}^{-1} \sin \dfrac{\pi}{5}$

10. $\text{Cos}^{-1} \cos \dfrac{\pi}{7}$

11. $\text{Tan}^{-1} \tan \dfrac{2\pi}{3}$

12. $\text{Cot}^{-1} \cot \dfrac{2\pi}{3}$

[ii]

13. $\sin \text{Arctan} \sqrt{3}$

14. $\sin \text{Arctan} \dfrac{\sqrt{3}}{3}$

15. $\cos \text{Arcsin} \dfrac{\sqrt{3}}{2}$

16. $\cos \text{Arcsin} \dfrac{\sqrt{2}}{2}$

17. $\tan \text{Cos}^{-1} \dfrac{\sqrt{2}}{2}$

18. $\tan \text{Cos}^{-1} \dfrac{\sqrt{3}}{2}$

19. $\text{Cos}^{-1} \sin \dfrac{\pi}{3}$

20. $\text{Cos}^{-1} \sin \pi$

21. $\text{Arcsin} \cos \dfrac{\pi}{6}$

22. $\text{Arcsin} \cos \dfrac{\pi}{4}$

23. $\text{Sin}^{-1} \tan \dfrac{\pi}{4}$

24. $\text{Sin}^{-1} \tan\left(-\dfrac{\pi}{4}\right)$

25. $\sin \text{Arctan} \dfrac{x}{2}$

26. $\sin \text{Arctan} \dfrac{a}{3}$

27. $\tan \text{Cos}^{-1} \dfrac{3}{x}$

28. $\cot \text{Sin}^{-1} \dfrac{5}{y}$

29. $\cot \text{Sin}^{-1} \dfrac{a}{b}$

30. $\tan \text{Cos}^{-1} \dfrac{p}{q}$

31. $\cos \text{Tan}^{-1} \dfrac{\sqrt{2}}{3}$

32. $\cos \text{Tan}^{-1} \dfrac{\sqrt{3}}{4}$

33. $\tan \text{Arcsin } 0.1$

34. $\tan \text{Arcsin } 0.2$

35. $\cot \text{Cos}^{-1}(-0.2)$

36. $\cot \text{Cos}^{-1}(-0.3)$

37. $\sin \text{Arccot } y$

38. $\sin \text{Arccot } x$

39. $\cos \text{Arctan } t$

40. $\sin \text{Arctan } t$

41. $\cot \text{Sin}^{-1} y$

42. $\tan \text{Cos}^{-1} y$

43. $\sin \text{Cos}^{-1} x$

44. $\cos \text{Sin}^{-1} x$

45. $\tan \left(\frac{1}{2} \text{Arcsin} \frac{4}{5} \right)$

46. $\tan \left(\frac{1}{2} \text{Arcsin} \frac{1}{2} \right)$

47. $\cos \left(\frac{1}{2} \text{Arcsin} \frac{1}{2} \right)$

48. $\cos \left(\frac{1}{2} \text{Arcsin} \frac{\sqrt{3}}{2} \right)$

49. $\sin \left(2 \text{Cos}^{-1} \frac{3}{5} \right)$

50. $\sin \left(2 \text{Cos}^{-1} \frac{1}{2} \right)$

51. $\cos \left(2 \text{Sin}^{-1} \frac{5}{13} \right)$

52. $\cos \left(2 \text{Cos}^{-1} \frac{4}{5} \right)$

53. $\sin \left(\text{Sin}^{-1} \frac{1}{2} + \text{Cos}^{-1} \frac{3}{5} \right)$

54. $\sin \left(\text{Sin}^{-1} \frac{1}{2} - \text{Cos}^{-1} \frac{4}{5} \right)$

55. $\cos \left(\text{Sin}^{-1} \frac{\sqrt{2}}{2} + \text{Cos}^{-1} \frac{3}{5} \right)$

56. $\cos \left(\text{Sin}^{-1} \frac{4}{5} - \text{Cos}^{-1} \frac{1}{2} \right)$

57. $\sin \left(\text{Sin}^{-1} x + \text{Cos}^{-1} y \right)$

58. $\sin \left(\text{Sin}^{-1} x - \text{Cos}^{-1} y \right)$

59. $\cos \left(\text{Sin}^{-1} x + \text{Cos}^{-1} y \right)$

60. $\cos \left(\text{Sin}^{-1} x - \text{Cos}^{-1} y \right)$

61. ▦ $\sin \left(\text{Sin}^{-1} 0.6032 + \text{Cos}^{-1} 0.4621 \right)$

62. ▦ $\cos \left(\text{Sin}^{-1} 0.7325 - \text{Cos}^{-1} 0.4838 \right)$

☆

63. An observer's eye is at point A, looking at a mural of height h, with the bottom of the mural y feet above the eye. The eye is x feet from the wall. Write an expression for θ in terms of x, y, and h.

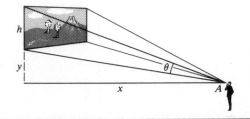

64. ▦ Evaluate the expression given in Exercise 63 when $x = 20$ ft, $y = 7$ ft, and $h = 25$ ft.

4.6 TRIGONOMETRIC EQUATIONS

[i] Simple Equations

When an equation contains a trigonometric expression with a variable such as sin x, it is called a *trigonometric equation*. To solve such an equation, we find all replacements for the variable that make the equation true.

Example 1 Solve $2 \sin x = 1$.

We first solve for sin x:

$$\sin x = \frac{1}{2}.$$

Now we note that the solutions are those numbers having a sine of $\frac{1}{2}$. We look for them. The unit circle is helpful. There are just two points on it for which the sine is $\frac{1}{2}$, as shown.

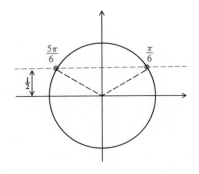

OBJECTIVES

You should be able to:

[i] Solve simple trigonometric equations not requiring the use of tables, finding all solutions, or all solutions in $[0, 2\pi)$.

[ii] Solve simple trigonometric equations requiring the use of tables, finding all solutions, or all solutions in $[0, 360°)$.

[iii] Solve trigonometric equations by factoring or using the quadratic formula.

1. Solve. Give answers in both degrees and radians.

$$2 \cos x = 1$$

They are the points for $\pi/6$ and $5\pi/6$. These numbers, plus any multiple of 2π, are the solutions

$$\frac{\pi}{6} + 2k\pi \quad \text{or} \quad \frac{5\pi}{6} + 2k\pi,$$

where k is any integer. In degrees, the solutions are

$$30° + k \cdot 360° \quad \text{or} \quad 150° + k \cdot 360°,$$

where k is any integer.

DO EXERCISE 1.

Example 2 Solve $4 \cos^2 x = 1$.

We first solve for $\cos x$:

$$\cos^2 x = \frac{1}{4}$$

$$|\cos x| = \frac{1}{2} \qquad \text{Taking the principal square root}$$

$$\cos x = \pm \frac{1}{2}.$$

2. Find all solutions in $[0, 2\pi)$. Give answers in both degrees and radians.

$$4 \sin^2 x = 1$$

Now we use the unit circle to find those numbers having a cosine of $\frac{1}{2}$ or $-\frac{1}{2}$. The solutions are $\pi/3$, $2\pi/3$, $4\pi/3$, $5\pi/3$, plus any multiple of 2π.

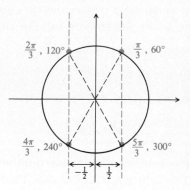

DO EXERCISE 2.

In solving trigonometric equations, it is usually sufficient to find just the solutions from 0 to 2π. We then remember that any multiple of 2π may be added to obtain the rest of the solutions.

The following example illustrates that when we look for solutions in the interval $[0, 2\pi)$ we must be cautious.

Example 3 Solve $2 \sin 2x = 1$ in the interval $[0, 2\pi)$.

We first solve for $\sin 2x$: $\sin 2x = \frac{1}{2}$.

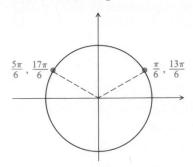

Using the unit circle we find points for which $\sin 2x = \frac{1}{2}$. However, since x is to run from 0 to 2π, we must let 2x run from 0 to 4π. These values of 2x are $\pi/6$, $5\pi/6$, $13\pi/6$, and $17\pi/6$. Thus the desired values of x in $[0, 2\pi)$ are half of these. Therefore,

$$x = \frac{\pi}{12}, \quad \frac{5\pi}{12}, \quad \frac{13\pi}{12}, \quad \frac{17\pi}{12}.$$

DO EXERCISE 3.

[ii] **Using Tables**

In solving some trigonometric equations, it is necessary to use tables. In such cases, we usually find answers in degrees.

Example 4 Solve $2 + \sin x = 2.5299$ in $[0, 360°)$.

We first solve for $\sin x$:

$$\sin x = 0.5299.$$

From the table we find the reference angle, $x = 32°$. Since $\sin x$ is positive, the solutions are to be found in the first and second quadrants. The solutions are

$$32° \quad \text{and} \quad 148°.$$

Example 5 Solve $\sin x - 1 = -1.5299$ in $[0, 360°)$.

We first solve for $\sin x$:

$$\sin x = -0.5299.$$

From the table we find the reference angle, $32°$. Since $\sin x$ is negative, the solutions are to be found in the third and fourth quadrants. The solutions are

$$212° \quad \text{and} \quad 328°.$$

DO EXERCISES 4 AND 5.

3. Find all solutions in $[0, 2\pi)$. Leave answers in terms of π.

$$2 \cos 2x = 1$$

Solve. Give answers in degrees.

4. $2 + \cos x = 2.7660$

5. $\cos x - 1 = -1.7660$

Solve.

6. $8 \cos^2 \theta + 2 \cos \theta = 1$

7. $2 \cos^2 \phi + \cos \phi = 0$

[iii] **Using Algebraic Techniques**

In solving trigonometric equations, we can expect to apply some algebra before concerning ourselves with the trigonometric part. In the next examples, we begin by factoring. The equations are reducible to quadratic.

Example 6 Solve $8 \cos^2 \theta - 2 \cos \theta = 1$.

Since we will be using the principle of zero products, we first obtain a 0 on one side of the equation.

$$8 \cos^2 \theta - 2 \cos \theta - 1 = 0$$
$$(4 \cos \theta + 1)(2 \cos \theta - 1) = 0 \qquad \text{Factoring}$$
$$4 \cos \theta + 1 = 0 \qquad \text{or} \quad 2 \cos \theta - 1 = 0 \qquad \begin{array}{l}\text{Principle of zero} \\ \text{products}\end{array}$$

$$\cos \theta = -\frac{1}{4} = -0.25 \quad \text{or} \qquad \cos \theta = \frac{1}{2}$$

$$\theta = 104°30', \ 255°30' \quad \text{or} \quad \theta = 60°, \ 300° \quad \left[\text{or } \frac{\pi}{3}, \frac{5\pi}{3} \right]$$

The solutions in $[0, 360°)$ are $104°30'$, $255°30'$, $60°$, and $300°$.

It may be helpful in accomplishing the algebraic part of solving trigonometric equations to make substitutions. If we use such an approach, the algebraic part of Example 6 would look like this:

$$8 \cos^2 \theta - 2 \cos \theta = 1.$$

Let $u = \cos \theta$.

$$8u^2 - 2u = 1$$
$$(4u + 1)(2u - 1) = 0$$
$$4u + 1 = 0 \qquad \text{or} \quad 2u - 1 = 0$$
$$u = -\frac{1}{4} \quad \text{or} \qquad u = \frac{1}{2}$$

Example 7 Solve $2 \sin^2 \phi + \sin \phi = 0$.

$$\sin \phi (2 \sin \phi + 1) = 0 \qquad \text{Factoring}$$
$$\sin \phi = 0 \qquad \text{or} \quad 2 \sin \phi + 1 = 0 \qquad \text{Principle of zero products}$$

$$\sin \phi = 0 \qquad \text{or} \qquad \sin \phi = -\frac{1}{2}$$

$$\phi = 0, \ \pi \quad \text{or} \qquad \phi = \frac{7\pi}{6}, \frac{11\pi}{6}$$

The solutions in $[0, 2\pi)$ are 0, π, $7\pi/6$, and $11\pi/6$.

DO EXERCISES 6 AND 7.

In case a trigonometric equation is quadratic but difficult or impossible to factor, we use the quadratic formula.

Example 8 Solve $10 \sin^2 x - 12 \sin x - 7 = 0$ in $[0, 360°)$.

It may help to make the substitution $u = \sin x$, in order to obtain $10u^2 - 12u - 7 = 0$. Then use the quadratic formula to find u, or $\sin x$.

$$\sin x = \frac{12 \pm \sqrt{144 + 280}}{20} \quad \text{Using the quadratic formula}$$

$$= \frac{12 \pm \sqrt{424}}{20} = \frac{12 \pm 2\sqrt{106}}{20} = \frac{6 \pm \sqrt{106}}{10}$$

$$= \frac{6 \pm 10.296}{10}$$

$$\sin x = 1.6296 \quad \text{or} \quad \sin x = -0.4296$$

Since sines are never greater than 1, the first of the equations on the preceding line has no solution. We look up the other number and find the reference angle to be $25°30'$ to the nearest $10'$. Thus the solutions are $205°30'$ and $334°30'$.

DO EXERCISE 8.

8. Solve using the quadratic formula.

$$10 \cos^2 x - 10 \cos x - 7 = 0$$

EXERCISE SET 4.6

[i], [ii] Solve, finding all solutions.

1. $\sin x = \dfrac{\sqrt{3}}{2}$

2. $\cos x = \dfrac{\sqrt{3}}{2}$

3. $\cos x = \dfrac{1}{\sqrt{2}}$

4. $\tan x = \sqrt{3}$

5. $\sin x = 0.3448$

6. $\cos x = 0.6406$

Solve, finding all solutions in $[0, 2\pi)$ or $[0°, 360°)$.

7. $\cos x = -0.5495$

8. $\sin x = -0.4279$

9. $2 \sin x + \sqrt{3} = 0$

10. $\sqrt{3} \tan x + 1 = 0$

11. $2 \tan x + 3 = 0$

12. $4 \sin x - 1 = 0$

[iii]

13. $4 \sin^2 x - 1 = 0$

14. $2 \cos^2 x = 1$

15. $\cot^2 x - 3 = 0$

16. $\csc^2 x - 4 = 0$

17. $2 \sin^2 x + \sin x = 1$

18. $2 \cos^2 x + 3 \cos x = -1$

19. $\cos^2 x + 2 \cos x = 3$

20. $2 \sin^2 x - \sin x = 3$

21. $4 \sin^3 x - \sin x = 0$

22. $2 \cos^2 x - \sqrt{3} \cos x = 0$

23. $2 \sin^2 \theta + 7 \sin \theta = 4$

24. $2 \sin^2 \theta - 5 \sin \theta + 2 = 0$

25. $6 \cos^2 \phi + 5 \cos \phi + 1 = 0$

26. $2 \sin^2 \phi + \sin \phi - 1 = 0$

27. $2 \sin t \cos t + 2 \sin t - \cos t - 1 = 0$

28. $2 \sin t \tan t + \tan t - 2 \sin t - 1 = 0$

29. $\cos 2x \sin x + \sin x = 0$

30. $\sin 2x \cos x - \cos x = 0$

☆ ───

Solve, restricting solutions to $[0, 2\pi)$ or $[0°, 360°)$ where sensible to do so.

31. $|\sin x| = \dfrac{\sqrt{3}}{2}$

32. $|\cos x| = \dfrac{1}{2}$

33. $\sqrt{\tan x} = \sqrt[4]{3}$

34. $12 \sin x - 7\sqrt{\sin x} + 1 = 0$

35. $16 \cos^4 x - 16 \cos^2 x + 3 = 0$

36. $\ln \cos x = 0$

37. $e^{\sin x} = 1$

38. $\sin \ln x = -1$

39. $e^{\ln \sin x} = 0$

40. Solve this system of equations for x and y:

$$x \cos \theta - y \sin \theta = 0,$$
$$x \sin \theta + y \cos \theta = 1.$$

41. ▦ Solve this system graphically:

$$y = 2 \sin \frac{x}{2},$$
$$y = \operatorname{Tan}^{-1} x.$$

42. Solve graphically: $\sin x = \tan \dfrac{x}{2}$.

──

OBJECTIVE

You should be able to:

[i] Solve trigonometric equations requiring the use of identities.

4.7 IDENTITIES IN SOLVING EQUATIONS

[i] When a trigonometric equation involves more than one function, we may use identities to put it in terms of a single function. This can usually be done in several ways. In the following three examples, we illustrate by solving the same equation three different ways.

Example 1 Solve the equation $\sin x + \cos x = 1$.

To express $\cos x$ in terms of $\sin x$, we use the identity

$$\sin^2 x + \cos^2 x \equiv 1.$$

From this we obtain $\cos x \equiv \pm \sqrt{1 - \sin^2 x}$. Now we substitute in the original equation:

$$\sin x \pm \sqrt{1 - \sin^2 x} = 1.$$

This is a radical equation. We get the radical alone on one side and then square both sides:

$$1 - \sin^2 x = 1 - 2 \sin x + \sin^2 x.$$

This simplifies to

$$-2 \sin^2 x + 2 \sin x = 0.$$

Now we factor:

$$-2 \sin x (\sin x - 1) = 0.$$

Using the principle of zero products, we have

$$-2 \sin x = 0 \quad \text{or} \quad \sin x - 1 = 0,$$
$$\sin x = 0 \quad \text{or} \quad \sin x = 1.$$

The values of x in $[0, 2\pi)$ satisfying these are

$$x = 0, \quad x = \pi, \quad \text{or} \quad x = \frac{\pi}{2}.$$

Now we check these in the original equation. We find that π does not check but the other values do. Thus the solutions are 0 and $\pi/2$.

CAUTION! **It is important to check when solving. Values are often obtained that are not solutions of the original equation.**

DO EXERCISE 1.

Example 2 Solve $\sin x + \cos x = 1$.

This time we will square both sides, obtaining

$$\sin^2 x + 2 \sin x \cos x + \cos^2 x = 1.$$

Since $\sin^2 x + \cos^2 x \equiv 1$, we can simplify to

$$2 \sin x \cos x = 0.$$

Now we use the identity $2 \sin x \cos x \equiv \sin 2x$, and obtain

$$\sin 2x = 0.$$

The values in $[0, 2\pi)$ satisfying this equation are 0, $\pi/2$, π, and $3\pi/2$. Only 0 and $\pi/2$ check.

DO EXERCISE 2.

Example 3 Solve $\sin x + \cos x = 1$.

Let us recall the identity $c \sin ax + d \cos ax \equiv A \sin (ax + b)$, where $A = \sqrt{c^2 + d^2}$ and b is a number whose cosine is c/A and whose sine is d/A. In this case $c = d = a = 1$, so $A = \sqrt{2}$ and $b = \pi/4$. Thus our equation is equivalent to

$$\sqrt{2} \sin \left(x + \frac{\pi}{4} \right) = 1.$$

Then

$$\sin \left(x + \frac{\pi}{4} \right) = \frac{1}{\sqrt{2}} = \frac{\sqrt{2}}{2}$$

$$x + \frac{\pi}{4} = \frac{\pi}{4} \quad \text{or} \quad x + \frac{\pi}{4} = \frac{3\pi}{4}$$

$$x = 0 \quad \text{or} \quad x = \frac{\pi}{2}.$$

Both values check.

DO EXERCISE 3.

1. Solve $\sin x - \cos x = 1$ as in Example 1.

2. Solve $\sin x - \cos x = 1$, as in Example 2.

3. Solve $\sin x - \cos x = 1$, as in Example 3.

4. Solve $\tan^2 x \cos x - \cos x = 0$.

Example 4 Solve $2\cos^2 x \tan x - \tan x = 0$.

First, we factor:

$$\tan x (2\cos^2 x - 1) = 0.$$

Now we use the identity $2\cos^2 x - 1 \equiv \cos 2x$:

$$\tan x \cos 2x = 0$$

$$\tan x = 0 \quad \text{or} \quad \cos 2x = 0 \qquad \text{Principle of zero products}$$

$$x = 0, \pi \quad \text{or} \qquad 2x = \frac{\pi}{2}, \frac{3\pi}{2}, \frac{5\pi}{2}, \frac{7\pi}{2}$$

$$x = 0, \pi \quad \text{or} \qquad x = \frac{\pi}{4}, \frac{3\pi}{4}, \frac{5\pi}{4}, \frac{7\pi}{4}.$$

All values check. The solutions in $[0, 2\pi)$ are 0, π, $\pi/4$, $3\pi/4$, $5\pi/4$, and $7\pi/4$.

DO EXERCISE 4.

Example 5 Solve $\cos 2x + \sin x = 1$.

We first use the identity $\cos 2x \equiv 1 - 2\sin^2 x$, to get

$$1 - 2\sin^2 x + \sin x = 1$$
$$-2\sin^2 x + \sin x = 0$$
$$\sin x (1 - 2\sin x) = 0 \qquad \text{Factoring}$$
$$\sin x = 0 \quad \text{or} \quad 1 - 2\sin x = 0 \qquad \text{Principle of zero products}$$

$$\sin x = 0 \quad \text{or} \qquad \sin x = \frac{1}{2}$$

$$x = 0, \pi \quad \text{or} \qquad x = \frac{\pi}{6}, \frac{5\pi}{6}.$$

5. Solve $\sin 2x + \cos x = 0$.

All values check. The solutions in $[0, 2\pi)$ are 0, π, $\pi/6$, and $5\pi/6$.

DO EXERCISE 5.

Example 6 Solve $\sin 5\theta \cos 2\theta - \cos 5\theta \sin 2\theta = \sqrt{2}/2$.

We use the identity

$$\sin(\alpha - \beta) \equiv \sin \alpha \cos \beta - \cos \alpha \sin \beta$$

to get $\sin 3\theta = \sqrt{2}/2$. Then

$$3\theta = \frac{\pi}{4}, \frac{3\pi}{4}, \frac{9\pi}{4}, \frac{11\pi}{4}, \frac{17\pi}{4}, \frac{19\pi}{4}$$

and

$$\theta = \frac{\pi}{12}, \frac{\pi}{4}, \frac{3\pi}{4}, \frac{11\pi}{12}, \frac{17\pi}{12}, \frac{19\pi}{12}.$$

Let us check $\pi/12$:

$$\sin \frac{5\pi}{12} \cos \frac{\pi}{6} - \cos \frac{5\pi}{12} \sin \frac{\pi}{6}.$$

We again use the difference identity to obtain $\sin(5\pi/12 - \pi/6)$. This is equal to sin $3\pi/12$ or sin $\pi/4$, and this is equal to $\sqrt{2}/2$. Hence $\pi/12$ checks. All of the above answers check.

DO EXERCISE 6.

Example 7 Solve $\tan^2 x + \sec x - 1 = 0$.

We use the identity $1 + \tan^2 x \equiv \sec^2 x$. Substituting, we get

$$\sec^2 x - 1 + \sec x - 1 = 0,$$

or

$$\sec^2 x + \sec x - 2 = 0$$
$$(\sec x + 2)(\sec x - 1) = 0 \qquad \text{Factoring}$$
$$\sec x = -2 \qquad \text{or} \qquad \sec x = 1 \qquad \text{Principle of zero products}$$
$$x = \frac{2}{3}\pi, \frac{4}{3}\pi \quad \text{or} \qquad x = 0.$$

All of these values check. The solutions in $[0, 2\pi)$ are 0, $2\pi/3$, and $4\pi/3$.

Example 8 Solve $\sin x = \sin (2x - \pi)$.

We can use the identity $\sin (y - \pi) \equiv -\sin y$ or the identity $\sin (\alpha - \beta) \equiv \sin \alpha \cos \beta - \cos \alpha \sin \beta$ with the right-hand side. We get

$$\sin x = -\sin 2x.$$

Next we use the identity $\sin 2x \equiv 2 \sin x \cos x$, and obtain

$$\sin x = -2 \sin x \cos x, \quad \text{or} \quad 2 \sin x \cos x + \sin x = 0$$
$$\sin x (2 \cos x + 1) = 0 \qquad \text{Factoring}$$

$$\sin x = 0 \quad \text{or} \quad \cos x = -\frac{1}{2} \qquad \text{Principle of zero products}$$

$$x = 0, \pi \quad \text{or} \qquad x = \frac{2\pi}{3}, \frac{4\pi}{3}.$$

All of these values check, so the solutions in $[0, 2\pi)$ are 0, π, $2\pi/3$, and $4\pi/3$.

DO EXERCISE 7.

6. Solve

$$\sin 3\theta \cos \theta - \cos 3\theta \sin \theta = \frac{1}{2}.$$

7. Solve $\cos x = \cos (\pi - 2x)$.

EXERCISE SET 4.7

[i] Find all solutions of the following equations in $[0, 2\pi)$.

1. $\tan x \sin x - \tan x = 0$

2. $2 \sin x \cos x + \sin x = 0$

3. $2 \sec x \tan x + 2 \sec x + \tan x + 1 = 0$

4. $2 \csc x \cos x - 4 \cos x - \csc x + 2 = 0$

5. $\sin 2x - \cos x = 0$

6. $\cos 2x - \sin x = 1$

7. $\sin 2x \sin x - \cos x = 0$

8. $\sin 2x \cos x - \sin x = 0$

9. $\sin 2x + 2 \sin x \cos x = 0$

10. $\cos 2x \sin x + \sin x = 0$

11. $\cos 2x \cos x + \sin 2x \sin x = 1$

12. $\sin 2x \sin x - \cos 2x \cos x = -\cos x$

13. $\sin 4x - 2 \sin 2x = 0$

14. $\sin 4x + 2 \sin 2x = 0$

15. $\sin 2x + 2\sin x - \cos x - 1 = 0$

16. $\sin 2x + \sin x + 2\cos x + 1 = 0$

17. $\sec^2 x = 4\tan^2 x$

18. $\sec^2 x - 2\tan^2 x = 0$

19. $\sec^2 x + 3\tan x - 11 = 0$

20. $\tan^2 x + 4 = 2\sec^2 x + \tan x$

21. $\cot x = \tan(2x - 3\pi)$

22. $\tan x = \cot(2x + \pi)$

23. $\cos(\pi - x) + \sin\left(x - \dfrac{\pi}{2}\right) = 1$

24. $\sin(\pi - x) + \cos\left(\dfrac{\pi}{2} - x\right) = 1$

25. $\dfrac{\cos^2 x - 1}{\sin\left(\dfrac{\pi}{2} - x\right) - 1} = \dfrac{\sqrt{2}}{2} + 1$

26. $\dfrac{\sin^2 x - 1}{\cos\left(\dfrac{\pi}{2} - x\right) + 1} = \dfrac{\sqrt{2}}{2} - 1$

27. $2\cos x + 2\sin x = \sqrt{6}$

28. $2\cos x + 2\sin x = \sqrt{2}$

29. $\sqrt{3}\cos x - \sin x = 1$

30. $\sqrt{2}\cos x - \sqrt{2}\sin x = 2$

☆ _____

Solve.

31. $\operatorname{Arccos} x = \operatorname{Arccos}\dfrac{3}{5} - \operatorname{Arcsin}\dfrac{4}{5}$

32. $\operatorname{Sin}^{-1} x = \operatorname{Tan}^{-1}\dfrac{1}{3} + \operatorname{Tan}^{-1}\dfrac{1}{2}$

33. Solve graphically: $\sin x - \cos x = \cot x$.

★ _____

34. Solve this system:

$$\sin x + \cos y = 1,$$
$$2\sin x = \sin 2y.$$

CHAPTER 4 REVIEW

[4.1, **i**]   Use the sum and difference formulas to write equivalent expressions. You need not simplify.

    1. $\cos\left(x + \dfrac{3\pi}{2}\right)$

    2. $\tan(45° - 30°)$

[4.1, **ii**]   **3.** Simplify $\cos 27° \cos 16° + \sin 27° \sin 16°$.

    4. Given $\tan \alpha = \sqrt{3}$, $\sin \beta = \sqrt{2}/2$, and α and β are between 0 and $\pi/2$, evaluate $\tan(\alpha - \beta)$ exactly.

[4.2, **ii**]   **5.** Find $\sin\dfrac{\pi}{8}$.

[4.2, **iii**]   **6.** Simplify $\dfrac{\sin 2\theta}{\sin^2 \theta}$.

[4.3, **i**]   **7.** Prove the identity: $\tan 2\theta \equiv \dfrac{2\tan \theta}{1 - \tan^2 \theta}$.

[4.4, **i**]   **8.** Find, in radians, all values of $\sin^{-1}\dfrac{1}{2}$.

    9. Use a table to find, in degrees, all values of $\cot^{-1} 0.1584$.

[4.4, **ii**]   **10.** Find $\operatorname{Sin}^{-1}\left(-\dfrac{\sqrt{2}}{2}\right)$.

[4.5, **ii**]   **11.** Evaluate or simplify $\tan\left(\operatorname{Arctan}\dfrac{7}{8}\right)$.

[4.1, **iii**]   **12.** Find the angle, to the nearest 10 minutes, from l_1 to l_2, given the equations

$$l_1{:}\, y = 2x - 4 \quad \text{and} \quad l_2{:}\, x - y = 2.$$

[4.6, **iii**]   **13.** Solve: $\sin^2 x - 7\sin x = 0$, solutions in $[0, 2\pi)$.

[4.3, **iii**]   **14.** Find an equivalent expression involving the sine function only: $6\sin 3x + 2\cos 3x$.

    15. Graph $y + 1 = 2\cos^2 x$.

    16. Graph $f(x) = 2\operatorname{Sin}^{-1}\left(x + \dfrac{\pi}{2}\right)$.

5

TRIANGLES, VECTORS, AND APPLICATIONS

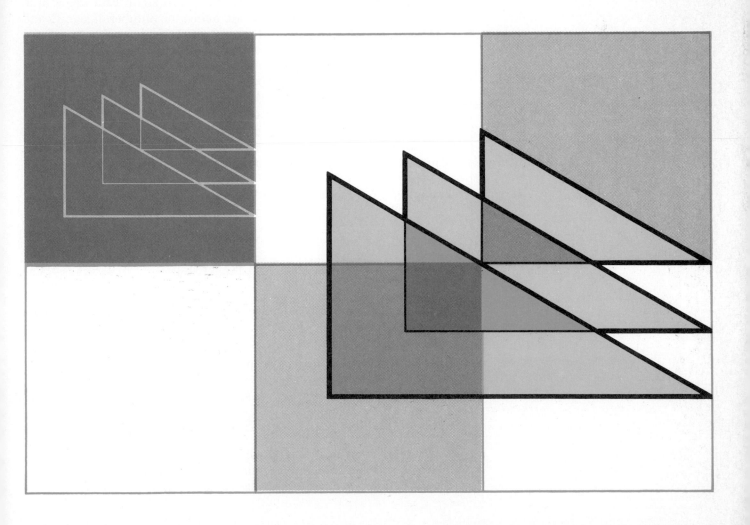

OBJECTIVES

You should be able to:

[i] Solve right triangles.
[ii] Solve applied problems involving the solution of right triangles.

1. Find the lengths a and b in this triangle. Assume that angle A is given to the nearest minute, and length AB to four-digit precision.

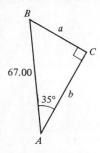

2. Solve this triangle. Use four-digit precision.

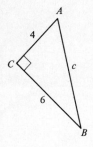

5.1 SOLVING RIGHT TRIANGLES AND APPLICATIONS

[i] Solving Triangles

In Section 2.1 the trigonometric functions were defined and the solving of right triangles was introduced. Tables of the trigonometric functions were considered in Chapter 3. We continue consideration of solving right triangles, a topic important in many applications of trigonometry. The word *trigonometry* actually originally meant "triangle measurement."

The precision to which we can find an angle, using a ratio of sides, such as the sine ratio, depends upon how precisely we know the lengths of the sides. The following table shows the relationship.

NUMBER OF DIGITS IN RATIO	PRECISION OF ANGLE MEASURE
4	To nearest minute
3	To nearest ten minutes
2	To nearest degree

The table can also be read in reverse. For example, if we know an angle to the nearest ten minutes, then the number of significant digits in a value obtained from the table is 3, and hence the number of significant digits obtained for the length of a side of a triangle is also 3.

When we have four-digit precision, but a number itself has fewer than four digits, we often indicate the precision by writing 0's after the decimal point, as in 18.00 or 7.000.

In some of the exercises of this section, the precision is not realistic. For example, one would scarcely consider the distance from Los Angeles to Chicago to the nearest thousandth of a mile. The exercises may use unwarranted precision, from the standpoint of reality, in order to provide valid practice in calculating.

Example 1 Find the length b in this triangle. Use four-digit precision.

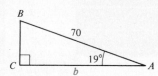

The known side is the hypotenuse. The side we seek is adjacent to the known angle. Thus we shall use the cosine: $\cos A = b/70$. We solve for b:

$$b = 70 \times \cos 19° = 70 \times 0.9455$$
$$b = 66.19.$$

DO EXERCISES 1 AND 2.

When we solve a triangle, we find the *measures* of its sides and angles not already known. We sometimes shorten this to saying that we "find the angles" or "find the sides."

[ii] Applications of Solving Triangles

In many applied problems, unknown parts of right triangles are to be found.

Example 2 *Finding cloud height.* A device for measuring cloud height at night consists of a vertical beam of light, which makes a spot on the clouds. The spot is viewed from a point 135 ft away. The angle of elevation is 67°40′. (The angle between the horizontal and a line of sight is called an *angle of elevation* or an *angle of depression,* the latter if the line of sight is below the horizontal.) Find the height of the clouds.

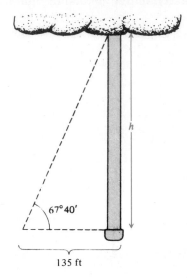

135 ft

From the drawing we have

$$\frac{h}{135} = \tan 67°40′$$

$$h = 135 \times \tan 67°40′$$

$$= 135 \times 2.434$$

$$= 329 \text{ ft.}$$

Note that distances are precise to three digits and the angle to the nearest ten minutes.

Following is a general procedure for solving problems involving triangles.

To solve a triangle problem:

1. **Draw a sketch of the problem situation.**
2. **Look for triangles and sketch them in.**
3. **Mark the known and unknown sides and angles.**
4. **Express the desired side or angle in terms of known trigonometric ratios. Then solve.**

DO EXERCISE 3.

3. The length of a guy wire to a pole is 37.7 ft. It makes an angle of 71°20′ with the ground, which is horizontal. How high above the ground is it attached to the pole?

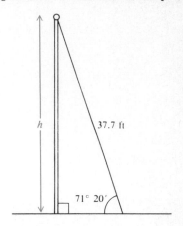

4. A downslope distance is measured to be 241.3 ft and the angle of depression α is measured to be 5°15′. Find the horizontal distance.

Calculators

A calculator would be most convenient for use with Example 2. The same is true for most of the examples and exercises of this chapter, and the use of a calculator is recommended, although not essential, except for certain exercises marked ▦.

If function values are obtained from a calculator, rather than Table 2, it should be kept in mind that the number of decimal places will be different; hence answers may vary because of rounding-error differences. The answers in the back of the book are computed using Table 2.

It is vital to note whether the calculator requires parts of degrees to be entered in tenths and hundredths, rather than minutes and seconds.

Example 3 (*Surveying.*) Horizontal distances must often be measured, even though terrain is not level. One way of doing it is as follows. Distance down a slope is measured with a surveyor's tape, and the distance d is measured by making a level sighting from A to a pole held vertically at B, or the angle α is measured by an instrument placed at A. Suppose that a slope distance L is measured to be 121.3 ft and the angle α is measured to be 3°25′. Find the horizontal distance from A to B.

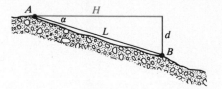

From the drawing we see that $H/L = \cos \alpha$. Thus $H = L \cos \alpha$, and in this case,

$$H = 121.3 \times 0.9982$$
$$= 121.1 \text{ ft.}$$

DO EXERCISE 4.

Example 4 In aerial navigation, directions are given in degrees, clockwise from north. Thus east is 90°, south is 180°, and so on. An airplane leaves an airport and travels for 100 mi in a direction 300°. How far north of the airport is the plane then? how far west?

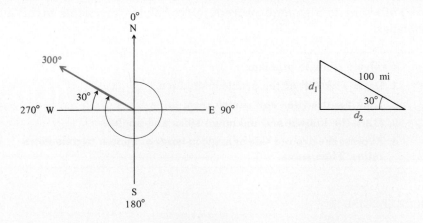

The direction of flight is as shown. In the triangle, d_1 is the northerly distance and d_2 is the westerly distance. Then

$$\frac{d_1}{100} = \sin 30° \quad\text{and}\quad \frac{d_2}{100} = \cos 30°,$$

$$d_1 = 100 \sin 30° = 100 \times 0.5$$
$$= 50 \text{ mi (to the nearest mile)},$$
$$d_2 = 100 \cos 30° = 100 \times 0.866$$
$$= 87 \text{ mi} \quad\text{(to the nearest mile)}.$$

Example 5 In surveying and some other applications, directions, or *bearings,* are given by reference to north or south using an acute angle. For example, N 40°W means 40° west of north and S 30°E means 30° east of south. A forest ranger at point A sights a fire directly south of him. A second ranger at a point B, 7 miles east, sights the fire at a bearing of S 27°20′W. How far from A is the fire?

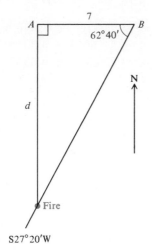

From the drawing we see that the desired distance d is part of a right triangle, as shown. We have

$$\frac{d}{7} = \tan 62°40′$$
$$d = 7 \tan 62°40′ = 7 \times 1.935$$
$$= 13.5 \text{ mi}.$$

DO EXERCISES 5 AND 6.

▦ **Example 6** From an observation tower two markers are viewed on the ground. The markers and the base of the tower are on a line and the observer's eye is 65.3 ft above the ground. The angles of depression to the markers are 53°10′ and 27°50′. How far is it from one marker to the other?

We first make a sketch. We look for right triangles and sketch them in. Then we mark the known information on the sketch. The distance we seek is d, which is $d_1 - d_2$.

5. An airplane flies 150 km from an airport in a direction of 115°. It is then how far east of the airport? how far south?

6. Directly east of a lookout station there is a small forest fire. The bearing of this fire from a station 12 km south of the first is N 57°10′ E. How far is the fire from the southerly lookout station?

7. From an airplane flying 7500 ft above level ground, one can see two towns directly to the east. The angles of depression to the towns are 5°10′ and 77°30′. How far apart are the towns, to the nearest mile?

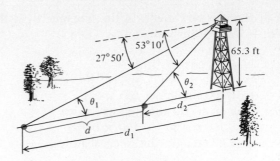

From the right triangles in the drawing, we have

$$\frac{d_1}{65.3} = \cot \theta_1 \quad \text{and} \quad \frac{d_2}{65.3} = \cot \theta_2.$$

Then $d_1 = 65.3 \cot 27°50′$ and $d_2 = 65.3 \cot 53°10′$. We could calculate these and subtract, but the use of a calculator is more efficient if we leave all calculations until last.

$$
\begin{aligned}
d = d_1 - d_2 &= 65.3 \cot 27°50′ - 65.3 \cot 53°10′ \\
&= 65.3(\cot 27°50′ - \cot 53°10′) \\
&= 65.3(\cot 27.83° - \cot 53.17°) \qquad \text{Converting} \\
&\qquad\qquad\qquad\qquad\qquad\qquad\qquad\quad \text{to decimals} \\
&= 65.3(1.894 - 0.7490) \\
&= 74.8 \text{ ft}
\end{aligned}
$$

DO EXERCISE 7.

EXERCISE SET 5.1

[i] In Exercises 1–18, standard lettering for a right triangle will be used: A, B, and C are the angles, C being the right angle. The sides opposite A, B, and C are a, b, and c, respectively. Solve the triangles, using three-digit precision.

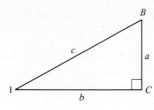

1. $A = 36°10′$, $a = 27.2$

2. $A = 87°40′$, $a = 9.73$

3. $B = 12°40′$, $b = 98.1$

4. $B = 69°50′$, $b = 127$

5. $A = 17°20′$, $b = 13.6$

6. $A = 78°40′$, $b = 1340$

7. $B = 23°10′$, $a = 0.0345$

8. $B = 69°20′$, $a = 0.00488$

9. $A = 47°30′$, $c = 48.3$

10. $A = 88°50′$, $c = 3950$

11. $B = 82°20′$, $c = 0.982$

12. $B = 56°30′$, $c = 0.0447$

13. $a = 12.0$, $b = 18.0$

14. $a = 10.0$, $b = 20.0$

15. $a = 16.0$, $c = 20.0$

16. $a = 15.0$, $c = 45.0$

17. $b = 1.80$, $c = 4.00$

18. $b = 100$, $c = 450$

[ii]

19. A guy wire to a pole makes an angle of 73°10′ with the level ground, and is 14.5 ft from the pole at the ground. How far above the ground is the wire attached to the pole?

20. A kite string makes an angle of 31°40′ with the (level) ground and 455 ft of string is out. How high is the kite?

21. A road rises 3 m per 100 horizontal m. What angle does it make with the horizontal?

22. A kite is 120 ft high when 670 ft of string is out. What angle does the kite make with the ground?

23. What is the angle of elevation of the sun when a 6-ft man casts a 10.3-ft shadow?

24. What is the angle of elevation of the sun when a 35-ft mast casts a 20-ft shadow?

25. From a balloon 2500 ft high, a command post is seen with an angle of depression of 7°40′. How far is it from a point on the ground below the balloon to the command post?

26. From a lighthouse 55 ft above sea level, the angle of depression to a small boat is 11°20′. How far from the foot of the lighthouse is the boat?

27. An airplane travels at 120 km/h for 2 hr in a direction of 243° from Chicago. At the end of this time, how far south of Chicago is the plane?

28. An airplane travels at 150 km/h for 2 hr in a direction of 138° from Omaha. At the end of this time, how far east of Omaha is the plane?

29. Ship A is due west of a lighthouse. Ship B is 12 km south of ship A. From ship B the bearing to the lighthouse is N 63°20′E. How far is ship A from the lighthouse?

30. Lookout station A is 15 km west of station B. The bearing from A to a fire directly south of B is S 37°50′E. How far is the fire from B?

31. From a balloon two km high, the angles of depression to two towns in line with the balloon are 81°20′ and 13°40′. How far apart are the towns?

32. From a balloon 1000 m high, the angles of depression to two artillery posts, in line with the balloon, are 11°50′ and 84°10′. How far apart are the artillery posts?

33. A weather balloon is directly west of two observing stations 10 km apart. The angles of elevation of the balloon from the two stations are 17°50° and 78°10′. How high is the balloon?

34. From two points south of a hill on level ground and 1000 ft apart, the angles of elevation of the hill are 12°20′ and 82°40′. How high is the hill?

35. Show that the area of a right triangle is $\frac{1}{4}c^2 \sin 2A$.

36. Show that the area of a right triangle is $bc \sin A/2$.

★

37. Find a formula for the distance to the horizon, as a function of the height of the observer above the earth. Calculate the distance to the horizon from an airplane at an altitude of 1000 ft.

38. In finding horizontal distance from slope distance (see Example 3) $H = L - C$, where C is a correction. Show that a good approximation to C is $d^2/2L$.

5.2 THE LAW OF SINES

[i] The trigonometric functions can be used to solve triangles that are not right triangles (oblique triangles). In order to solve oblique triangles we need to derive some properties, one of which is called the *law of sines*.

We shall consider any oblique triangle. It may or may not have an obtuse angle. We consider both cases, but the derivations are essentially the same.

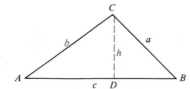 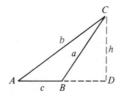

The triangles are lettered in the standard way, with angles A, B, and C and the sides opposite them a, b, and c, respectively. We have

OBJECTIVES

You should be able to:

[i] Use the law of sines to solve any triangle, given a side and two angles.

[ii] Use the law of sines to solve triangles, given two sides and an angle opposite one of them, finding two solutions when they exist, and recognizing when a solution does not exist.

drawn an altitude from vertex C. It has length h. In either triangle we now have, from triangle ADC,

$$\frac{h}{b} = \sin A \qquad \text{or} \qquad h = b \sin A.$$

From triangle DBC on the left we have $h/a = \sin B$, or $h = a \sin B$. On the right we have $h/a = \sin(180° - B) = \sin B$. So in either kind of triangle we now have

$$h = b \sin A \qquad \text{and} \qquad h = a \sin B.$$

Thus it follows that

$$b \sin A = a \sin B.$$

We divide by $\sin A \sin B$ to obtain

$$\frac{a}{\sin A} = \frac{b}{\sin B}.$$

There is no danger of dividing by 0 here because we are dealing with triangles whose angles are never 0° or 180°.

If we were to consider an altitude from vertex A in the triangles shown above, the same argument would give us

$$\frac{b}{\sin B} = \frac{c}{\sin C}.$$

We combine these results to obtain the law of sines, which holds for right triangles as well as oblique triangles.

THEOREM 1

___

The law of sines. In any triangle ABC,

$$\frac{a}{\sin A} = \frac{b}{\sin B} = \frac{c}{\sin C}$$

(the sides are proportional to the sines of the opposite angles).

___

Solving Triangles (AAS)

When two angles and a side of any triangle are known, the law of sines can be used to solve the triangle.

▦ **Example 1** In triangle ABC, $a = 4.56$, $A = 43°$, and $C = 57°$. Solve the triangle.

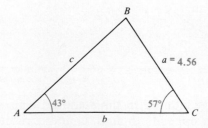

We first draw a sketch. We find B, as follows:

$$B = 180° - (43° + 57°) = 80°.$$

We can now find the other two sides, using the law of sines:

$$\frac{c}{\sin C} = \frac{a}{\sin A}$$

$$c = \frac{a \sin C}{\sin A} = \frac{4.56 \sin 57°}{\sin 43°}$$

$$= \frac{4.56 \times 0.8387}{0.6820} = 5.61,$$

$$\frac{b}{\sin B} = \frac{a}{\sin A}$$

$$b = \frac{a \sin B}{\sin A} = \frac{4.56 \sin 80°}{0.6820}$$

$$= \frac{4.56 \times 0.9848}{0.6820} = 6.58.$$

We have now found the unknown parts of the triangle: $B = 80°$, $c = 5.61$, and $b = 6.58$.

DO EXERCISE 1.

[ii] The Ambiguous Case (SSA)

When two sides of a triangle and an angle opposite one of them are known, the law of sines can be used to solve the triangle. However, there may be more than one solution. Thus this is known as the ambiguous case. Suppose a, b, and A are given. Then the various possibilities are as shown in the four cases below.

Case I

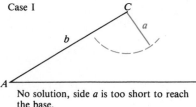

No solution, side a is too short to reach the base.

Case II

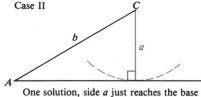

One solution, side a just reaches the base and is perpendicular to it.

Case III

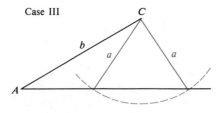

Two solutions, an arc of radius a meets the base at two points.

Case IV

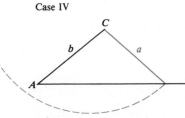

$a \geq b$. One solution, an arc of radius a meets the base at just one point, other than A.

The following examples correspond to the four possibilities just described.

1. Solve this triangle.

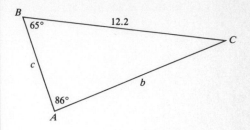

2. Solve this triangle.

$$a = 40,\ b = 12,\ B = 57°$$

3. Solve this triangle.

$$a = 3,\ b = 4,\ A = 48°35'$$

Example 2 (*Case I.*) In triangle ABC, $a = 15$, $b = 25$, and $A = 47°$. Solve the triangle.

We look for B:

$$\frac{a}{\sin A} = \frac{b}{\sin B}.$$

Then

$$\sin B = \frac{b \sin A}{a} = \frac{25 \sin 47°}{15} = \frac{25 \times 0.7314}{15} = 1.219.$$

Since there is no angle having a sine greater than 1, there is no solution.

DO EXERCISE 2.

Example 3 (*Case II.*) In triangle ABC, $a = 12$, $b = 5$, and $B = 24°38'$. Solve the triangle.

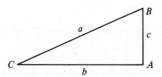

We look for A:

$$\frac{a}{\sin A} = \frac{b}{\sin B}$$

$$\sin A = \frac{a \sin B}{b} = \frac{12 \sin 24°38'}{5}$$

$$= \frac{12 \times 0.4168}{5} = 1.000$$

$$A = 90°.$$

Then $C = 90° - 24°38' = 65°22'$. Since $c/a = \cos B$,

$$c = a \cos B = 12 \times 0.9090 = 10.9.$$

DO EXERCISE 3.

Example 4 (*Case III.*) In triangle ABC, $a = 20$, $b = 15$, and $B = 30°$. Solve the triangle.

We look for A:

$$\frac{a}{\sin A} = \frac{b}{\sin B}$$

$$\sin A = \frac{a \sin B}{b} = \frac{20 \sin 30°}{15}$$

$$\sin A = \frac{20 \times 0.5}{15} = 0.6667.$$

There are two angles less than 180° having a sine of 0.6667. They are 42° and 138°, to the nearest degree. This gives us two possible solutions.

Possible solution 1. $A = 42°$. Then $C = 180° - (30° + 42°) = 108°$.

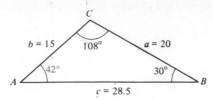

We now find c:

$$\frac{c}{\sin C} = \frac{b}{\sin B}$$

$$c = \frac{b \sin C}{\sin B} = \frac{15 \sin 108°}{\sin 30°}$$

$$= \frac{15 \times 0.9511}{0.5} = 28.5.$$

These parts make a triangle, as shown; hence we have a solution.

Possible solution 2. $A = 138°$. Then $C = 12°$.

We now find c:

$$c = \frac{b \sin C}{\sin B} = \frac{15 \sin 12°}{\sin 30°} = \frac{15 \times 0.2079}{0.5} = 6.2.$$

These parts make a triangle; hence we have a second solution.

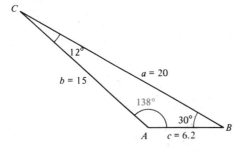

DO EXERCISE 4.

Example 5 (*Case IV.*) In triangle ABC, $a = 25$, $b = 10$, and $A = 42°$. Solve the triangle.

We look for B:

$$\frac{b}{\sin B} = \frac{a}{\sin A}$$

$$\sin B = \frac{b \sin A}{a} = \frac{10 \sin 42°}{25}$$

$$= \frac{10 \times 0.6691}{25} = 0.2676.$$

Then $B = 15°30'$ or $B = 164°30'$. Since $a > b$, we know there is only one solution. If we had not noticed this, we could tell it now. An angle

4. In triangle ABC, $a = 25$, $b = 20$, and $B = 33°$. Solve the triangle.

192 TRIANGLES, VECTORS, AND APPLICATIONS

5. In triangle ABC, $b = 20$, $c = 10$, and $B = 38°$. Solve the triangle.

of $164°30'$ cannot be an angle of this triangle because it already has an angle of $42°$, and these two would total more than $180°$.

$$C = 180° - (42° + 15°30') = 122°30',$$

$$\frac{c}{\sin C} = \frac{a}{\sin A},$$

$$c = \frac{a \sin C}{\sin A} = \frac{25 \sin 122°30'}{\sin 42°} \qquad \text{Solving for } c$$

$$c = \frac{25 \times 0.8434}{0.6691} = 31.5$$

DO EXERCISE 5.

EXERCISE SET 5.2

[i] In Exercises 1–12, solve the triangle ABC.

1. $A = 133°$, $B = 30°$, $b = 18$ **2.** $B = 120°$, $C = 30°$, $a = 16$ **3.** $B = 38°$, $C = 21°$, $b = 24$

4. $A = 131°$, $C = 23°$, $b = 10$ **5.** $A = 68°30'$, $C = 42°40'$, $c = 23.5$ **6.** $B = 118°20'$, $C = 45°40'$, $b = 42.1$

[ii]

7. $A = 36°$, $a = 24$, $b = 34$ **8.** $C = 43°$, $c = 28$, $b = 27$ **9.** $A = 116°20'$, $a = 17.2$, $c = 13.5$

10. $A = 47°50'$, $a = 28.3$, $b = 18.2$ **11.** $C = 61°10'$, $c = 30.3$, $b = 24.2$ **12.** $B = 58°40'$, $a = 25.1$, $b = 32.6$

[i], [ii]

13. Points A and B are on opposite sides of a lunar crater. Point C is 50 meters from A. The measure of $\angle BAC$ is determined to be $112°$ and the measure of $\angle ACB$ is determined to be $42°$. What is the width of the crater?

14. A guy wire to a pole makes a $71°$ angle with level ground. At a point 25 ft farther from the pole than the guy wire, the angle of elevation of the top of the pole is $37°$. How long is the guy wire?

15. A pole leans away from the sun at an angle of $7°$ to the vertical. When the angle of elevation of the sun is $51°$, the pole casts a shadow 47 ft long on level ground. How long is the pole?

16. A vertical pole stands by a road that is inclined $10°$ to the horizontal. When the angle of elevation of the sun is $23°$, the pole casts a shadow 38 ft long directly downhill along the road. How long is the pole?

17. A reconnaissance airplane leaves its airport on the east coast of the United States and flies in a direction of $085°$. Because of bad weather it returns to another airport 230 km to the north of its home base. For the return it flies in a direction of $283°$. What was the total distance it flew?

18. Lookout station B is 10.2 km east of station A. The bearing of a fire from A is S $10°40'$ W. The bearing of the fire from B is S $31°20'$ W. How far is the fire from A? from B?

19. A boat leaves a lighthouse A and sails 5.1 km. At this time it is sighted from lighthouse B, 7.2 km west of A. The bearing of the boat from B is N $65°10'$ E. How far is the boat from B?

20. An airplane leaves airport A and flies 200 km. At this time its bearing from airport B, 250 km to the west, is $120°$. How far is the airplane from B?

☆

21. Prove that the area of a parallelogram is the product of two sides and the sine of the included angle.

22. Prove that the area of a quadrilateral is half the product of the lengths of its diagonals and the sine of an angle between the diagonals.

★

23. When two objects, such as ships, airplanes, or runners, move in straight-line paths, if the distance between them is decreasing and if the bearing from one of them to the other is constant, they will collide. ("Constant bearing means collision," as mariners put it.) Prove that this statement is true.

5.3 THE LAW OF COSINES

[i] A second property of triangles important in solving oblique triangles is called the *law of cosines*. To derive this property we consider any triangle ABC placed on a coordinate system. We place the origin at one of the vertices, say C, and the positive half of the x-axis along one of the sides, say CB. Then the coordinates of B are $(a, 0)$, and the coordinates of A are $(b \cos C, b \sin C)$.

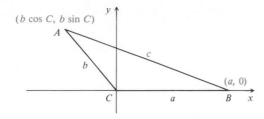

We will next use the distance formula to determine c^2:

$$c^2 = (b \cos C - a)^2 + (b \sin C - 0)^2.$$

Now we multiply and simplify:

$$c^2 = b^2 \cos^2 C - 2ab \cos C + a^2 + b^2 \sin^2 C$$
$$c^2 = a^2 + b^2(\sin^2 C + \cos^2 C) - 2ab \cos C$$
$$c^2 = a^2 + b^2 - 2ab \cos C.$$

Had we placed the origin at one of the other vertices, we would have obtained

$$a^2 = b^2 + c^2 - 2bc \cos A$$

or

$$b^2 = a^2 + c^2 - 2ac \cos B.$$

This result can be summarized as follows.

THEOREM 2

The law of cosines. In any triangle ABC,

$$a^2 = b^2 + c^2 - 2bc \cos A,$$
$$b^2 = a^2 + c^2 - 2ac \cos B, \quad \text{and}$$
$$c^2 = a^2 + b^2 - 2ab \cos C.$$

(In any triangle, the square of a side is the sum of the squares of the other two sides, minus twice the product of those sides and the cosine of the included angle.)

Only one of the above formulas need be memorized. The other two can be obtained by a change of letters.

Solving Triangles (SAS)

When two sides of a triangle and the included angle are known, we can use the law of cosines to find the third side. The law of cosines or the law of sines can then be used to finish solving the triangle.

OBJECTIVES

You should be able to:

[i] Use the law of cosines, with the law of sines, to solve any triangle, given two sides and the included angle.

[ii] Use the law of cosines to solve any triangle, given three sides.

1. In triangle ABC, $b = 18$, $c = 28$, and $A = 122°$. Solve the triangle.

Example 1 In triangle ABC, $a = 24$, $c = 32$, and $B = 115°$. Solve the triangle.

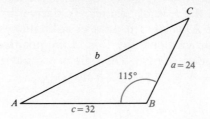

We first find the third side. From the law of cosines,

$$b^2 = a^2 + c^2 - 2ac \cos B$$
$$= 24^2 + 32^2 - 2 \cdot 24 \cdot 32(-0.4226)$$
$$= 2249.$$

Then $b = \sqrt{2249} = 47.4$. Next, we use the law of sines to find a second angle:

$$\frac{a}{\sin A} = \frac{b}{\sin B} \quad \text{and} \quad \sin A = \frac{a \sin B}{b}$$

$$\sin A = \frac{24 \sin 115°}{47.4} = \frac{24 \times 0.9063}{47.4} = 0.4589.$$

Thus

$$A = 27°20'.$$

The third angle is now easy to find:

$$C = 180° - (115° + 27°20') = 37°40'.$$

DO EXERCISE 1.

[ii] Solving Triangles (SSS)

When all three sides of a triangle are known, the law of cosines can be used to solve the triangle.

Example 2 In triangle ABC, $a = 18$, $b = 25$, and $c = 12$. Solve the triangle.

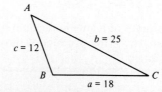

Let us find angle B. We select the formula from the law of cosines that contains $\cos B$; in other words, $b^2 = a^2 + c^2 - 2ac \cos B$. We solve this for $\cos B$ and substitute:

$$\cos B = \frac{a^2 + c^2 - b^2}{2ac} = \frac{18^2 + 12^2 - 25^2}{2 \cdot 18 \cdot 12} = -0.3634.$$

Then $B = 111°20'$. Similarly, we shall find angle A:*

$$a^2 = b^2 + c^2 - 2bc \cos A$$

$$\cos A = \frac{b^2 + c^2 - a^2}{2bc} = \frac{25^2 + 12^2 - 18^2}{2 \cdot 25 \cdot 12} = 0.7417.$$

Thus $A = 42°10'$. Then $C = 180° - (111°20' + 42°10') = 26°30'$.

DO EXERCISE 2.

2. In triangle ABC, $a = 25$, $b = 10$, and $c = 20$. Solve the triangle.

EXERCISE SET 5.3

In Exercises 1–12, solve the triangles.

[i]

1. $A = 30°$, $b = 12$, $c = 24$ **2.** $C = 60°$, $a = 15$, $b = 12$ **3.** $A = 133°$, $b = 12$, $c = 15$

4. $B = 116°$, $a = 31$, $c = 25$ **5.** $B = 72°40'$, $c = 16$, $a = 78$ **6.** $A = 24°30'$, $b = 68$, $c = 14$

[ii]

7. $a = 12$, $b = 14$, $c = 20$ **8.** $a = 22$, $b = 22$, $c = 35$ **9.** $a = 3.3$, $b = 2.7$, $c = 2.8$

10. $a = 16$, $b = 20$, $c = 32$ **11.** $a = 2.2$, $b = 4.1$, $c = 2.4$ **12.** $a = 3.6$, $b = 6.2$, $c = 4.1$

[i], [ii]

13. A ship leaves a harbor and sails 15 nautical mi east. It then sails 18 nautical mi in a direction of S 27°E. How far is it, then, from the harbor and in what direction?

14. An airplane leaves an airport and flies west 147 km. It then flies 200 km in a direction of 220°. How far is it, then, from the airport and in what direction?

15. Two ships leave harbor at the same time. The first sails N 15°W at 25 knots (a knot is one nautical mile per hour). The second sails N 32°E at 20 knots. After 2 hr, how far apart are the ships?

16. Two airplanes leave an airport at the same time. The first flies 150 km/h in a direction of 320°. The second flies 200 km/h in a direction of 200°. After 3 hr, how far apart are the planes?

17. A hill is inclined 5° to the horizontal. A 45-ft pole stands at the top of the hill. How long a rope will it take to reach from the top of the pole to a point 35 ft downhill from the base of the pole?

18. A hill is inclined 15° to the horizontal. A 40-ft pole stands at the top of the hill. How long a rope will it take to reach from the top of the pole to a point 68 ft downhill from the base of the pole?

19. A piece of wire 5.5 m long is bent into a triangular shape. One side is 1.5 m long and another is 2 m long. Find the angles of the triangle.

20. A triangular lot has sides 120 ft long, 150 ft long, and 100 ft long. Find the angles of the lot.

21. A slow-pitch softball diamond is a square 60 ft on a side. The pitcher's mound is 46 ft from home. How far is it from the pitcher's mound to first base?

22. A baseball diamond is a square 90 ft on a side. The pitcher's mound is 60.5 ft from home. How far does the pitcher have to run to cover first?

23. The longer base of an isosceles trapezoid measures 14 ft. The nonparallel sides measure 10 ft, and the base angles measure 80°.

a) Find the length of a diagonal.

b) Find the area.

24. An isosceles triangle has a vertex angle of 38° and this angle is included by two sides, each measuring 20 ft. Find the area of the triangle.

25. A field in the shape of a parallelogram has sides that measure 50 yd and 70 yd. One angle of the field measures 78°. Find the area of the field.

26. An aircraft takes off to fly a 180-mile trip. After flying 75 miles it is 10 miles off course. How much should the heading be corrected to then fly straight to the destination, assuming no wind correction?

---

*The law of sines could be used at this point.

27. From the top of a hill 20 ft above the surface of a lake, a tree across the lake is sighted. The angle of elevation is 11°. From the same position, the top of the tree is seen by reflection in the lake. The angle of depression is 14°. What is the height of the tree?

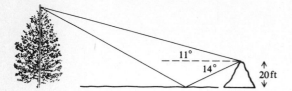

28. A bridge is being built across a canyon. The length of the bridge is 5045 ft. From the deepest point in the canyon, the angles of elevation of the ends of the bridge are 78° and 72°. How deep is the canyon?

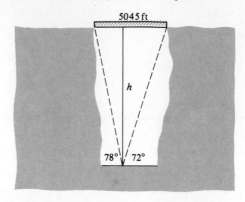

29. Find a formula for the area of an isosceles triangle in terms of the congruent sides and their included angle. Under what conditions will the area of a triangle with fixed congruent sides be a maximum?

30. (*Surveying.*) In surveying, a series of bearings and distances is called a *traverse*. In the following, measurements are taken as shown.

a) Compute the bearing BC.

b) Compute the distance and bearing AC.

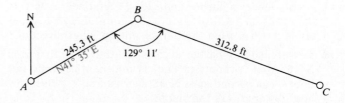

31. Show that in any triangle ABC,

$$a^2 + b^2 + c^2 = 2(bc \cos A + ac \cos B + ab \cos C).$$

32. Show that in any triangle ABC,

$$\frac{\cos A}{a} + \frac{\cos B}{b} + \frac{\cos C}{c} = \frac{a^2 + b^2 + c^2}{2abc}.$$

OBJECTIVES

You should be able to:

[i] Given two vectors, find their sum, or resultant.

[ii] Solve applied problems involving finding sums of vectors.

5.4 VECTORS

In many applications there arise certain quantities in which a direction is specified. Any such quantity having a *direction* and a *magnitude* is called a *vector quantity,* or *vector.* Here are some examples of vector quantities.

Displacement. An object moves a certain distance in a certain direction.

A train travels 100 mi to the northeast.

A person takes 5 steps to the west.

A batter hits a ball 100 m along the left-field foul line.

Velocity.   An object travels at a certain speed in a certain direction.

The wind is blowing 15 mph from the northwest.

An airplane is traveling 450 km/h in a direction of 243°.

Force.   A push or pull is exerted on an object in a certain direction.

A 15-kg force is exerted downward on the handle of a jack.

A 25-lb upward force is required to lift a box.

A wagon is being pulled up a 30° incline, requiring an effort of 200 kg.

We shall represent vectors abstractly by directed line segments, or arrows. The length is chosen, according to some scale, to represent the magnitude of the vector, and the direction of the arrow represents the direction of the vector. For example, if we let 1 cm represent 5 km/h, then a 15-km/h wind from the northwest would be represented by an arrow 3 cm long, as shown.

[i] Vector Addition

To "add" vectors, we find a single vector that would have the same effect as the vectors combined. We shall illustrate vector addition for displacements, but it is done the same way for any vector quantities. Suppose a person takes 4 steps east and then 3 steps north. He will then be 5 steps from the starting point in the direction shown. The *sum* of the two vectors is the vector 5 steps in magnitude and in the direction shown. The sum is also called the *resultant* of the two vectors.

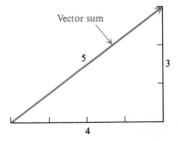

In general, if we have two vectors **a** and **b**,* we can add them as in the above example. That is, we place the tail of one arrow at the head of the other and then find the vector that forms the third side of a

---

*It is usual to denote vectors using boldface type, and sometimes in handwriting to write a bar over the letters.

1. Two forces of 5.0 kg and 14.0 kg act at right angles to each other. Find the resultant, specifying the angle that it makes with the larger force.

triangle. Or, we can place the tails of the arrows together, complete a parallelogram, and find the diagonal of the parallelogram.

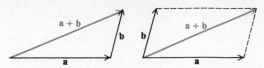

Example 1 Forces of 15 kg and 25 kg act on an object at right angles to each other. Find their *sum*, or *resultant*, giving the angle that it makes with the larger force.

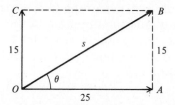

We make a drawing, this time a rectangle, using s for the length of vector **OB**. Since OAB is a right triangle, we have

$$\tan \theta = \frac{15}{25} = 0.6.$$

Thus θ, the angle the resultant makes with the larger force, is 31°. Now $s/15 = \csc \theta$, or $s = 15 \csc \theta = 15 \times 1.942 = 29.1$. Thus the resultant **OB** has a magnitude of 29.1 kg and makes an angle of 31° with the larger force.

DO EXERCISE 1.

[ii] Applications

Example 2 An airplane heads in a direction of 100° at 180-km/h airspeed while a wind is blowing 40 km/h from 220°. Find the speed of the airplane over the ground and the direction of its track over the ground.

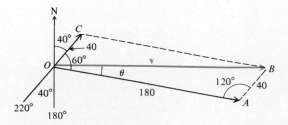

We first make a drawing. The wind is represented by **OC** and the velocity vector of the airplane by **OA**. The resultant velocity is **v**, the sum of the two vectors. We denote the length of **v** by |**v**|. By the law of cosines in $\triangle OAB$, we have

$$|\mathbf{v}|^2 = 40^2 + 180^2 - 2 \cdot 40 \cdot 180 \cos 120°$$
$$= 41{,}200.$$

Thus |**v**| is 203 km/h. By the law of sines in the same triangle,

$$\frac{40}{\sin \theta} = \frac{203}{\sin 120°},$$

or

$$\sin \theta = \frac{40 \sin 120°}{203} = 0.1706.$$

Thus $\theta = 10°$ to the nearest degree. Therefore the ground speed of the airplane is 203 km/h and its track is in a direction of 90°.

DO EXERCISE 2.

2. An airplane heads in a direction of 90° at 140-km/h airspeed. The wind is 50 km/h from 210°. Find the direction and speed of the airplane over the ground.

EXERCISE SET 5.4

[i] In Exercises 1–8, magnitudes of vectors **a** and **b** are given and the angle between the vectors θ. Find the resultant, giving the direction by specifying to the nearest degree the angle it makes with vector **a**.

1. $|\mathbf{a}| = 45$, $|\mathbf{b}| = 35$, $\theta = 90°$
2. $|\mathbf{a}| = 54$, $|\mathbf{b}| = 43$, $\theta = 90°$
3. $|\mathbf{a}| = 10$, $|\mathbf{b}| = 12$, $\theta = 67°$
4. $|\mathbf{a}| = 25$, $|\mathbf{b}| = 30$, $\theta = 75°$
5. $|\mathbf{a}| = 20$, $|\mathbf{b}| = 20$, $\theta = 117°$
6. $|\mathbf{a}| = 30$, $|\mathbf{b}| = 30$, $\theta = 123°$
7. $|\mathbf{a}| = 23$, $|\mathbf{b}| = 47$, $\theta = 27°$
8. $|\mathbf{a}| = 32$, $|\mathbf{b}| = 74$, $\theta = 72°$

[ii]

9. Two forces of 5 kg and 12 kg act on an object at right angles. Find the magnitude of the resultant and the angle it makes with the smaller force.

10. Two forces of 30 kg and 40 kg act on an object at right angles. Find the magnitude of the resultant and the angle it makes with the smaller force.

11. Forces of 420 kg and 300 kg act on an object. The angle between the forces is 50°. Find the resultant, giving the angle it makes with the larger force.

12. Forces of 410 kg and 600 kg act on an object. The angle between the forces is 47°. Find the resultant, giving the angle it makes with the smaller force.

13. A balloon is rising 12 ft/sec while a wind is blowing 18 ft/sec. Find the speed of the balloon and the angle it makes with the horizontal.

14. A balloon is rising 10 ft/sec while a wind is blowing 5 ft/sec. Find the speed of the balloon and the angle it makes with the horizontal.

15. A boat heads 35°, propelled by a force of 750 lb. A wind from 320° exerts a force of 150 lb on the boat. How large is the resultant force and in what direction is the boat moving?

16. A boat heads 220°, propelled by a 650-lb force. A wind from 080° exerts a force of 100 lb on the boat. How large is the resultant force, and in what direction is the boat moving?

17. A ship sails N 80°E for 120 nautical mi, then S 20°W for 200 nautical mi. How far is it then from the starting point, and in what direction?

18. An airplane flies 032° for 210 km, then 280° for 170 km. How far is it, then, from the starting point, and in what direction?

19. A motorboat has a speed of 15 km/h. It crosses a river whose current has a speed of 3 km/h. In order to cross the river at right angles, the boat should be pointed in what direction?

20. An airplane has an airspeed of 150 km/h. It is to make a flight in a direction of 080° while there is a 25-km/h wind from 350°. What should be the airplane's heading?

OBJECTIVES

You should be able to:

[i] Resolve vectors into components.
[ii] Add vectors using components.
[iii] Change from rectangular to polar notation for vectors.
[iv] Change from polar to rectangular notation for vectors.
[v] Do simple manipulations in vector algebra.

5.5 VECTORS AND COORDINATES

[i] Components of Vectors

Given a vector, it is often convenient to reverse the addition procedure, that is, to find two vectors whose sum is the given vector. Usually the two vectors we seek will be perpendicular. The two vectors we find are called *components* of the given vector.

Example 1 A certain vector **a** has a magnitude of 130 and is inclined 40° with the horizontal. Resolve the vector into horizontal and vertical components.

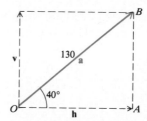

We first make a drawing showing horizontal and vertical vectors whose sum is the given vector **a**. From $\triangle OAB$, we see

$$|\mathbf{h}| = 130 \cos 40° = 99.6$$

and

$$|\mathbf{v}| = 130 \sin 40° = 83.6.$$

These are the components we seek.

Example 2 An airplane is flying at 200 km/h in a direction of 305°. Find the westerly and northerly components of its velocity.

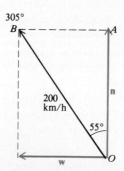

We first make a drawing showing westerly and northerly vectors whose sum is the given velocity. From $\triangle OAB$, we see that

$$|\mathbf{n}| = 200 \cos 55° = 115 \text{ km/h},$$
$$|\mathbf{w}| = 200 \sin 55° = 164 \text{ km/h}.$$

Example 3 A ten-pound block is sitting on a 30° incline. Find the components of the weight parallel and perpendicular to the incline.

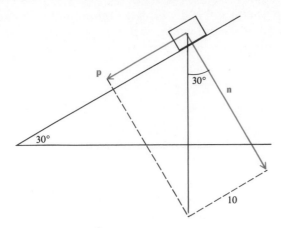

We make a drawing showing the components. From the drawing we see that

$$|\mathbf{p}| = 10 \sin 30° = 5 \, \text{lb},$$
$$|\mathbf{n}| = 10 \cos 30° = 8.67 \, \text{lb}.$$

DO EXERCISES 1 AND 2.

[ii] Adding Vectors Using Components

Given a vector, we can resolve it into components. Given components of a vector, we can find the vector. We simply find the vector sum of the components.

Example 4 A vector **v** has a westerly component of 12 and a southerly component of 16. Find the vector.

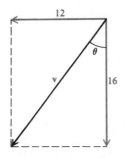

From the drawing we see that $\tan \theta = \frac{12}{16} = 0.75$. Thus $\theta = 36°50'$. Then $|\mathbf{v}| = 12 \csc 36°50' = 20$. Thus the vector has a magnitude of 20 and a direction of S 36°50′W.

DO EXERCISE 3.

1. A vector of magnitude 100 points southeast. Resolve the vector into easterly and southerly components.

2. Two boys are pulling on a rope attached to a tree with a force of 45 lb in a direction S 35°W. Find the westerly and southerly components of the force.

3. A wind has an easterly component (*from* the east) of 10 km/h and a southerly component (*from* the south) of 16 km/h. Find the magnitude and direction of the wind.

4. One vector has components of 7 up and 9 to the left. A second vector has components of 3 down and 12 to the left. Find the sum, expressing it

a) in terms of components;
b) giving magnitude and direction.

When we have two vectors resolved into components, we can add the two vectors by adding the components.

Example 5 A vector **v** has a westerly component of 3 and a northerly component of 5. A second vector **w** has an easterly component of 8 and a northerly component of 4. Find the components of the sum. Find the sum.

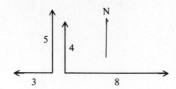

Adding the east–west components, we obtain 5 east. Adding the north–south components, we obtain 9 north. The components of **v** + **w** are 5 east and 9 north. We now add these components:

Add the vectors, giving answers as ordered pairs.

5. (7, 2) and (5, −10)

$$\tan \theta = \frac{9}{5} = 1.8,$$
$$\theta = 61°,$$
$$|\mathbf{v} + \mathbf{w}| = 9 \csc 61° = 10.3.$$

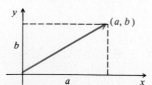

The sum **v** + **w** has a magnitude of 10.3 and a direction of N 29°E.

DO EXERCISE 4.

Analytic Representation of Vectors

If we place a coordinate system so that the origin is at the tail of an arrow representing a vector, we say that the vector is in *standard position*. Then if we know the coordinates of the other end of the vector, we know the vector. The coordinates will be the x-component and the y-component of the vector. Thus we can consider an ordered pair (a, b) to be a vector. When vectors are given in this form, it is easy to add them. We simply add the respective components.

6. (−2, 7) and (14, −3)

Example 6 Find **u** + **v**, where **u** = (3, −7) and **v** = (4, 2).

The sum is found by adding the x-components and then adding the y-components: **u** + **v** = (3, −7) + (4, 2) = (7, −5).

DO EXERCISES 5 AND 6.

[iii] **Polar Notation**

Vectors can also be specified by giving their length and direction. When this is done, we say that we have polar notation for the vector. An example of polar notation is (15, 260°). The angle is measured from the positive half of the x-axis counterclockwise. In Section 5.6, we consider a coordinate system (called a *polar coordinate system*) in which polar notation is very natural.

Example 7 Find polar notation for the vector **v**, where **v** = (7, −2).

From the diagram, we see the reference angle φ.

$$\tan \phi = \frac{-2}{7} = -0.2857$$

$$\phi = -16°,$$
$$|\mathbf{v}| = \sqrt{(-2)^2 + 7^2} = 7.28.$$

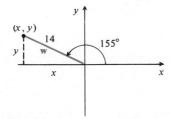

Thus polar notation for **v** is

$$(7.28, 344°) \quad \text{or} \quad (7.28, -16°).$$

DO EXERCISE 7.

[iv] Given polar notation for a vector we can also find the ordered pair notation (rectangular notation).

Example 8 Find rectangular notation for the vector **w**, where **w** = (14, 155°).

From the drawing we have

$$x = 14 \cos 155° = -12.7,$$
$$y = 14 \sin 155° = 5.92.$$

Thus rectangular notation for **w** is (−12.7, 5.92).

DO EXERCISE 8.

7. Find polar notation for the vector (−8, 3).

8. Find rectangular notation for the vector (15, 341°).

9. For $\mathbf{u} = (-3, 5)$ and $\mathbf{v} = (4, 2)$, find the following.

a) $\mathbf{u} + \mathbf{v}$

b) $\mathbf{v} - \mathbf{u}$

c) $5\mathbf{u} - 2\mathbf{v}$

d) $|3\mathbf{u} + 2\mathbf{v}|$

[v] Properties of Vectors and Vector Algebra

With vectors represented in rectangular notation (a, b), it is easy to determine many of their properties. We consider the set of all ordered pairs of real numbers (a, b). This is the set of all vectors on a plane. Consider the sum of any two vectors.

$$\mathbf{u} = (a, b), \qquad \mathbf{v} = (c, d),$$
$$\mathbf{u} + \mathbf{v} = (a + c, b + d), \qquad \mathbf{v} + \mathbf{u} = (c + a, d + b).$$

We have just shown that addition of vectors is commutative. Other properties are likewise not difficult to show. The system of vectors is not a field, but it does have some of the familiar field properties.

PROPERTIES OF VECTORS

COMMUTATIVITY. Addition of vectors is commutative.

ASSOCIATIVITY. Addition of vectors is associative.

IDENTITY. There is an additive identity, the vector $(0, 0)$, or $\mathbf{0}$.

INVERSES. Every vector (a, b) has an additive inverse $(-a, -b)$.

SUBTRACTION. We define $\mathbf{u} - \mathbf{v}$ to be the vector which when added to $\mathbf{v}$ gives $\mathbf{u}$. It follows that if $\mathbf{u} = (a, b)$ and $\mathbf{v} = (c, d)$, then $\mathbf{u} - \mathbf{v} = (a - c, b - d)$. Thus subtraction is always possible. Also, $\mathbf{u} - \mathbf{v} = \mathbf{u} + (-\mathbf{v})$.

LENGTH, OR ABSOLUTE VALUE. If $\mathbf{u} = (a, b)$, then it follows that $|\mathbf{u}| = \sqrt{a^2 + b^2}$.

A vector can be multiplied by a real number. A real number in this context is called a *scalar*. For example, $2\mathbf{v}$ is a vector in the same direction as $\mathbf{v}$ but twice as long. Analytically, we define scalar multiplication as follows.

DEFINITION

Scalar multiplication. If r is any real number and $\mathbf{v} = (a, b)$, then $r\mathbf{v} = (ar, br)$.

Example 9 Do the following calculations, where $\mathbf{u} = (4, 3)$ and $\mathbf{v} = (-5, 8)$: (a) $\mathbf{u} + \mathbf{v}$; (b) $\mathbf{u} - \mathbf{v}$; (c) $3\mathbf{u} - 4\mathbf{v}$; (d) $|3\mathbf{u} - 4\mathbf{v}|$.

a) $\mathbf{u} + \mathbf{v} = (4, 3) + (-5, 8) = (-1, 11)$

b) $\mathbf{u} - \mathbf{v} = \mathbf{u} + (-\mathbf{v}) = (4, 3) + (5, -8) = (9, -5)$

c) $3\mathbf{u} - 4\mathbf{v} = 3(4, 3) - 4(-5, 8) = (12, 9) - (-20, 32)$
$$= (32, -23)$$

d) $|3\mathbf{u} - 4\mathbf{v}| = \sqrt{32^2 + (-23)^2} = \sqrt{1553} \approx 39.41$

DO EXERCISE 9.

EXERCISE SET 5.5

[ii] In Exercises 1–14, rectangular notation for vectors is given. In Exercises 1–6, add, giving rectangular notation for the answers.

1. $(3, 7) + (2, 9)$

2. $(5, -7) + (-3, 2)$

3. $(17, 7.6) + (-12.2, 6.1)$

4. $(-15.2, 37.1) + (7.9, -17.8)$

5. $(-650, -750) + (-12, 324)$

6. $(-354, -973) + (-75, 256)$

[iii] In Exercises 7–14, find polar notation.

7. $(3, 4)$

8. $(4, 3)$

9. $(10, -15)$

10. $(17, -10)$

11. $(-3, -4)$

12. $(-4, -3)$

13. $(-10, 15)$

14. $(-17, 10)$

[iv] In Exercises 15–22, find rectangular notation.

15. $(4, 30°)$

16. $(8, 60°)$

17. $(10, 235°)$

18. $(15, 210°)$

19. $(20, 330°)$

20. $(20, 200°)$

21. $(100, -45°)$

22. $(150, -60°)$

[i]

23. A vector **u** with magnitude 150 is inclined upward 52° from the horizontal. Find the horizontal and vertical components of **u**.

24. A vector **u** with magnitude 170 is inclined downward 63° from the horizontal. Find the horizontal and vertical components of **u**.

25. An airplane is flying 220° at 250 km/h. Find the southerly and westerly components of its velocity **v**.

26. A wind is blowing from 310° at 25 mph. Find the southerly and easterly components of the wind velocity **v**.

[ii]

27. A force **f** has a westerly component of 25 kg and a southerly component of 35 kg. Find the magnitude and direction of **f**.

28. A force **f** has an upward component of 65 kg and a component to the left of 90 kg. Find the magnitude and direction of **f**.

29. Vector **u** has a westerly component of 15 and a northerly component of 22. Vector **v** has a northerly component of 6 and an easterly component of 8. Find (a) the components of **u** + **v**; (b) the magnitude and direction of **u** + **v**.

30. Vector **u** has a component of 18 upward and a component of 12 to the left. Vector **v** has a component of 5 downward and a component of 35 to the right. Find (a) the components of **u** + **v**; (b) the magnitude and direction of **u** + **v**.

[v] In Exercises 31–40, do the calculations for the following vectors: **u** = (3, 4), **v** = (5, 12), and **w** = (−6, 8).

31. $3\mathbf{u} + 2\mathbf{v}$

32. $3\mathbf{v} - 2\mathbf{w}$

33. $(\mathbf{u} + \mathbf{v}) - \mathbf{w}$

34. $\mathbf{u} - (\mathbf{v} + \mathbf{w})$

35. $|\mathbf{u}| + |\mathbf{v}|$

36. $|\mathbf{u}| - |\mathbf{v}|$

37. $|\mathbf{u} + \mathbf{v}|$

38. $|\mathbf{u} - \mathbf{v}|$

39. $2|\mathbf{u} + \mathbf{v}|$

40. $2|\mathbf{u}| + 2|\mathbf{v}|$

☆

41. If **PQ** is any vector, what is **PQ** + **QP**?

42. The *inner product* of vectors **u** • **v** is a scalar, defined as follows: $\mathbf{u} \cdot \mathbf{v} = |\mathbf{u}||\mathbf{v}| \cos \theta$, where θ is the angle between the vectors. Show that vectors **u** and **v** are perpendicular if and only if **u** • **v** = 0.

★

43. Show that scalar multiplication is distributive over vector addition.

44. Let **u** = (3, 4). Find a vector that has the same direction as **u** but length 1.

45. Prove that for any vectors **u** and **v**, $|\mathbf{u} + \mathbf{v}| \le |\mathbf{u}| + |\mathbf{v}|$.

OBJECTIVES

You should be able to:

[i] Plot points, given their polar co-ordinates, and determine polar coordinates of points on a graph.

[ii] Convert from rectangular to polar coordinates and from polar to rectangular coordinates.

[iii] Graph simple polar equations.

5.6 POLAR COORDINATES

[i] In graphing we locate a point with an ordered pair of numbers (a, b). We may consider any such ordered pair to be a vector. From our work with vectors we know that we could also locate a point with a vector, given a length and a direction. When we use rectangular notation to locate points, we describe the coordinate system as *rectangular*.* When we use polar notation, we describe the coordinate system as *polar*. As this diagram shows, any point has rectangular coordinates (x, y) and has polar coordinates (r, θ). On a polar graph the origin is called the *pole* and the positive half of the x-axis is called the *polar axis*.

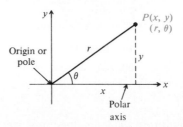

To plot points on a polar graph, we usually locate θ first, then move a distance r from the pole in that direction. If r is negative, we move in the opposite direction. Polar-coordinate graph paper, shown below, facilitates plotting. Point B illustrates that θ may be in radians. Point E illustrates that the polar coordinates of a point are not unique.

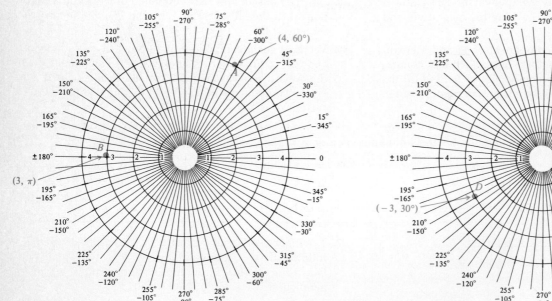

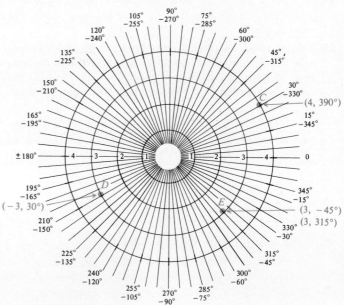

*Also called *Cartesian*, after the French mathematician René Descartes.

DO EXERCISES 1–4.

[ii] Polar and Rectangular Equations

Some curves have simpler equations in polar coordinates than in rectangular coordinates. For others the reverse is true. The following relationships between the two kinds of coordinates allow us to convert an equation from rectangular to polar coordinates:

$$x = r \cos \theta, \qquad y = r \sin \theta.$$

Example 1 Convert to a polar equation: $x^2 + y^2 = 25$.

$$
\begin{aligned}
(r \cos \theta)^2 + (r \sin \theta)^2 &= 25 \qquad \text{Substituting for x and y} \\
r^2 \cos^2 \theta + r^2 \sin^2 \theta &= 25 \\
r^2(\cos^2 \theta + \sin^2 \theta) &= 25 \\
r^2 &= 25 \\
r &= 5
\end{aligned}
$$

Example 1 illustrates that the polar equation of a circle centered at the origin is much simpler than the rectangular equation.

Example 2 Convert to a polar equation: $2x - y = 5$.

$$
\begin{aligned}
2(r \cos \theta) - (r \sin \theta) &= 5 \\
r(2 \cos \theta - \sin \theta) &= 5
\end{aligned}
$$

DO EXERCISES 5 AND 6.

The relationships that we need to convert from polar equations to rectangular equations can easily be determined from the triangle in the diagram on p. 206. They are as follows:

$$r = \pm\sqrt{x^2 + y^2},$$

$$\sin \theta = \frac{\pm y}{\sqrt{x^2 + y^2}},$$

$$\cos \theta = \frac{\pm x}{\sqrt{x^2 + y^2}},$$

$$\tan \theta = \frac{y}{x}$$

To convert to rectangular notation, we substitute as needed from either of the two lists, choosing the + or − signs as appropriate in each case.

Example 3 Convert to a rectangular (Cartesian) equation: $r = 4$.

$$
\begin{aligned}
+\sqrt{x^2 + y^2} &= 4 \qquad \text{Substituting for r} \\
x^2 + y^2 &= 16 \qquad \text{Squaring}
\end{aligned}
$$

In squaring like this, we must be careful not to introduce solutions of the equation that are not already present. In this example we did not, because the graph is a circle of radius 4 centered at the origin, in both cases.

1. Plot the following points.
   a) $(3, 60°)$ d) $(5, -60°)$
   b) $(0, 10°)$ e) $(2, 3\pi/2)$
   c) $(-5, 120°)$

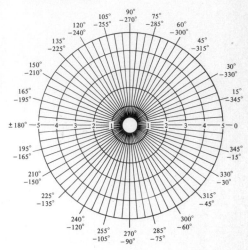

2. Find polar coordinates of each of these points. Give two answers for each.

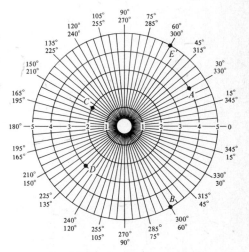

3. Find polar coordinates of these points.
   a) $(3, 3)$ c) $(-3, 3\sqrt{3})$
   b) $(0, -4)$ d) $(2\sqrt{3}, -2)$

4. Find rectangular coordinates of these points.
   a) $(5, 30°)$ c) $(-5, 45°)$
   b) $(10, \pi/3)$ d) $(-8, -5\pi/6)$

Convert to polar equations.

5. $2x + 5y = 9$

6. $x^2 + y^2 + 8x = 0$

Convert to Cartesian equations.

7. $r = 7$

Example 4 Convert to a rectangular equation: $r \cos \theta = 6$.

From our first list we know that $x = r \cos \theta$, so we have $x = 6$.

Example 5 Convert to a rectangular equation: $r = 2 \cos \theta + 3 \sin \theta$.

Let us first multiply on both sides by r:

$$r^2 = 2r \cos \theta + 3r \sin \theta$$
$$x^2 + y^2 = 2x + 3y \quad \text{Substituting}$$

DO EXERCISES 7–9.

8. $r \sin \theta = 5$

[iii] Graphing Polar Equations

To graph a polar equation, we usually make a table of values, choosing values of θ and calculating corresponding values of r. We plot the points and then complete the graph, as in the rectangular case. A difference arises in the polar case, because as θ increases sufficiently, points will begin to repeat and the curve will be traced again and again. When such a point is reached, the curve is complete.

9. $r - 3 \cos \theta = 5 \sin \theta$

Example 6 Graph $r = \cos \theta$.

We first make a table of values.

θ	0°	30°	45°	60°	90°	120°	135°	150°	180°
r	1	0.866	0.707	0.5	0	−0.5	−0.707	−0.866	−1

We plot these points and note that the last point is the same as the first. (Points having negative r are plotted in color.)

10. Graph $r = 1 - \sin \theta$.

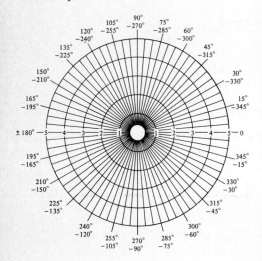

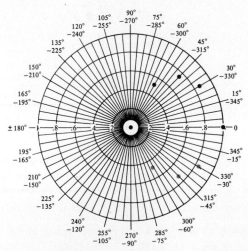

We try another point, $(-0.866, 210°)$, and find that it has already been plotted. Since the repeating has started, we plot no more points, but draw the graph, which is a circle as shown.

DO EXERCISE 10.

EXERCISE SET 5.6

[i] Use polar coordinate paper. Graph the following points.

1. $(4, 30°)$ 2. $(5, 45°)$ 3. $(0, 37°)$ 4. $(0, 48°)$

5. $(-6, 150°)$ 6. $(-5, 135°)$ 7. $(-8, 210°)$ 8. $(-5, 270°)$

9. $(3, -30°)$ 10. $(6, -45°)$ 11. $(7, -315°)$ 12. $(4, -270°)$

13. $(-3, -30°)$ 14. $(-6, -45°)$ 15. $(-3.2, 27°)$ 16. $(-6.8, 34°)$

17. $\left(6, \dfrac{\pi}{4}\right)$ 18. $\left(5, \dfrac{\pi}{6}\right)$ 19. $\left(4, \dfrac{3\pi}{2}\right)$ 20. $\left(3, \dfrac{3\pi}{4}\right)$

21. $\left(-6, \dfrac{\pi}{4}\right)$ 22. $\left(-5, \dfrac{\pi}{6}\right)$ 23. $\left(-4, -\dfrac{3\pi}{2}\right)$ 24. $\left(-3, -\dfrac{3\pi}{4}\right)$

[ii] Find polar coordinates of the following points.

25. $(4, 4)$ 26. $(5, 5)$ 27. $(0, 5)$ 28. $(0, -3)$

29. $(4, 0)$ 30. $(-5, 0)$ 31. $(3, 3\sqrt{3})$ 32. $(-3, -3\sqrt{3})$

33. $(\sqrt{3}, 1)$ 34. $(-\sqrt{3}, 1)$ 35. $(3\sqrt{3}, 3)$ 36. $(4\sqrt{3}, -4)$

Find Cartesian coordinates of the following points.

37. $(4, 45°)$ 38. $(5, 60°)$ 39. $(0, 23°)$ 40. $(0, -34°)$

41. $(-3, 45°)$ 42. $(-5, 30°)$ 43. $(6, -60°)$ 44. $(3, -120°)$

45. $\left(10, \dfrac{\pi}{6}\right)$ 46. $\left(12, \dfrac{3\pi}{4}\right)$ 47. $\left(-5, \dfrac{5\pi}{6}\right)$ 48. $\left(-6, \dfrac{3\pi}{4}\right)$

Convert to polar equations.

49. $3x + 4y = 5$ 50. $5x + 3y = 4$ 51. $x = 5$ 52. $y = 4$

53. $x^2 + y^2 = 36$ 54. $x^2 + y^2 = 16$ 55. $x^2 - 4y^2 = 4$ 56. $x^2 - 5y^2 = 5$

Convert to rectangular equations.

57. $r = 5$ 58. $r = 8$ 59. $\theta = \dfrac{\pi}{4}$ 60. $\theta = \dfrac{3\pi}{4}$

61. $r \sin\theta = 2$ 62. $r \cos\theta = 5$ 63. $r = 4\cos\theta$ 64. $r = -3\sin\theta$

65. $r - r\sin\theta = 2$ 66. $r + r\cos\theta = 3$ 67. $r - 2\cos\theta = 3\sin\theta$ 68. $r + 5\sin\theta = 7\cos\theta$

[iii] Graph.

69. $r = 4\cos\theta$ 70. $r = 4\sin\theta$ 71. $r = 1 - \cos\theta$

72. $r = 1 - \sin\theta$ 73. $r = \sin 2\theta$ 74. $r = \sin 3\theta$

★

75. Convert to a rectangular equation: $r = \sec^2 \dfrac{\theta}{2}$.

76. The center of a regular hexagon is at the origin and one vertex is the point $(4, 0°)$. Find the coordinates of the other vertices.

77. Graph $r = 2\cos 3\theta$.

78. Graph $r = 3\sin 2\theta$.

5.7 FORCES IN EQUILIBRIUM

[i] When several forces act through the same point on an object, their vector sum must be **0** in order for a balance to occur. When a balance occurs, then the object is either stationary or moving in a straight line without acceleration. The fact that the vector sum must be **0** for a balance and vice versa allows us to solve many applied problems involving forces. In these problems we usually resolve vectors

OBJECTIVE

You should be able to:

[i] Solve applied problems in which several forces acting at a point are in equilibrium.

1. A weight of 300 lb is supported by two ropes at a point A. One rope is horizontal and the other makes an angle of 30° with the horizontal. Find the forces (of tension) in the two ropes.

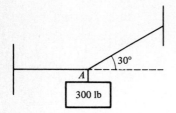

into components. For a balance, the sum of the components must be **0** separately.*

Example 1 A 200-lb sign is hanging from the end of a horizontal hinged boom, supported by a cable from the end of the boom and inclined 35° with the horizontal. Find the force (tension) in the cable and the force (compression) in the boom.

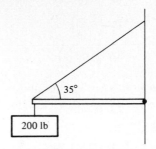

At the end of the boom we have three forces acting at a point: the weight of the sign acting down, the cable pulling up and to the right, and the boom pushing to the left. We draw a force diagram showing these.

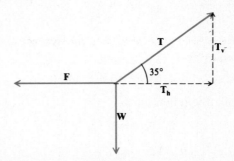

We have called the tension in the cable **T**, and the compression in the boom **F** and the weight **W**. We resolve the tension in the cable into horizontal and vertical components. We have a balance, so $\mathbf{F} + \mathbf{T} + \mathbf{W} = \mathbf{0}$. In fact, the sum of the horizontal components must be **0**, and the sum of the vertical components must also be **0**. Thus we know that $|\mathbf{T_v}| = 200$ lb and that $|\mathbf{F}| = |\mathbf{T_h}|$. Then we have

$$|\mathbf{T_v}| = 200 = |\mathbf{T}| \sin 35°$$

$$|\mathbf{T}| = \frac{|\mathbf{T_v}|}{\sin 35°} = \frac{200}{0.5736} = 349 \text{ lb.}$$

Then $|\mathbf{F}| = |\mathbf{T_h}| = 349 \cos 35° = 286$ lb. The answer is that the tension in the cable is 349 lb and the compression in the boom is 286 lb.

DO EXERCISE 1.

_____

*For a balance, the rotational tendency of the forces must also be zero. This will always be the case if the forces are concurrent (act through the same point). In this section we consider only applications in which forces are concurrent.

Example 2 A block weighing 100 kg rests on a 25° incline. Find the components of its weight perpendicular and parallel to the incline.

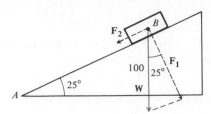

The weight **W** is 100 kg acting downward as shown. We draw $\mathbf{F_1}$ and $\mathbf{F_2}$, the components. The angle at B has the same measure as the angle at A because their sides are respectively perpendicular. Thus we have

$$|\mathbf{F_1}| = 100 \cos 25° = 90.6 \text{ kg} \qquad \text{and} \qquad |\mathbf{F_2}| = 100 \sin 25° = 42.3 \text{ kg}.$$

DO EXERCISE 2.

In a situation like that of Example 2, the force necessary to hold the block on the incline is a force the size of $\mathbf{F_2}$ but opposite in direction. The force with which the block pushes on the incline is $\mathbf{F_1}$.

Example 3 A 500-kg block is suspended by two ropes as shown. Find the tension in each rope.

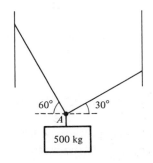

At point A there are three forces acting: the block is pulling down, and the two ropes are pulling upward and outward. We draw a force diagram showing these.

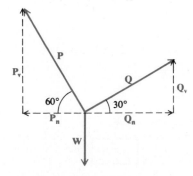

2. A block weighing 100 kg rests on a 30° incline. Find the components of its weight parallel to and perpendicular to the incline.

3. Two ropes support a 1000-kg weight, as shown. Find the tension in each rope.

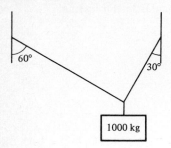

1000 kg

We have called the forces exerted by the ropes **P** and **Q**. Their horizontal components have magnitudes $|\mathbf{P}| \cos 60°$ and $|\mathbf{Q}| \cos 30°$. Since there is a balance, these must be the same:

$$|\mathbf{P}| \cos 60° = |\mathbf{Q}| \cos 30°.$$

The vertical components in the ropes have magnitudes $|\mathbf{P}| \sin 60°$ and $|\mathbf{Q}| \sin 30°$. Since there is a balance, these must total 500 lb, the weight of the block:

$$|\mathbf{P}| \sin 60° + |\mathbf{Q}| \sin 30° = 500.$$

We now have two equations with unknowns $|\mathbf{P}|$ and $|\mathbf{Q}|$:

$$|\mathbf{P}| \cos 60° - |\mathbf{Q}| \cos 30° = 0, \tag{1}$$
$$|\mathbf{P}| \sin 60° + |\mathbf{Q}| \sin 30° = 500. \tag{2}$$

To solve, we begin by solving equation (1) for P:

$$|\mathbf{P}| = |\mathbf{Q}| \frac{\cos 30°}{\cos 60°}. \tag{3}$$

Next, we substitute in equation (2):

$$|\mathbf{Q}| \frac{\cos 30° \sin 60°}{\cos 60°} + |\mathbf{Q}| \sin 30° = 500$$

$$|\mathbf{Q}| \frac{\cos 30° \sin 60°}{\cos 60°} + |\mathbf{Q}| \frac{\sin 30° \cos 60°}{\cos 60°} = 500$$

$$|\mathbf{Q}| (\cos 30° \sin 60° + \sin 30° \cos 60°) = 500 \cos 60°$$

$$|\mathbf{Q}| (\sin 90°) = 500 \cos 60° \quad \text{Using an identity}$$

$$|\mathbf{Q}| = 500 \cdot \frac{1}{2} = 250 \text{ kg}.$$

Then substituting in equation (3), we obtain

$$|\mathbf{P}| = 250 \frac{\cos 30°}{\cos 60°} = 250 \frac{\sqrt{3}/2}{1/2} = 250 \sqrt{3} = 433 \text{ kg}.$$

DO EXERCISE 3.

EXERCISE SET 5.7

[i]

1. A 150-lb sign is hanging from the end of a hinged boom, supported by a cable inclined 42° with the horizontal. Find the tension in the cable and the compression in the boom.

2. A 300-lb sign is hanging from the end of a hinged boom, supported by a cable inclined 51° from the horizontal. Find the tension in the cable and the compression in the boom (see the figure for Exercise 1).

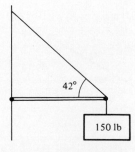

42°

150 lb

3. A weight of 200 kg is supported by a frame made of two rods and hinged at *A*, *B*, and *C* (see the figure below). Find the forces exerted by the two rods.

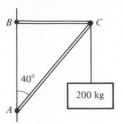

4. A weight of 300 kg is supported by the hinged frame of Exercise 3, but where the angle shown is 50°. Find the forces exerted by the two rods.

5. The force due to air movement on an airplane wing in flight is 2800 lb, acting at an angle of 28° with the vertical. What is the lift (vertical component) and what is the drag (horizontal component)?

6. The force due to air movement on an airplane wing in flight is 3500 lb, acting at an angle of 32° with the vertical. What is the lift (vertical component) and what is the drag (horizontal component)?

7. A 100-kg block of ice rests on a 37° incline. What force parallel to the incline is necessary to keep it from sliding down?

8. What force is necessary to pull a 6500-kg truck up a 7° incline?

9. The moon's gravity is $\frac{1}{6}$ that of the earth. To simulate the moon's gravity an incline is used, as shown in the figure to the right. The astronaut is suspended in a sling and is free to move horizontally. The component of his weight perpendicular to the incline should be $\frac{1}{6}$ of his weight. What must the angle θ be?

10. To simulate gravity on a planet having $\frac{3}{8}$ of the earth's gravity, what should the angle θ be? (See the figure and explanation for Exercise 9.)

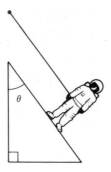

11. A 400-kg block is suspended by two ropes as shown. Find the tension in each rope.

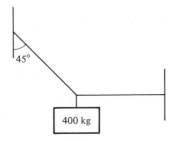

12. A 1000-lb block is suspended by two ropes as shown. Find the tension in each rope.

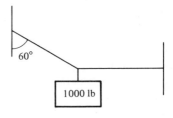

13. A 2000-kg block is suspended by two ropes as shown. Find the tension in each rope.

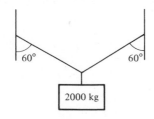

14. A 1500-lb block is suspended by two ropes as shown. Find the tension in each rope.

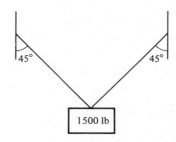

15. A 2500-kg block is suspended by two ropes as shown. Find the tension in each rope.

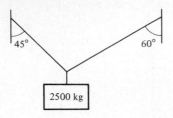

16. A 2000-kg block is suspended by two ropes as shown. Find the tension in each rope.

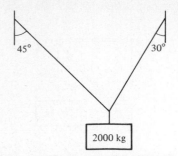

CHAPTER 5 REVIEW

In Exercises 1 and 2 solve the triangles. Angle C is a right angle. Use three-digit precision.

[5.1, i]   **1.** $a = 7.3$, $c = 8.6$ **2.** $a = 30.5$, $B = 51°10'$.

3. One leg of a right triangle bears east. The hypotenuse is 734 cm long and bears N 57°20′E. Find the perimeter.

[5.2, i]   **4.** Solve triangle ABC, where $B = 118°20'$, $C = 27°30'$, and $b = 0.974$.

5. In an isosceles triangle, the base angles each measure 52°20′ and the base is 513 cm long. Find the lengths of the other two sides.

[5.3, i]   **6.** In triangle ABC, $a = 3.7$, $c = 4.9$, and $B = 135°$. Find b.

[5.1, ii] **7.** An observer's eye is 6 ft above the floor. A mural is being viewed. The bottom of the mural is at floor level. The observer looks downward 13° to see the bottom and upward 17° to see the top. How tall is the mural?

[5.2, ii] **8.** A parallelogram has sides of lengths 3.21 cm and 7.85 cm. One of its angles measures 147°. Find the area of the parallelogram.

[5.5, ii] **9.** A car is moving north at 45 mph. A ball is thrown from the car at 37 mph in a direction of N 45°E. Find the speed and direction of the ball over the ground.

[5.7, i]   **10.** A block weighing 150 lb rests on an inclined plane. The plane makes an angle of 45° with the ground. Find the components of the weight parallel and perpendicular to the plane.

[5.5, iv] **11.** Find rectangular notation for the vector $(20, -200°)$.

[5.5, iii] **12.** Find polar notation for the vector $(-2, 3)$.

[5.6, ii] **13.** Convert to a polar equation: $x^2 + y^2 + 2x - 3y = 0$.

[5.5, ii] **14.** Vector **u** has a component of 12 lb upward and a component of 18 lb to the right. Vector **v** has components of 35 lb downward and 35 lb to the left. Find the components of **u** + **v** and the magnitude and direction of **u** + **v**.

[5.5, v]   **15.** Do the following calculations for **u** = (4, 3) and **v** = (−3, 4).
a) 4**u** − 3**v**
b) |**u** + **v**|

[5.7, i]   **16.** A crane is supporting a 500-kg steel beam. The boom of the crane is inclined 30° from the vertical. Find the compression in the boom of the crane.

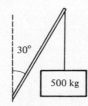

17. A parallelogram has sides of lengths 3.42 and 6.97. Its area is 18.4. Find the sizes of its angles.

6

IMAGINARY AND COMPLEX NUMBERS

OBJECTIVES

You should be able to:

[i] Express imaginary numbers (square roots of negative numbers) in terms of i, and simplify.

[ii] Add, subtract, and multiply complex numbers, expressing the answer as $a + bi$. Also factor sums of squares.

[iii] Determine whether a complex number is a solution of an equation.

[iv] Use the fact that for equality of complex numbers, the real parts must be the same and the imaginary parts must be the same.

Express in terms of i.

1. $\sqrt{-6}$

2. $-\sqrt{-10}$

3. $\sqrt{-4}$

4. $-\sqrt{-25}$

5. Simplify $\sqrt{-5}\sqrt{-2}$.

Simplify.

6. $\dfrac{\sqrt{-22}}{\sqrt{-2}}$

7. $\dfrac{\sqrt{-21}}{\sqrt{3}}$

8. $\sqrt{-16} + \sqrt{-9}$

9. $\sqrt{-25} - \sqrt{-4}$

10. $\sqrt{-17} + \sqrt{-9}$

6.1 IMAGINARY AND COMPLEX NUMBERS

[i] Imaginary Numbers

Negative numbers do not have square roots in the system of real numbers. Certain equations have no solutions. A new kind of number, called *imaginary*, was invented so that negative numbers would have square roots and certain equations would have solutions. These numbers were devised, starting with an imaginary unit, named i, with the agreement that $i^2 = -1$ or $i = \sqrt{-1}$. All other imaginary numbers can be expressed as a product of i and a real number.

Example 1 Express $\sqrt{-5}$ and $-\sqrt{-7}$ in terms of i.

$$\sqrt{-5} = \sqrt{-1 \cdot 5} = \sqrt{-1}\sqrt{5} = i\sqrt{5},$$
$$-\sqrt{-7} = -\sqrt{-1 \cdot 7} = -\sqrt{-1}\sqrt{7} = -i\sqrt{7}$$

It is also to be understood that the imaginary unit obeys the familiar laws of real numbers, such as the commutative and associative laws.

Example 2 Simplify $\sqrt{-3}\sqrt{-7}$.

Important. We first express the two imaginary numbers in terms of i:

$$\sqrt{-3}\sqrt{-7} = i\sqrt{3} \cdot i\sqrt{7}.$$

Now, rearranging and combining, we have

$$i^2\sqrt{3}\sqrt{7} = -1 \cdot \sqrt{21}$$
$$= -\sqrt{21}.$$

Had we not expressed imaginary numbers in terms of i at the outset, we would have obtained $\sqrt{21}$ instead of $-\sqrt{21}$.

> *Important:* **All imaginary numbers must be expressed in terms of i before simplifying.**

DO EXERCISES 1–5.

Example 3 Simplify $-\sqrt{20}/\sqrt{-5}$.

$$\frac{-\sqrt{20}}{\sqrt{-5}} = \frac{-\sqrt{20}}{i\sqrt{5}} \cdot \frac{i}{i} = \frac{-i\sqrt{20}}{i^2\sqrt{5}}$$
$$= \frac{-i}{-1}\sqrt{\frac{20}{5}} = 2i$$

Example 4 Simplify $\sqrt{-9} + \sqrt{-25}$.

$$\sqrt{-9} + \sqrt{-25} = i\sqrt{9} + i\sqrt{25}$$
$$= 3i + 5i = (3 + 5)i$$
$$= 8i$$

DO EXERCISES 6–10.

Powers of i

Let us look at the powers of i:

$$i^2 = -1, \qquad\qquad i^5 = i^4 \cdot i = 1 \cdot i = i,$$
$$i^3 = i^2 \cdot i = -1 \cdot i = -i, \qquad i^6 = i^5 \cdot i = i \cdot i = -1,$$
$$i^4 = i^2 \cdot i^2 = -1(-1) = 1, \qquad i^7 = i^6 \cdot i = -1 \cdot i = -i.$$

The first four powers of i are all different, but thereafter there is a repeating pattern, in cycles of four. Note that $i^4 = 1$ and that all powers of i^4, such as i^8, i^{16}, and so on, are 1. To find a higher power we express it in terms of the nearest power of 4 less than the given one.

Example 5 Simplify i^{17} and i^{23}.

$$i^{17} = i^{16} \cdot i$$

Since i^{16} is a power of i^4 and $i^4 = 1$, we know $i^{16} = 1$ so

$$i^{17} = 1 \cdot i = i,$$
$$i^{23} = i^{20} \cdot i^3.$$

Since i^{20} is a power of i^4, $i^{20} = 1$, so $i^{23} = 1 \cdot i^3 = 1 \cdot (-i) = -i$.

DO EXERCISES 11–13.

[ii] Complex Numbers

The equation $x^2 + 1 = 0$ has no solution in real numbers, but it has the imaginary solutions i and $-i$. There are still rather simple-looking equations that do not have either real or imaginary solutions. For example, $x^2 - 2x + 2 = 0$ does not. If we allow sums of real and imaginary numbers, however, this equation and many others have solutions, as we shall show.

In order that more equations will have solutions, we invent a new system of numbers called the *system of complex numbers.** A complex number is a sum of a real number and an imaginary number.

DEFINITION

The set of complex numbers consists of all numbers $a + bi$, where a and b are real numbers.

For the complex number $a + bi$ we say that the *real part* is a and the *imaginary part* is bi.

We also agree that the familiar properties of real numbers (the *field* properties) hold also for complex numbers†. We list them.

Simplify.

11. i^{25}

12. i^{18}

13. i^{31}

---

*One may wonder why we do not invent a system of imaginary numbers. That would not make sense because products of imaginary numbers are not necessarily imaginary. Consider, for example, $i^2 = -1$.
†In a more rigorous treatment, addition and multiplication are defined for complex numbers. It is then proved that the field properties hold.

Simplify.

14. $(8 - i) + (4 + 2i)$

15. $(9 + 2i) - (4 + 3i)$

16. $(2 + 4i)(3 + i)$

17. $(4 + 5i) + (4 - 5i)$

18. $2i(4 + 3i)$

19. $(5 + 6i) - (5 + 3i)$

Factor.

20. $x^2 + 4$

21. $9 + y^2$

> **COMMUTATIVITY.** Addition and multiplication are commutative.
>
> **ASSOCIATIVITY.** Addition and multiplication are associative.
>
> **DISTRIBUTIVITY.** Multiplication is distributive over addition and also over subtraction.
>
> **IDENTITIES.** The additive identity is $0 + 0i$, or 0. The multiplicative identity is $1 + 0i$, or 1.
>
> **ADDITIVE INVERSES.** Every complex number $a + bi$ has the additive inverse $-a - bi$.
>
> **MULTIPLICATIVE INVERSES.** Every nonzero complex number has a multiplicative inverse, or reciprocal.

The complex-numbers system is an extension of the real-number system. Any number $a + bi$, where a and b are real numbers, is a complex number. The number b can be 0, in which case we have $a + 0i$, which simplifies to the real number a. Thus the complex numbers include all of the real numbers. They also include all of the imaginary numbers, because any imaginary number bi is equal to $0 + bi$.

Calculations

Since in the system of complex numbers the field properties hold, calculations are much the same as for real numbers. The primary difference is that one must remember that $i^2 = -1$.

Example 6 Simplify $(8 + 6i) + (3 + 2i)$.

$$(8 + 6i) + (3 + 2i) = 8 + 3 + 6i + 2i$$
$$= 11 + 8i$$

Example 7 Simplify $(1 + 2i)(1 + 3i)$.

$$(1 + 2i)(1 + 3i) = 1 + 6i^2 + 2i + 3i$$
$$= 1 - 6 + 2i + 3i \qquad i^2 = -1$$
$$= -5 + 5i$$

Example 8 Simplify $(3 + 2i) - (5 - 2i)$.

$$(3 + 2i) - (5 - 2i) = (3 + 2i) - 5 + 2i$$
$$= 3 - 5 + 2i + 2i$$
$$= -2 + 4i$$

DO EXERCISES 14–19.

In the field of real numbers, a sum of two squares cannot be factored. In the system of complex numbers a sum of squares is always factorable.

Example 9 Factor $x^2 + y^2$.

$$(x + yi)(x - yi)$$

A check by multiplying will show that this is correct.

DO EXERCISES 20 AND 21.

[iii] **Solutions of Equations**

In the system of complex numbers, a great many equations have solutions. In fact, any equation $P(x) = 0$, where $P(x)$ is a nonconstant polynomial, has a solution.

Example 10 Determine whether $1 + i$ is a solution of $x^2 - 2x + 2 = 0$.

$$x^2 - 2x + 2 = 0$$

$(1 + i)^2 - 2(1 + i) + 2$	0 Substituting
$1 + 2i + i^2 - 2 - 2i + 2$	
$1 + 2i - 1 - 2 - 2i + 2$	
0	

The number $1 + i$ is a solution.

DO EXERCISE 22.

[iv] **Equality for Complex Numbers**

An equation $a + bi = c + di$ will be true if and only if the real parts are the same and the imaginary parts are the same. In other words, we have the following.

$$a + bi = c + di \quad \text{if and only if} \quad a = c \text{ and } b = d.$$

Example 11 Suppose that $3x + yi = 5x + 1 + 2i$. Find x and y.

We equate the real parts: $3x = 5x + 1$. Solving this equation, we obtain $x = -\frac{1}{2}$. We equate the imaginary parts: $yi = 2i$. Thus $y = 2$.

DO EXERCISE 23.

22. Determine whether $1 - i$ is a solution of $x^2 - 2x + 2 = 0$.

23. Given that $3x + 1 + (y + 2)i = 2x + 2yi$, find x and y.

EXERCISE SET 6.1

[i] In Exercises 1–6, express in terms of i.

1. $\sqrt{-15}$ **2.** $\sqrt{-17}$ **3.** $\sqrt{-16}$
4. $\sqrt{-25}$ **5.** $-\sqrt{-12}$ **6.** $-\sqrt{-20}$

In Exercises 7–20, simplify. Leave answers in terms of i in Exercises 7–10.

7. $\sqrt{-16} + \sqrt{-25}$ **8.** $\sqrt{-36} - \sqrt{-4}$ **9.** $\sqrt{-7} - \sqrt{-10}$
10. $\sqrt{-5} + \sqrt{-7}$ **11.** $\sqrt{-5}\sqrt{-11}$ **12.** $\sqrt{-7}\sqrt{-8}$
13. $-\sqrt{-4}\sqrt{-5}$ **14.** $-\sqrt{-9}\sqrt{-7}$ **15.** $\dfrac{-\sqrt{5}}{\sqrt{-2}}$ **16.** $\dfrac{\sqrt{-7}}{-\sqrt{5}}$
17. $\dfrac{\sqrt{-9}}{-\sqrt{4}}$ **18.** $\dfrac{-\sqrt{25}}{\sqrt{-16}}$ **19.** $\dfrac{-\sqrt{-36}}{\sqrt{-9}}$ **20.** $\dfrac{\sqrt{-25}}{-\sqrt{-16}}$

[ii] In Exercises 21–38, simplify.

21. $(2 + 3i) + (4 + 2i)$ **22.** $(5 - 2i) + (6 + 3i)$ **23.** $(4 + 3i) + (4 - 3i)$ **24.** $(2 + 3i) + (-2 - 3i)$
25. $(8 + 11i) - (6 + 7i)$ **26.** $(9 - 5i) - (4 + 2i)$ **27.** $2i - (4 + 3i)$ **28.** $3i - (5 + 2i)$

29. $(1 + 2i)(1 + 3i)$ **30.** $(1 + 4i)(1 - 3i)$ **31.** $(1 + 2i)(1 - 3i)$ **32.** $(2 + 3i)(2 - 3i)$

33. $3i(4 + 2i)$ **34.** $5i(3 - 4i)$ **35.** $(2 + 3i)^2$ **36.** $(3 - 2i)^2$

37. i^{13} **38.** i^{72}

Factor.

39. $4x^2 + 25y^2$ **40.** $16a^2 + 49b^2$

[iii]

41. Determine whether $1 + 2i$ is a solution of $x^2 - 2x + 5 = 0$.

42. Determine whether $1 - 2i$ is a solution of $x^2 - 2x + 5 = 0$.

[iv] In Exercises 43 and 44, solve for x and y.

43. $4x + 7i = -6 + yi$ **44.** $-4 + (x + y)i = 2x - 5y + 5i$

☆ _____

45. A function f is defined as follows: $f(z) = z^2 - 4z + i$. Find $f(3 + i)$.

46. A function g is defined as follows: $g(z) = 2z^2 + z - 2i$. Find $g(2 - i)$.

47. Show that the general rule for radicals, in real numbers, $\sqrt{a \cdot b} = \sqrt{a} \cdot \sqrt{b}$, does not hold for complex numbers.

48. Show that the general rule for radicals, in real numbers, $\sqrt{\dfrac{a}{b}} = \dfrac{\sqrt{a}}{\sqrt{b}}$ does not hold for complex numbers.

★ _____

49. Solve $z^2 = -2i$. (Hint: Let $z = a + bi$.)

50. Solve $\dfrac{z^2}{4} = i$. (Hint: Let $z = a + bi$.)

OBJECTIVES

You should be able to:

[i] Find the conjugate of a complex number and divide complex numbers.

[ii] Find the reciprocal of a complex number and express it in the form $a + bi$.

[iii] Given a polynomial in z, find a polynomial in $\bar{z}$ that is its conjugate.

Find the conjugate of each number.

1. $7 + 2i$ **2.** $6 - 4i$

3. $-5i$ **4.** $3i$

5. -3 **6.** 8

6.2 CONJUGATES AND DIVISION

[i] **Conjugates**

We shall define the conjugate of a complex number as follows.

DEFINITION

The *conjugate* of a complex number $a + bi$ is $a - bi$, and the conjugate of $a - bi$ is $a + bi$.

We illustrate:

The conjugate of $3 + 4i$ is $3 - 4i$.
The conjugate of $5 - 7i$ is $5 + 7i$.
The conjugate of $5i$ is $-5i$.
The conjugate of 6 is 6.

DO EXERCISES 1–6.

Division

Fractional notation is useful for division of complex numbers. We also use the notion of conjugates.

Example 1 Divide $4 + 5i$ by $1 + 4i$.

We shall write fractional notation and then multiply by 1.

$$\frac{4 + 5i}{1 + 4i} = \frac{4 + 5i}{1 + 4i} \cdot \frac{1 - 4i}{1 - 4i}$$ Note that $1 - 4i$ is the conjugate of the divisor.

$$= \frac{(4 + 5i)(1 - 4i)}{1^2 - 4^2i^2}$$

$$= \frac{24 - 11i}{1 + 16} = \frac{24 - 11i}{17}$$ $i^2 = -1$

$$= \frac{24}{17} - \frac{11}{17}i$$

The procedure of the last example allows us always to find a quotient of two numbers and express it in the form $a + bi$. This is true because the product of a number and its conjugate is always a real number (we will prove this later), giving us a real-number denominator.

DO EXERCISES 7 AND 8.

[ii] Reciprocals

We can find the reciprocal, or multiplicative inverse, of a complex number by division. The reciprocal of a complex number z is $1/z$.

Example 2 Find the reciprocal of $2 - 3i$ and express it in the form $a + bi$.

a) The reciprocal of $2 - 3i$ is $\dfrac{1}{2 - 3i}$.

b) We can express it in the form $a + bi$ as follows:

$$\frac{1}{2 - 3i} = \frac{1}{2 - 3i} \cdot \frac{2 + 3i}{2 + 3i} = \frac{2 + 3i}{2^2 - 3^2i^2}$$

$$= \frac{2 + 3i}{4 + 9} = \frac{2}{13} + \frac{3}{13}i.$$

DO EXERCISE 9.

[iii] Properties of Conjugates

We may use a single letter for a complex number. For example, we could shorten $a + bi$ to z. To denote the conjugate of a number, we use a bar. The conjugate of z is $\bar{z}$. Or, the conjugate of $a + bi$ is $\overline{a + bi}$. Of course, by definition of conjugates, $\overline{a + bi} = a - bi$ and $\overline{a - bi} = a + bi$. We have already noted that the product of a number and its conjugate is always a real number. Let us state this formally and prove it.

THEOREM 1

For any complex number z, $z \cdot \bar{z}$ is a real number.

Divide.

7. $\dfrac{1 + 3i}{3 + 2i}$

8. $\dfrac{2 + i}{3 - 2i}$

9. Find the reciprocal of $3 + 4i$ and express it in the form $a + bi$.

10. Compare $\overline{(3 + 2i) + (4 - 5i)}$ and $\overline{(3 + 2i)} + \overline{(4 - 5i)}$.

Proof. Let $z = a + bi$. Then

$$z \cdot \bar{z} = (a + bi)(a - bi)$$
$$= a^2 - b^2 i^2$$
$$= a^2 + b^2.$$

Since a and b are real numbers, so is $a^2 + b^2$. Thus $z \cdot \bar{z}$ is real.

The sum of a number and its conjugate is also always real. We state this as the second property.

THEOREM 2

For any complex number z, $z + \bar{z}$ is a real number.

Proof. Let $z = a + bi$. Then

$$z + \bar{z} = (a + bi) + (a - bi)$$
$$= 2a.$$

Since a is a real number, $2a$ is real. Thus $z + \bar{z}$ is real.

Now we consider the conjugate of a sum and compare it with the sum of the conjugates.

Example 3 Compare $\overline{(2 + 4i) + (5 + i)}$ and $\overline{(2 + 4i)} + \overline{(5 + i)}$.

a) $\overline{(2 + 4i) + (5 + i)} = \overline{7 + 5i}$ Adding the complex numbers
$$= 7 - 5i \quad \text{Taking the conjugate}$$

b) $\overline{(2 + 4i)} + \overline{(5 + i)} = (2 - 4i) + (5 - i)$ Taking conjugates
$$= 7 - 5i \quad \text{Adding}$$

DO EXERCISE 10.

Taking the conjugate of a sum gives the same result as adding the conjugates. Let us state and prove this.

THEOREM 3

For any complex numbers z and w, $\overline{z + w} = \bar{z} + \bar{w}$.

Proof. Let $z = a + bi$ and $w = c + di$. Then

$$\overline{z + w} = \overline{(a + bi) + (c + di)}$$
$$= \overline{(a + c) + (b + d)i},$$

by adding. We now take the conjugate and obtain $(a + c) - (b + d)i$. Now

$$\bar{z} + \bar{w} = \overline{(a + bi)} + \overline{(c + di)}$$
$$= (a - bi) + (c - di),$$

taking the conjugates. We will now add to obtain $(a + c) - (b + d)i$, the same result as before. Thus $\overline{z + w} = \bar{z} + \bar{w}$.

Let us next consider the conjugate of a product.

Example 4 Compare $\overline{(3 + 2i)(4 - 5i)}$ and $\overline{(3 + 2i)} \cdot \overline{(4 - 5i)}$.

$$\overline{(3 + 2i)(4 - 5i)} = \overline{22 - 7i} \qquad \text{Multiplying}$$
$$= 22 + 7i, \qquad \text{Taking the conjugate}$$
$$\overline{(3 + 2i)} \cdot \overline{(4 - 5i)} = (3 - 2i)(4 + 5i) \qquad \text{Taking conjugates}$$
$$= 22 + 7i \qquad \text{Multiplying}$$

DO EXERCISE 11.

The conjugate of a product is the product of the conjugates. This is our next result.

THEOREM 4

For any complex numbers z and w, $\overline{z \cdot w} = \overline{z} \cdot \overline{w}$.

Proof. Let $z = a + bi$ and $w = c + di$. Then

$$\overline{z \cdot w} = \overline{(a + bi)(c + di)}$$
$$= \overline{(ac - bd) + (bc + ad)i}. \qquad \text{Multiplying}$$

Taking the conjugate, we obtain $(ac - bd) - (bc + ad)i$. Now

$$\overline{z} \cdot \overline{w} = \overline{(a + bi)} \cdot \overline{(c + di)} = (a - bi)(c - di),$$

taking conjugates. By multiplication we obtain $(ac - bd) - (bc + ad)i$, the same result as before. Thus $\overline{z \cdot w} = \overline{z} \cdot \overline{w}$.

Let us now consider conjugates of powers, using the preceding result.

Example 5 Show that for any complex number z, $\overline{z^2} = \overline{z}^2$.

$$\overline{z^2} = \overline{z \cdot z} \qquad \text{By the definition of exponents}$$
$$= \overline{z} \cdot \overline{z} \qquad \text{By Theorem 4}$$
$$= \overline{z}^2 \qquad \text{By the definition of exponents}$$

DO EXERCISE 12.

We now state our next result.

THEOREM 5

For any complex number z, $\overline{z^n} = \overline{z}^n$. In other words, the conjugate of a power is the power of the conjugate. We understand that the exponent is a natural number.

The conjugate of a real number $a + 0i$ is $a - 0i$, and both are equal to a. Thus a real number is its own conjugate. We state this as our next result.

THEOREM 6

If z is a real number, then $\overline{z} = z$.

11. Compare $\overline{(2 + 5i)(1 + 3i)}$ and $\overline{(2 + 5i)} \cdot \overline{(1 + 3i)}$.

12. Show that for any complex number z, $\overline{z^3} = \overline{z}^3$.

Find a polynomial in $\bar{z}$ that is the conjugate of each of the following.

13. $5z^3 + 4z^2 - 2z + 1$

14. $7z^5 - 3z^3 + 8z^2 + z$

Conjugates of Polynomials

Given a polynomial in z, where z is a variable for a complex number, we can find its conjugate in terms of $\bar{z}$.

Example 6 Find a polynomial in $\bar{z}$ that is the conjugate of $3z^2 + 2z - 1$.

We write an expression for the conjugate and then use the properties of conjugates.

$$\overline{3z^2 + 2z - 1} = \overline{3z^2} + \overline{2z} - \overline{1} \quad \text{By Theorem 3}$$
$$= \overline{3}\,\overline{z^2} + \overline{2} \cdot \overline{z} - \overline{1} \quad \text{By Theorem 4}$$
$$= 3\overline{z^2} + 2\overline{z} - 1 \quad \text{The conjugate of a real number is the number itself.}$$
$$= 3\bar{z}^2 + 2\bar{z} - 1 \quad \text{By Theorem 5}$$

DO EXERCISES 13 AND 14.

EXERCISE SET 6.2

[i] In Exercises 1–16, simplify.

1. $\dfrac{4 + 3i}{1 - i}$ **2.** $\dfrac{2 - 3i}{5 - 4i}$ **3.** $\dfrac{\sqrt{2} + i}{\sqrt{2} - i}$ **4.** $\dfrac{\sqrt{3} + i}{\sqrt{3} - i}$

5. $\dfrac{3 + 2i}{i}$ **6.** $\dfrac{2 + 3i}{i}$ **7.** $\dfrac{i}{2 + i}$ **8.** $\dfrac{3}{5 - 11i}$

9. $\dfrac{1 - i}{(1 + i)^2}$ **10.** $\dfrac{1 + i}{(1 - i)^2}$ **11.** $\dfrac{3 - 4i}{(2 + i)(3 - 2i)}$ **12.** $\dfrac{(4 - i)(5 + i)}{(6 - 5i)(7 - 2i)}$

13. $\dfrac{1 + i}{1 - i} \cdot \dfrac{2 - i}{1 - i}$ **14.** $\dfrac{1 - i}{1 + i} \cdot \dfrac{2 + i}{1 + i}$ **15.** $\dfrac{3 + 2i}{1 - i} + \dfrac{6 + 2i}{1 - i}$ **16.** $\dfrac{4 - 2i}{1 + i} + \dfrac{2 - 5i}{1 + i}$

[ii] In Exercises 17–24, find the reciprocal and express it in the form $a + bi$.

17. $4 + 3i$ **18.** $4 - 3i$ **19.** $5 - 2i$ **20.** $2 + 5i$

21. i **22.** $-i$ **23.** $-4i$ **24.** $5i$

[iii] In Exercises 25–28, find a polynomial in $\bar{z}$ that is the conjugate.

25. $3z^5 - 4z^2 + 3z - 5$ **26.** $7z^4 + 5z^3 - 12z$

27. $4z^7 - 3z^5 + 4z$ **28.** $5z^{10} - 7z^8 + 13z^2 - 4$

☆

29. Solve $z + 6\bar{z} = 7$.   **30.** Solve $5z - 4\bar{z} = 7 + 8i$.

31. Let $z = a + bi$. Find $\frac{1}{2}(z + \bar{z})$.   **32.** Let $z = a + bi$. Find $\frac{1}{2}i(\bar{z} - z)$.

33. Find $f(3 + i)$, where $f(z) = \dfrac{\bar{z}}{z - 1}$.   **34.** Show that for any complex number z, $z + \bar{z}$ is real.

★

35. Let $z = a + bi$.

Find a general expression for $\dfrac{1}{z}$.

36. Let $z = a + bi$ and $w = c + di$.

Find a general expression for $\dfrac{w}{z}$.

37. State and prove a theorem about the conjugate of a polynomial.

6.3 EQUATIONS AND COMPLEX NUMBERS

[i] Linear Equations

Linear equations with complex coefficients are solved in the same way as equations with real-number coefficients. The steps used in solving depend on the field properties in each case.

Example 1 Solve $3ix + 4 - 5i = (1 + i)x + 2i$.

$$3ix - (1 + i)x = 2i - (4 - 5i) \qquad \text{Adding } -(1 + i)x \text{ and } -(4 - 5i)$$

$$(-1 + 2i)x = -4 + 7i \qquad \text{Simplifying}$$

$$x = \frac{-4 + 7i}{-1 + 2i} \qquad \text{Dividing}$$

$$x = \frac{-4 + 7i}{-1 + 2i} \cdot \frac{-1 - 2i}{-1 - 2i} \qquad \text{Simplifying}$$

$$x = \frac{18 + i}{5} = \frac{18}{5} + \frac{1}{5}i$$

DO EXERCISE 1.

[ii] Quadratic Equations

The quadratic formula for equations with real-number coefficients is usually derived by completing the square. A glance at that derivation shows that all the steps can also be done with complex-number coefficients with one possible exception. We need to know that every nonzero complex number has two square roots that are additive inverses of each other. This is the case, as will be shown later. Thus the quadratic formula holds for equations with complex coefficients.

THEOREM 7

Any equation $ax^2 + bx + c = 0$, where $a \neq 0$ and a, b, and c are complex numbers, has solutions $\dfrac{-b \pm \sqrt{b^2 - 4ac}}{2a}$.

Example 2 Solve $x^2 + (1 + i)x - 2i = 0$.

We note that $a = 1$, $b = 1 + i$, and $c = -2i$. Thus

$$x = \frac{-(1 + i) \pm \sqrt{(1 + i)^2 - 4 \cdot 1 \cdot (-2i)}}{2 \cdot 1} \qquad \text{Substituting in the formula}$$

$$= \frac{-1 - i \pm \sqrt{10i}}{2}. \qquad \text{Simplifying}$$

DO EXERCISE 2.

In Example 1 we have not attempted to evaluate $\sqrt{10i}$. That will be done later (Example 6). Let us say, however, that $10i$ has two square roots that are additive inverses of each other. The same is true of every

1. Solve.

$$3 - 4i + 2ix = 3i - (1 - i)x$$

2. Solve.

$$2x^2 + (1 - i)x + 2i = 0$$

3. Solve.

$$5x^2 + 6x + 5 = 0$$

4. Find an equation having i and $1 + i$ as solutions.

5. Find an equation having 2, i, and $-i$ as solutions.

nonzero complex number. There is no problem of negative numbers not having square roots, because in the system of complex numbers there are no positive or negative numbers. Whenever we speak of positive or negative numbers, we will be referring to the system of real numbers.

[iii] Quadratic Equations with Real Coefficients

Since any real number a is $a + 0i$, we can consider real numbers to be special kinds of complex numbers. If a quadratic equation has real coefficients, the quadratic formula always gives solutions in the system of complex numbers. Recall that in using the formula, we take the square root of $b^2 - 4ac$. If this *discriminant* is negative, the solutions will have nonzero imaginary parts.

Example 3 Solve $x^2 + 2x + 5 = 0$.

We note that $a = 1$, $b = 2$, and $c = 5$. Thus

$$x = \frac{-2 \pm \sqrt{2^2 - 4 \cdot 1 \cdot 5}}{2 \cdot 1} = \frac{-2 \pm \sqrt{-16}}{2} = \frac{-2 \pm 4i}{2}.$$

The solutions are $-1 + 2i$ and $-1 - 2i$.

It is easy to see from Example 3 that when there are nonzero imaginary parts, an equation has two solutions that are conjugates of each other.

DO EXERCISE 3.

[iv] Writing Equations with Specified Solutions

The principle of zero products for real numbers states that a product is 0 if and only if at least one of the factors is 0. This principle also holds for complex numbers, as we will show in Section 6.4. Since the principle holds for complex numbers we can write equations having specified solutions.

Example 4 Find an equation having the numbers 1, i, and $-i$ as solutions.

The factors we use will be $x - 1$, $x - i$, and $x + i$. Next we set the product of these factors equal to 0:

$$(x - 1)(x - i)(x + i) = 0.$$

Now we multiply and simplify:

$$(x - 1)(x^2 - i^2) = 0$$
$$(x - 1)(x^2 + 1) = 0$$
$$x^3 - x^2 + x - 1 = 0.$$

Example 5 Find an equation having -1, i, and $1 + i$ as solutions.

$$(x + 1)(x - i)[x - (1 + i)] = 0$$
$$(x^2 + x - ix - i)(x - 1 - i) = 0$$
$$x^3 - 2ix^2 - ix - 2x - 1 + i = 0$$

DO EXERCISES 4 AND 5.

[v] Finding Square Roots

In Example 2 we used the quadratic formula, but did not evaluate $\sqrt{10i}$. We wish to find complex numbers z for which $z^2 = 10i$. We do so in the following example.

Example 6 Solve $z^2 = 10i$.

Let $z = x + yi$. Then we have $(x + yi)^2 = 10i$, or $x^2 + 2xyi + y^2i^2 = 10i$, or $x^2 - y^2 + 2xyi = 0 + 10i$. Since the real parts must be the same and the imaginary parts the same, $x^2 - y^2 = 0$ and $2xy = 10$. Here we have two equations in real numbers. We can find a solution by solving the second equation for y and substituting in the first. Since

$$y = \frac{5}{x},$$

then

$$x^2 - \left(\frac{5}{x}\right)^2 = 0$$

$$x^2 - \frac{25}{x^2} = 0 \quad \text{or} \quad x^4 - 25 = 0.$$

We now have an equation reducible to quadratic:

$$(x^2 + 5)(x^2 - 5) = 0 \qquad \text{Factoring}$$
$$x^2 + 5 = 0 \quad \text{or} \quad x^2 - 5 = 0. \qquad \text{Principle of zero products}$$

The real solutions are $\sqrt{5}$ and $-\sqrt{5}$. Now, going back to $2xy = 10$, we see that if $x = \sqrt{5}$, then $y = \sqrt{5}$, and if $x = -\sqrt{5}$, then $y = -\sqrt{5}$. Thus the solutions of our equation (and the square roots of 10i) are $\sqrt{5} + \sqrt{5}i$ and $-\sqrt{5} - \sqrt{5}i$.

DO EXERCISE 6.

6. Solve.

$$z^2 = 3 - 4i$$

EXERCISE SET 6.3

[i] Solve.

1. $(3 + i)x + i = 5i$

2. $(2 + i)x - i = 5 + i$

3. $2ix + 5 - 4i = (2 + 3i)x - 2i$

4. $5ix + 3 + 2i = (3 - 2i)x + 3i$

5. $(1 + 2i)x + 3 - 2i = 4 - 5i + 3ix$

6. $(1 - 2i)x + 2 - 3i = 5 - 4i + 2x$

7. $(5 + i)x + 1 - 3i = (2 - 3i)x + 2 - i$

8. $(5 - i)x + 2 - 3i = (3 - 2i)x + 3 - i$

[ii] In Exercises 9–14, solve using the quadratic formula. You need not evaluate square roots.

9. $x^2 + (1 - i)x + i = 0$

10. $x^2 + (1 + i)x - i = 0$

11. $2x^2 + ix + 1 = 0$

12. $2x^2 - ix + 1 = 0$

13. $3x^2 + (1 + 2i)x + 1 - i = 0$

14. $3x^2 + (1 - 2i)x + 1 + i = 0$

[iii] Solve.

15. $x^2 - 2x + 5 = 0$

16. $x^2 - 4x + 5 = 0$

17. $x^2 - 4x + 13 = 0$

18. $x^2 - 6x + 13 = 0$

19. $x^2 + 3x + 4 = 0$

20. $3x^2 + x + 2 = 0$

[iv] Find an equation having the specified solutions.

21. $2i, -2i$ **22.** $3i, -3i$ **23.** $1 + i, 1 - i$ **24.** $2 + i, 2 - i$

25. $2 + 3i, 2 - 3i$ **26.** $4 + 3i, 4 - 3i$ **27.** $3, i$ **28.** $5, i$

29. $1, 3i, -3i$ **30.** $1, 2i, -2i$ **31.** $2, 1 + i, i$ **32.** $3, 1 - i, i$

33. $i, 2i, -i$ **34.** $i, -2i, -i$

[v] Solve.

35. $z^2 = 4i$ **36.** $z^2 = -4i$ **37.** $z^2 = 3 + 4i$ **38.** $z^2 = 5 - 12i$

☆ ───

Solve. (*Hint*: Factor first.)

39. $x^3 - 8 = 0$ **40.** $x^3 + 8 = 0$

Solve.

41. $2x + 3y = 7 - 7i,$ **42.** $3x + 4y = 11 + 9i,$
$\quad\ \ 3x - 2y = 4 + 9i$ $\quad\ \ 2x - y = -5i$

★ ───

43. Solve for x and y: $\sin^2 x + i \cos^2 y = 1 + \dfrac{i}{2}$.

OBJECTIVES

You should be able to:

[i] Graph a complex number $a + bi$ and graph the sum of two numbers.

[ii] Given rectangular, or binomial, notation for a complex number, find polar notation.

[iii] Given polar notation for a complex number, find binomial notation.

[iv] Use polar notation to multiply and divide complex numbers.

6.4 GRAPHICAL REPRESENTATION AND POLAR NOTATION

[i] Graphs

The real numbers can be graphed on a line. Complex numbers are graphed on a plane. We graph a complex number $a + bi$ in the same way that we graph an ordered pair of real numbers (a, b). In place of an x-axis we have a *real axis*, and in place of a y-axis we have an *imaginary axis*.

Examples Graph the following.

1. $3 + 2i$

2. $-4 + 5i$

3. $-5 - 4i$

See Fig. 6.1.

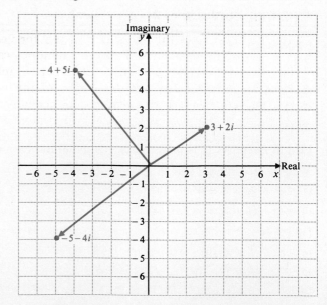

Figure 6.1

Horizontal distances correspond to the real part of a number. Vertical distances correspond to the imaginary part. Graphs of the numbers are shown above as vectors. The horizontal component of the vector for $a + bi$ is a and the vertical component is b. Complex numbers are sometimes used in the study of vectors.

DO EXERCISE 1.

Adding complex numbers is like adding vectors using components. For example, to add $3 + 2i$ and $5 + 4i$ we add the real parts and the imaginary parts to obtain $8 + 6i$. Graphically, then, the sum of two complex numbers looks like a vector sum. It is the diagonal of a parallelogram.

Example 4 Show graphically $2 + 2i$ and $3 - i$. Show also their sum.

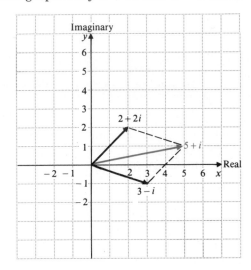

DO EXERCISE 2.

[ii] Polar Notation for Complex Numbers

In Chapter 5 we studied polar notation for vectors, in which we use the length of a vector and the angle that it makes with the x-axis. In a similar way we develop polar notation for complex numbers. From the diagram we can see that the length of the vector for $a + bi$ is $\sqrt{a^2 + b^2}$. Note that this quantity is a nonnegative real number. It is called the *absolute value* of $a + bi$, and is denoted $|a + bi|$.

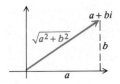

DEFINITION

The *absolute value* of a complex number $a + bi$ is denoted $|a + bi|$ and is defined to be $\sqrt{a^2 + b^2}$.

---

1. Graph these complex numbers.

   a) $5 - 3i$
   b) $-3 + 4i$
   c) $-5 - 2i$
   d) $5 + 5i$

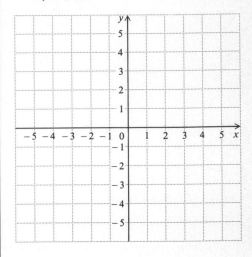

2. Graph each pair of complex numbers and graph their sum.

   a) $2 + 3i$, $1 - 5i$
   b) $-5 + 2i$, $-1 - 4i$

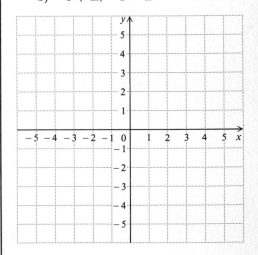

3. Find the following absolute values.

a) $|4 - 3i|$

b) $|-12 - 5i|$

4. Write rectangular notation for

$$\sqrt{2}\,(\cos 315° + i \sin 315°).$$

5. Write binomial notation for

$$2 \operatorname{cis}\left(-\frac{\pi}{6}\right).$$

Example 5 Find $|3 + 4i|$.

$$|3 + 4i| = \sqrt{3^2 + 4^2} = \sqrt{9 + 16} = 5$$

DO EXERCISE 3.

Now let us consider any complex number $a + bi$. Suppose that its absolute value is r. Let us also suppose that the angle that the vector makes with the real axis is θ. As this diagram shows, we have

$$a = r \cos \theta \quad \text{and} \quad b = r \sin \theta.$$

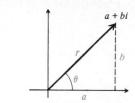

Thus

$$a + bi = r \cos \theta + ir \sin \theta$$
$$= r(\cos \theta + i \sin \theta).$$

This is polar notation for $a + bi$. The angle θ is called the *argument*.

Polar notation for complex numbers is also called *trigonometric notation*, and is often abbreviated *r cis θ*.

[ii], [iii] **Change of Notation**

To change from polar notation to *binomial*, or *rectangular*, notation $a + bi$, we proceed as in Section 5.6, recalling that $a = r \cos \theta$ and $b = r \sin \theta$.

Example 6 Write binomial notation for $2(\cos 120° + i \sin 120°)$.

$$a = 2 \cos 120° = -1,$$
$$b = 2 \sin 120° = \sqrt{3}$$

Thus $2(\cos 120° + i \sin 120°) = -1 + i\sqrt{3}$.

Example 7 Write binomial notation for $\sqrt{8} \operatorname{cis} \frac{7\pi}{4}$.

$$a = \sqrt{8} \cos \frac{7\pi}{4} = \sqrt{8} \cdot \frac{1}{\sqrt{2}} = 2,$$

$$b = \sqrt{8} \sin \frac{7\pi}{4} = \sqrt{8} \cdot \frac{-1}{\sqrt{2}} = -2$$

Thus $\sqrt{8} \operatorname{cis} \frac{7\pi}{4} = 2 - 2i$.

DO EXERCISES 4 AND 5.

To change from binomial notation to polar notation, we remember that $r = \sqrt{a^2 + b^2}$ and θ is an angle for which $\sin \theta = b/r$ and $\cos \theta = a/r$.

Example 8 Find polar notation for $1 + i$.

We note that $a = 1$ and $b = 1$. Then

$$r = \sqrt{1^2 + 1^2} = \sqrt{2},$$

$$\sin \theta = \frac{1}{\sqrt{2}} \quad \text{and} \quad \cos \theta = \frac{1}{\sqrt{2}}.$$

Thus $\theta = \pi/4$, or $45°$, and we have

$$1 + i = \sqrt{2} \operatorname{cis} \frac{\pi}{4} \quad \text{or} \quad 1 + i = \sqrt{2} \operatorname{cis} 45°.$$

Example 9 Find polar notation for $\sqrt{3} - i$.

$$r = \sqrt{(\sqrt{3})^2 + (-1)^2} = 2,$$

$$\sin \theta = -\frac{1}{2} \quad \text{and} \quad \cos \theta = \frac{\sqrt{3}}{2}.$$

Thus $\theta = 11\pi/6$, or $330°$, and we have

$$\sqrt{3} - i = 2 \operatorname{cis} \frac{11\pi}{6} = 2 \operatorname{cis} 330°.$$

In changing to polar notation, note that there are many angles satisfying the given conditions. We ordinarily choose the smallest positive angle.

DO EXERCISES 6 AND 7.

[iv] Multiplication and Polar Notation

Multiplication of complex numbers is somewhat easier to do with polar notation than with rectangular notation. We simply multiply the absolute values and add the arguments. Let us state this more formally and prove it.

THEOREM 8

For any complex numbers $r_1 \operatorname{cis} \theta_1$ and $r_2 \operatorname{cis} \theta_2$,

$$(r_1 \operatorname{cis} \theta_1)(r_2 \operatorname{cis} \theta_2) = r_1 \cdot r_2 \operatorname{cis} (\theta_1 + \theta_2).$$

Proof. Let us first multiply $a_1 + b_1 i$ by $a_2 + b_2 i$:

$$(a_1 + b_1 i)(a_2 + b_2 i) = (a_1 a_2 - b_1 b_2) + (a_2 b_1 + a_1 b_2)i.$$

Recall that

$$a_1 = r_1 \cos \theta_1, \qquad b_1 = r_1 \sin \theta_1,$$

and

$$a_2 = r_2 \cos \theta_2, \qquad b_2 = r_2 \sin \theta_2.$$

6. Write polar notation for $1 - i$.

7. Write polar notation for

$$-3\sqrt{2} - 3\sqrt{2}i.$$

We substitute these in the product above, to obtain

$$r_1(\cos \theta_1 + i \sin \theta_1) \cdot r_2(\cos \theta_2 + i \sin \theta_2)$$
$$= (r_1 r_2 \cos \theta_1 \cos \theta_2 - r_1 r_2 \sin \theta_1 \sin \theta_2)$$
$$+ (r_1 r_2 \sin \theta_1 \cos \theta_2 + r_1 r_2 \cos \theta_1 \sin \theta_2)i.$$

This simplifies to

$$r_1 r_2(\cos \theta_1 \cos \theta_2 - \sin \theta_1 \sin \theta_2) + r_1 r_2(\sin \theta_1 \cos \theta_2 + \cos \theta_1 \sin \theta_2)i.$$

Now, using identities for sums of angles, we simplify, obtaining

$$r_1 r_2 \cos (\theta_1 + \theta_2) + r_1 r_2 \sin (\theta_1 + \theta_2)i,$$

or

$$r_1 r_2 \operatorname{cis} (\theta_1 + \theta_2),$$

which was to be shown.

To divide complex numbers we do the reverse of the above. We state that fact, but will omit the proof.

THEOREM 9

For any complex numbers $r_1 \operatorname{cis} \theta_1$ and $r_2 \operatorname{cis} \theta_2$, $(r_2 \neq 0)$,

$$\frac{r_1 \operatorname{cis} \theta_1}{r_2 \operatorname{cis} \theta_2} = \frac{r_1}{r_2} \operatorname{cis} (\theta_1 - \theta_2).$$

Example 10 Find the product of $3 \operatorname{cis} 40°$ and $7 \operatorname{cis} 20°$.

$$3 \operatorname{cis} 40° \cdot 7 \operatorname{cis} 20° = 3 \cdot 7 \operatorname{cis} (40° + 20°) = 21 \operatorname{cis} 60°$$

Example 11 Find the product of $2 \operatorname{cis} \pi$ and $3 \operatorname{cis} \left(-\dfrac{\pi}{2}\right)$.

$$2 \operatorname{cis} \pi \cdot 3 \operatorname{cis} \left(-\frac{\pi}{2}\right) = 2 \cdot 3 \operatorname{cis} \left(\pi - \frac{\pi}{2}\right) = 6 \operatorname{cis} \frac{\pi}{2}.$$

Example 12 Convert to polar notation and multiply: $(1 + i)(\sqrt{3} - i)$.

We first find polar notation (see Examples 8 and 9):

$$1 + i = \sqrt{2} \operatorname{cis} 45°, \qquad \sqrt{3} - i = 2 \operatorname{cis} 330°.$$

We now multiply, using Theorem 8:

$$(\sqrt{2} \operatorname{cis} 45°)(2 \operatorname{cis} 330°) = 2 \cdot \sqrt{2} \operatorname{cis} 375° \quad \text{or} \quad 2\sqrt{2} \operatorname{cis} 15°.$$

Example 13 Divide $2 \operatorname{cis} \pi$ by $4 \operatorname{cis} \dfrac{\pi}{2}$.

$$\frac{2 \operatorname{cis} \pi}{4 \operatorname{cis} \dfrac{\pi}{2}} = \frac{2}{4} \operatorname{cis} \left(\pi - \frac{\pi}{2}\right)$$

$$= \frac{1}{2} \operatorname{cis} \frac{\pi}{2}$$

Example 14 Convert to polar notation and divide: $(1 + i)/(1 - i)$.

We first convert to polar notation:

$$1 + i = \sqrt{2}\ \text{cis}\ 45°, \qquad \text{See Example 8.}$$
$$1 - i = \sqrt{2}\ \text{cis}\ 315°.$$

We now divide, using Theorem 9:

$$\frac{1 + i}{1 - i} = \frac{\sqrt{2}\ \text{cis}\ 45°}{\sqrt{2}\ \text{cis}\ 315°} = 1 \cdot \text{cis}\ (45° - 315°)$$

$$= 1 \cdot \text{cis}\ (-270°) \quad \text{or} \quad 1 \cdot \text{cis}\ 90°.$$

DO EXERCISES 8–12.

The Principle of Zero Products

In the system of complex numbers, when we multiply by 0 the result is 0. This is half of the principle of zero products. We have yet to show that the converse is true: that if a product is zero, at least one of the factors must be zero. With polar notation that is easy to do. The number 0 has polar notation $0\ \text{cis}\ \theta$, where θ can be any angle. Now consider a product $r_1\ \text{cis}\ \theta_1 \cdot r_2\ \text{cis}\ \theta_2$, where neither factor is 0. This means that $r_1 \neq 0$ and $r_2 \neq 0$. Since these are nonzero real numbers, their product is not zero, and the product of the complex numbers $r_1 r_2\ \text{cis}\ (\theta_1 + \theta_2)$ is not zero. Thus the principle of zero products holds in complex numbers.

Multiply.

8. $5\ \text{cis}\ 25° \cdot 4\ \text{cis}\ 30°$

9. $8\ \text{cis}\ \pi \cdot \dfrac{1}{2}\ \text{cis}\ \dfrac{\pi}{4}$

10. Convert to polar notation and multiply.

$$(1 + i)(2 + 2i)$$

11. Divide.

$$10\ \text{cis}\ \frac{\pi}{2} \div 5\ \text{cis}\ \frac{\pi}{4}$$

12. Convert to polar notation and divide.

$$\frac{\sqrt{3} - i}{1 + i}$$

EXERCISE SET 6.4

[i] In Exercises 1–8, graph each pair of complex numbers and their sum.

1. $3 + 2i,\ 2 - 5i$ **2.** $4 + 3i,\ 3 - 4i$

3. $-5 + 3i,\ -2 - 3i$ **4.** $-4 + 2i,\ -3 - 4i$

5. $2 - 3i,\ -5 + 4i$ **6.** $3 - 2i,\ -5 + 5i$

7. $-2 - 5i,\ 5 + 3i$ **8.** $-3 - 4i,\ 6 + 3i$

[iii] In Exercises 9–16, find rectangular notation.

9. $3(\cos 30° + i \sin 30°)$ **10.** $6(\cos 150° + i \sin 150°)$

11. $10\ \text{cis}\ 270°$ **12.** $5\ \text{cis}\ (-60°)$

13. $\sqrt{8}\left(\cos \dfrac{\pi}{4} + i \sin \dfrac{\pi}{4}\right)$ **14.** $5\left(\cos \dfrac{\pi}{3} + i \sin \dfrac{\pi}{3}\right)$

15. $\sqrt{8}\ \text{cis}\ \dfrac{5\pi}{4}$ **16.** $\sqrt{8}\ \text{cis}\left(-\dfrac{\pi}{4}\right)$

[ii] In Exercises 17–22, find polar notation.

17. $1 - i$ **18.** $\sqrt{3} + i$

19. $10\sqrt{3} - 10i$ **20.** $-10\sqrt{3} + 10i$

21. -5 **22.** $-5i$

[iv] In Exercises 23–30, convert to polar notation and then multiply or divide.

23. $(1 - i)(2 + 2i)$

24. $(1 + i\sqrt{3})(1 + i)$

25. $(2\sqrt{3} + 2i)(2i)$

26. $(3\sqrt{3} - 3i)(2i)$

27. $\dfrac{1 - i}{1 + i}$

28. $\dfrac{1 - i}{\sqrt{3} - i}$

29. $\dfrac{2\sqrt{3} - 2i}{1 + \sqrt{3}i}$

30. $\dfrac{3 - 3\sqrt{3}i}{\sqrt{3} - i}$

☆ ———————————————————————————————

31. Show that for any complex number z, $|z| = |-z|$. (*Hint:* Let $z = a + bi$.)

32. Show that for any complex number z, $|z| = |\bar{z}|$. (*Hint:* Let $z = a + bi$.)

33. Show that for any complex number z, $|z\bar{z}| = |z^2|$.

34. Show that for any complex number z, $|z^2| = |z|^2$.

35. Show that for any complex numbers z and w, $|z \cdot w| = |z| \cdot |w|$. (*Hint:* Let $z = r_1 \operatorname{cis} \theta_1$ and $w = r_2 \operatorname{cis} \theta_2$.)

36. Show that for any complex number z and any nonzero complex number w, $\left|\dfrac{z}{w}\right| = \dfrac{|z|}{|w|}$. (Use the hint for Exercise 35.)

37. On a complex plane, graph $|z| = 1$.

38. On a complex plane, graph $z + \bar{z} = 3$.

★ ———————————————————————————————

39. Find polar notation for $(\cos \theta + i \sin \theta)^{-1}$.

═══════════════════════════════════════

OBJECTIVES

You should be able to:

[i] Use DeMoivre's theorem to raise complex numbers to powers.

[ii] Find the nth roots of a complex number.

1. Find $(1 - i)^{10}$.

2. Find $(\sqrt{3} + i)^4$.

6.5 DeMOIVRE'S THEOREM

[i] Powers of Complex Numbers

An important theorem about powers and roots of complex numbers is named for the French mathematician DeMoivre (1667–1754). Let us consider a number $r \operatorname{cis} \theta$ and its square:

$$(r \operatorname{cis} \theta)^2 = (r \operatorname{cis} \theta)(r \operatorname{cis} \theta) = r \cdot r \operatorname{cis} (\theta + \theta) = r^2 \operatorname{cis} 2\theta.$$

Similarly, we see that

$$(r \operatorname{cis} \theta)^3 = r \cdot r \cdot r \operatorname{cis} (\theta + \theta + \theta) = r^3 \operatorname{cis} 3\theta.$$

The generalization of this is DeMoivre's theorem.

THEOREM 10
———————————————————————————————

DeMoivre's theorem. **For any complex number $r \operatorname{cis} \theta$ and any natural number n, $(r \operatorname{cis} \theta)^n = r^n \operatorname{cis} n\theta$.**

———————————————————————————————

Example 1 Find $(1 + i)^9$.

We first find polar notation: $1 + i = \sqrt{2} \operatorname{cis} 45°$. Then

$$\begin{aligned}
(1 + i)^9 &= (\sqrt{2} \operatorname{cis} 45°)^9 \\
&= \sqrt{2}^9 \operatorname{cis} 9 \cdot 45° \\
&= 2^{9/2} \operatorname{cis} 405° \\
&= 16\sqrt{2} \operatorname{cis} 45°.
\end{aligned}$$

405° has the same terminal side as 45°.

DO EXERCISES 1 AND 2.

[ii] Roots of Complex Numbers

As we shall see, every nonzero complex number has two square roots. A number has three cube roots, four fourth roots, and so on. In general a nonzero complex number has n different nth roots. They can be found by the formula that we now state and prove.

THEOREM 11

The nth roots of a complex number r cis θ are given by

$$r^{1/n} \text{ cis} \left(\frac{\theta}{n} + k \cdot \frac{360°}{n} \right),$$

where $k = 0, 1, 2, \ldots, n - 1$.

We show that this formula gives us n different roots, using DeMoivre's theorem. We take the expression for the nth roots and raise it to the nth power, to show that we get r cis θ:

$$\left[r^{1/n} \text{ cis} \left(\frac{\theta}{n} + k \cdot \frac{360°}{n} \right) \right]^n = (r^{1/n})^n \text{ cis} \left(\frac{\theta}{n} \cdot n + k \cdot n \cdot \frac{360°}{n} \right)$$

$$= r \text{ cis } (\theta + k \cdot 360°) = r \text{ cis } \theta.$$

Thus we know that the formula gives us nth roots for any natural number k. Next we show that there are at least n different roots. To see this, consider substituting 0, 1, 2, and so on, for k. When $k = n$ the cycle begins to repeat, but from 0 to $n - 1$ the angles obtained and their sines and cosines are all different. There cannot be more than n different nth roots.

Example 2 Find the square roots of $2 + 2\sqrt{3}i$.

We first find polar notation: $2 + 2\sqrt{3}i = 4$ cis $60°$. Then

$$(4 \text{ cis } 60°)^{1/2} = 4^{1/2} \text{ cis} \left(\frac{60°}{2} + k \cdot \frac{360°}{2} \right), \quad k = 0, 1,$$

$$= 2 \text{ cis} \left(30° + k \cdot \frac{360°}{2} \right), \quad k = 0, 1.$$

Thus the roots are 2 cis 30° and 2 cis 210°, or

$$\sqrt{3} + i \quad \text{and} \quad -\sqrt{3} - i.$$

DO EXERCISE 3.

In Example 2 it should be noted that the two square roots of a number were additive inverses of each other. The same is true of the square roots of any complex number. To see this, let us find the square roots of any complex number r cis θ.

$$(r \text{ cis } \theta)^{1/2} = r^{1/2} \text{ cis} \left(\frac{\theta}{2} + k \cdot \frac{360°}{2} \right), \quad k = 0, 1,$$

$$= r^{1/2} \text{ cis} \frac{\theta}{2} \quad \text{or} \quad r^{1/2} \text{ cis} \left(\frac{\theta}{2} + 180° \right)$$

3. Find the square roots.

a) $2i$

b) $10i$

4. Find and graph the cube roots of −1.

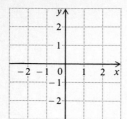

Now let us look at the two numbers on a graph. They lie on a line, so if one number has binomial notation $a + bi$, the other has binomial notation $-a - bi$. Hence their sum is 0 and they are additive inverses of each other.

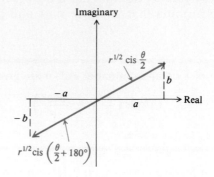

Example 3 Find the cube roots of 1. Locate them on a graph.

$$1 = 1 \text{ cis } 0°$$

$$(1 \text{ cis } 0°)^{1/3} = 1^{1/3} \text{ cis } \left(\frac{0°}{3} + k \cdot \frac{360°}{3} \right), \quad k = 0, 1, 2.$$

The roots are 1 cis 0°, 1 cis 120°, and 1 cis 240°, or

$$1, \quad -\frac{1}{2} + \frac{\sqrt{3}}{2} i, \quad \text{and} \quad -\frac{1}{2} - \frac{\sqrt{3}}{2} i.$$

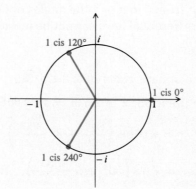

Note in the example that the graphs of the cube roots lie equally spaced about a circle. The same is true of the nth roots of any complex number.

DO EXERCISE 4.

EXERCISE SET 6.5

[i] In Exercises 1–6, raise the number to the power and write polar notation for the answer.

1. $\left(2 \text{ cis } \frac{\pi}{3} \right)^3$

2. $\left(3 \text{ cis } \frac{\pi}{2} \right)^4$

3. $\left(2 \text{ cis } \frac{\pi}{6} \right)^6$

4. $\left(2 \text{ cis } \frac{\pi}{5} \right)^5$

5. $(1 + i)^6$

6. $(1 - i)^6$

In Exercises 7–14, raise the number to the power and write rectangular notation for the answer.

7. $(2 \operatorname{cis} 240°)^4$

8. $(2 \operatorname{cis} 120°)^4$

9. $(1 + \sqrt{3}i)^4$

10. $(-\sqrt{3} + i)^6$

11. $\left(\dfrac{1}{\sqrt{2}} + \dfrac{1}{\sqrt{2}}i\right)^{10}$

12. $\left(\dfrac{1}{\sqrt{2}} - \dfrac{1}{\sqrt{2}}i\right)^{12}$

13. $\left(\dfrac{\sqrt{3}}{2} + \dfrac{1}{2}i\right)^{12}$

14. $\left(\dfrac{\sqrt{3}}{2} - \dfrac{1}{2}i\right)^{14}$

[ii]

15. Solve $x^2 = -1 + \sqrt{3}i$. That is, find the square roots of $-1 + \sqrt{3}i$.

16. Solve $x^2 = -\sqrt{3} - i$. That is, find the square roots of $-\sqrt{3} - i$.

17. Solve $x^3 = i$.

18. ▦ Solve $x^3 = 68.4321$.

19. Find and graph the fourth roots of 16.

20. Find and graph the fourth roots of i.

☆ ───────────────────────────────

In Exercises 21 and 22, solve. Evaluate square roots.

21. $x^2 + (1 - i)x + i = 0$

22. $3x^2 + (1 + 2i)x + 1 - i = 0$

CHAPTER 6 REVIEW

Simplify.

[6.1, ii]   **1.** $(2 - 2i)(3 + 4i)$

2. $(3 - 5i) - (2 - i)$

3. $(6 + 2i) + (-4 - 3i)$

4. $\dfrac{2 - 3i}{1 - 3i}$

[6.1, iv]   **5.** Solve for x and y: $4x + 2i = 8 - (2 + y)i$.

[6.2, iii]   **6.** Find a polynomial in $\bar{z}$ that is the conjugate of $3z^3 + z - 7$.

[6.3, iv]   **7.** Find an equation having the solutions $1 - 2i$, $1 + 2i$.

Solve.

[6.3, iii]   **8.** $5x^2 - 4x + 1 = 0$

[6.3, ii]   **9.** $x^2 + 3ix - 1 = 0$

[6.3, v]   **10.** Find the square roots of $4i$.

[6.4, i]   **11.** Graph the pair of complex numbers $-3 - 2i$, $4 + 7i$, and their sum.

[6.4, iii]   **12.** Find rectangular notation for $2(\cos 135° + i \sin 135°)$.

[6.4, ii]   **13.** Find polar notation for $1 + i$.

[6.4, iv]   **14.** Find the product of $7 \operatorname{cis} 18°$ and $10 \operatorname{cis} 32°$.

[6.5, ii]   **15.** Find the cube roots of $1 + i$.

16. Let $f(z) = 3z^2 - 2\bar{z} + 5i$. Find $f(2 - i)$.

17. Solve $3z + 2\bar{z} = 5 + 2i$.

18. Solve
$2x - 3y = 7 + 7i,$
$3x + 2y = 4 - 9i.$

[6.5, ii]   **19.** Solve $x^3 - 27 = 0$.

20. Graph $|z - (1 + i)| = 1$.

7

EXPONENTIAL AND LOGARITHMIC FUNCTIONS

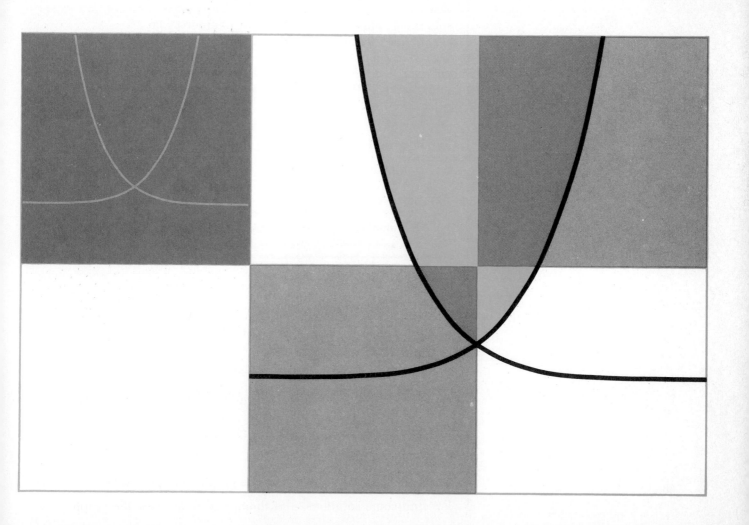

OBJECTIVES

You should be able to:

[i] Graph exponential and logarithmic equations.

[ii] Given an exponential equation of the type $a^b = c$, write an equivalent logarithmic equation $\log_a c = b$, and vice versa.

[iii] Solve equations like $\log_3 9 = x$, $\log_x 9 = 2$, and $\log_3 x = 2$.

[iv] Simplify expressions like $a^{\log_a x}$ and $\log_a a^x$.

7.1 EXPONENTIAL AND LOGARITHMIC FUNCTIONS

Irrational Exponents

We have defined exponential notation for cases in which the exponent is a rational number. For example $x^{2.34}$, or $x^{234/100}$, means to take the 100th root of x and raise the result to the 234th power. We now consider irrational exponents, such as π or $\sqrt{2}$.

Let us consider 2^π. We know that π has an unending decimal representation,

$$3.1415926535\ldots.$$

Now consider this sequence of numbers:

$$3, \quad 3.1, \quad 3.14, \quad 3.141, \quad 3.1415, \quad 3.14159, \ldots.$$

Each of these numbers is an approximation to π, the more decimal places the better the approximation. Let us use these (rational) numbers to form a sequence as follows:

$$2^3, \quad 2^{3.1}, \quad 2^{3.14}, \quad 2^{3.141}, \quad 2^{3.1415}, \quad 2^{3.14159}, \ldots.$$

Each of the numbers in this sequence is already defined, the exponent being rational. The numbers in this sequence get closer and closer to some real number. We define that number to be 2^π.

We can define exponential notation for any irrational exponent in a similar way. Thus any exponential expression a^x, $a > 0$, now has meaning, whether the exponent is rational or irrational. The usual laws of exponents still hold, in case exponents are irrational. We will not prove that fact here, however.

Exponential Functions

Exponential functions are defined using exponential notation.

DEFINITION

The function $f(x) = a^x$, where *a* is some positive constant different from 1, is called the *exponential function, base a.*

CAUTION! Here are some exponential functions:

$$f(x) = 2^x, \qquad g(x) = 3^x, \qquad h(x) = (0.178)^x.$$

Note that the variable is the exponent. The following are *not* exponential functions:

$$f(x) = x^2, \qquad g(x) = x^3, \qquad h(x) = x^{0.178}.$$

Note that the variable is not the exponent.

---

[i] **Graphs of Exponential and Logarithmic Functions**

Exponential Functions

Example 1 Graph $y = 2^x$. Use the graph to approximate 2^π.

We find some solutions, plot them, and then draw the graph.

x	0	1	2	3	-1	-2	-3
y	1	2	4	8	$\frac{1}{2}$	$\frac{1}{4}$	$\frac{1}{8}$

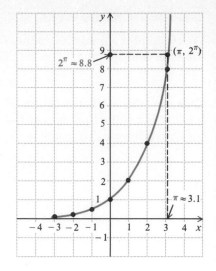

1. Use the graph of $y = 2^x$.

   a) Is this function increasing or decreasing?
   b) What is the domain?
   c) What is the range?
   d) What is the y-intercept?
   e) Use the graph to approximate $2^{\sqrt{3}}$ ($\sqrt{3} \approx 1.732$).

▦ (Note that as x increases, the function values increase. Check this on your calculator. As x decreases, the function values decrease toward 0. Check this on your calculator.)

To approximate 2^π, we locate π on the x-axis, at about 3.1. Then we find the corresponding function value. It is about 8.8.

DO EXERCISES 1 AND 2.

Let us now look at some other exponential functions. We will make comparisons, using transformations.

Example 2 Graph $y = 4^x$.

We could plot points and connect them, but let us be more clever. We note that $4^x = (2^2)^x = 2^{2x}$. Thus the function we wish to graph is

$$y = 2^{2x}.$$

Compare this with $y = 2^x$, graphed above. The graph of $y = 2^{2x}$ is a compression, in the x-direction toward the y-axis, of the graph of $y = 2^x$. Knowing this allows us to graph $y = 2^{2x}$ at once. Each point on the graph of 2^x is moved half the distance to the y-axis.

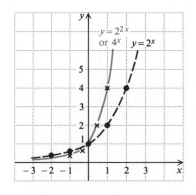

DO EXERCISES 3–5.

2. Graph $y = 4^x$, and compare it with the graph of $y = 2^x$.

   a) Is the function increasing or decreasing?
   b) What is the domain of $y = 4^x$?
   c) What is the range?
   d) What is the y-intercept?
   e) Use the graph to approximate $4^{\sqrt{2}}$ ($\sqrt{2} \approx 1.414$).
   f) Which function increases faster, 2^x or 4^x?

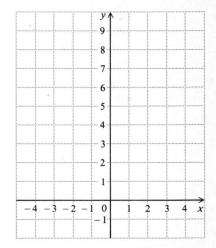

3. Graph $y = 8^x$. Use graph paper. [*Hint:* $8^x = (2^3)^x = 2^{3x}$, hence this is a horizontal compression of the graph of $y = 2^x$.]

4. Graph $y = 1^x$. Use graph paper.

5. Graph $y = \left(\frac{1}{2}\right)^x$. Use graph paper.

6. Graph $y = \left(\dfrac{1}{3}\right)^x$.

$\left[\textit{Hint: } \left(\dfrac{1}{3}\right)^x = 3^{-x}. \right]$

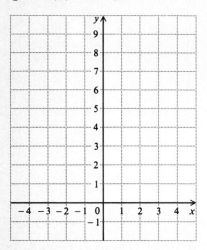

Example 3 Graph $y = (\tfrac{1}{2})^x$.

We could plot points and connect them, but again let us be more clever. We note that $(\tfrac{1}{2})^x = 1/2^x = 2^{-x}$. Thus the function we wish to graph is

$$y = 2^{-x}.$$

Compare this with the graph of $y = 2^x$ in Example 1. The graph of $y = 2^{-x}$ is a reflection, across the y-axis, of the graph of $y = 2^x$. Knowing this allows us to graph $y = 2^{-x}$ at once.

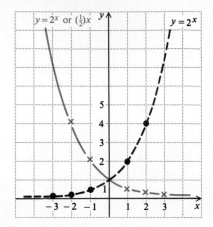

DO EXERCISE 6.

The preceding examples and exercises illustrate exponential functions of various bases. If $a = 1$, then $f(x) = a^x = 1^x = 1$ and the graph is a horizontal line. This is why we exclude 1 as a base for an exponential function. We summarize.

THEOREM 1

1. **When $a > 1$, the function $f(x) = a^x$ is an increasing function. The greater the value of a, the faster the function increases.**

2. **When $0 < a < 1$, the function $f(x) = a^x$ is a decreasing function. The greater the value of a, the more slowly the function decreases.**

It should be noted that for any value of a, the y-intercept is $(0, 1)$.

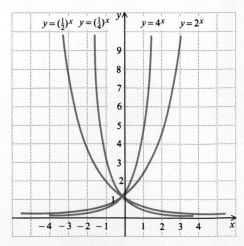

Logarithmic Functions

The inverse of an exponential function, for $a > 0$ and $a \neq 1$, is a function. It is called a *logarithmic* or *logarithm function*. Thus one way to describe a logarithm function is to interchange variables in $y = a^x$:

$$x = a^y.$$

The most useful and interesting logarithmic functions are those for which $a > 1$. The graph of such a function is a reflection of $y = a^x$ across the line $y = x$, as shown below. Note that the domain of a logarithm function is the set of all positive real numbers.

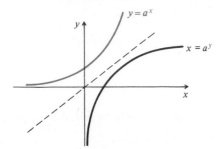

For logarithm functions we use the notation $\log_a (x)$ or $\log_a x$.* That is, we use the symbol $\log_a x$ to denote the second coordinates of a function $x = a^y$. In other words, a logarithmic function can be described as $y = \log_a x$.

DEFINITION

The following are equivalent:

1. $x = a^y$; and
2. $y = \log_a x$ (read "y equals the log, base a, of x").

Thus $\log_a x$ represents the exponent in the equation $x = a^y$, so the logarithm, base a, of a number x is the power to which a is raised to get x.

Example 4 Graph $y = \log_3 x$.

The equation $y = \log_3 x$ is equivalent to $x = 3^y$. The graph is a reflection of $y = 3^x$ across the line $y = x$. We make a table of values for $y = 3^x$ and then interchange x and y.

For $y = 3^x$:

x	0	1	2	-1	-2
y	1	3	9	$\frac{1}{3}$	$\frac{1}{9}$

For $y = \log_3 x$ (or $x = 3^y$):

x	1	3	9	$\frac{1}{3}$	$\frac{1}{9}$
y	0	1	2	-1	-2

*The parentheses in $\log_a (x)$ are like those in $f(x)$. In the case of logarithm functions we usually omit the parentheses.

7. Graph $y = \log_2 x$.
What is the domain of this function? What is the range?

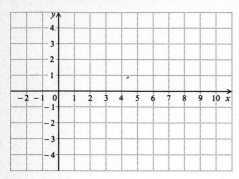

8. Graph $y = \log_4 x$.
What is the domain of this function? What is the range?

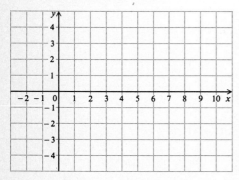

Write equivalent logarithmic equations.

9. $6^0 = 1$

10. $10^{-3} = 0.001$

11. $16^{1/4} = 2$

12. $\left(\dfrac{6}{5}\right)^{-2} = \dfrac{25}{36}$

Write equivalent exponential equations.

13. $\log_2 32 = 5$

14. $\log_{10} 1000 = 3$

15. $\log_{10} 0.01 = -2$

16. $\log_{\sqrt5} 5 = 2$

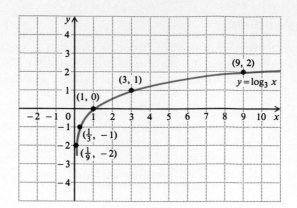

The graph of $y = \log_a x$, for any a, has the x-intercept $(1, 0)$.

DO EXERCISES 7 AND 8.

[ii] Converting Exponential and Logarithmic Equations

It is important to be able to convert from an exponential equation to a logarithmic equation.

Examples Convert to logarithmic equations.

5. $8 = 2^x \rightarrow x = \log_2 8$ It helps, in such conversions, to remember that the *logarithm is the exponent.*

6. $y^{-1} = 4 \rightarrow -1 = \log_y 4$

7. $a^b = c \rightarrow b = \log_a c$

DO EXERCISES 9–12.

It is also important to be able to convert from a logarithmic equation to an exponential equation.

Examples Convert to exponential equations.

8. $y = \log_3 5 \rightarrow 3^y = 5$ Again, it helps to remember that the *logarithm is the exponent.*

9. $-2 = \log_a 7 \rightarrow a^{-2} = 7$

10. $a = \log_b d \rightarrow b^a = d$

DO EXERCISES 13–16.

[iii] Solving Logarithmic Equations

Certain equations containing logarithmic notation can be solved by first converting to exponential notation.

Examples

11. Solve $\log_2 x = -3$.

$\log_2 x = -3$ is equivalent to $2^{-3} = x$. So $x = \frac{1}{8}$.

12. Solve $\log_{27} 3 = x$.

$\log_{27} 3 = x$ is equivalent to $27^x = 3$. Since $27^{1/3} = 3$ we have $x = \frac{1}{3}$.

13. Solve $\log_x 4 = \frac{1}{2}$.

$\log_x 4 = \frac{1}{2}$ is equivalent to $x^{1/2} = 4$. Since $(x^{1/2})^2 = 4^2$, we have $x = 16$.

DO EXERCISES 17–21.

[iv] Simplifying the Expressions $a^{\log_a x}$ and $\log_a a^x$

The exponential and logarithm functions are inverses of each other. Let us recall an important fact about functions and their inverses (Section 1.7). If the domains are suitable, then for any x,

$$f\big(f^{-1}(x)\big) = x$$

and

$$f^{-1}\big(f(x)\big) = x.$$

We apply this fact to exponential and logarithm functions. Suppose f is the exponential function, base a:

$$f(x) = a^x.$$

Then f^{-1} is the logarithm function, base a:

$$f^{-1}(x) = \log_a x.$$

Now let us find $f\big(f^{-1}(x)\big)$:

$$f\big(f^{-1}(x)\big) = a^{f^{-1}(x)} = a^{\log_a x} = x.$$

Thus for any suitable base a, $a^{\log_a x} = x$ for any positive number x (negative numbers and 0 do not have logarithms).

Next, let us find $f^{-1}\big(f(x)\big)$:

$$f^{-1}\big(f(x)\big) = \log_a f(x) = \log_a a^x = x.$$

Thus for any suitable base a, $\log_a a^x = x$ for any number x whatever.

These facts are important in simplification and should be learned well.

THEOREM 2

For any number a, suitable as a logarithm base,

1. $a^{\log_a x} = x$, for any positive number x; and

2. $\log_a a^x = x$, for any number x.

Examples Simplify.

14. $2^{\log_2 5} = 5$

15. $10^{\log_{10} t} = t$

16. $\log_e e^{-3} = -3$

17. $\log_{10} 10^{5.6} = 5.6$

DO EXERCISES 22–27.

Solve.

17. $\log_{10} x = 4$

18. $\log_x 81 = 4$

19. $\log_2 16 = x$

20. $\log_3 3 = x$

21. $\log_5 \dfrac{1}{25} = x$

Simplify.

22. $4^{\log_4 3}$

23. $7^{\log_7 \pi}$

24. $b^{\log_b 42}$

25. $\log_5 5^{37}$

26. $\log_e e^M$

27. $\log_{10} 10^{3.2}$

EXERCISE SET 7.1

[i] Graph.

1. a) $y = 4^x$ b) $y = \left(\frac{1}{4}\right)^x$ c) $y = \log_4 x$

2. a) $y = 3^x$ b) $y = \left(\frac{2}{3}\right)^x$ c) $y = \log_3 x$

3. $y = 5^x$ **4.** $y = 2^x$ **5.** $y = \log_2 x$ **6.** $y = \log_{10} x$

[ii] Write equivalent exponential equations.

7. $\log_2 32 = 5$ **8.** $\log_{10} 1000 = 3$ **9.** $\log_{10} 0.01 = -2$

10. $\log_{\sqrt{5}} 5 = 2$ **11.** $\log_6 6 = 1$ **12.** $\log_b M = N$

Write equivalent logarithmic equations.

13. $6^0 = 1$ **14.** $10^{-3} = 0.001$ **15.** $\left(\frac{6}{5}\right)^{-2} = \frac{25}{36}$ **16.** $5^4 = 625$

17. $5^{-2} = \frac{1}{25}$ **18.** $8^{1/3} = 2$ **19.** $e^{0.08} = 1.0833$ **20.** $10^{0.4771} = 3$

[iii] Solve.

21. $\log_{10} x = 4$ **22.** $\log_3 x = 2$ **23.** $\log_x \frac{1}{32} = 5$ **24.** $\log_8 x = \frac{1}{3}$

Solve for x.

25. $\log_2 16 = x$ **26.** $\log_3 3 = x$ **27.** $\log_4 2 = x$ **28.** $\log_2 64 = x$

29. $\log_{10} 10^2 = x$ **30.** $\log_3 3^4 = x$ **31.** $\log_\pi \pi = x$ **32.** $\log_a a = x$

33. $\log_{10} 0.001 = x$ **34.** $\log_{10} 1000 = x$

[iv] Simplify.

35. $3^{\log_3 4x}$ **36.** $5^{\log_5 (4x-5)}$ **37.** $\log_Q Q^{\sqrt{5}}$ **38.** $\log_e e^{|x-4|}$

☆ ──

Graph.

39. $y = \log_2 (x + 3)$ **40.** $y = \log_3 (x - 2)$ **41.** $y = 2^x - 1$ **42.** $y = 2^{x-3}$

43. $f(x) = 2^{|x|}$ **44.** $f(x) = \log_3 |x|$ **45.** $f(x) = 2^x + 2^{-x}$ **46.** $f(x) = 2^{-(x-1)}$

What is the domain of each function?

47. $f(x) = 3^x$ **48.** $f(x) = \log_{10} x$ **49.** $f(x) = \log_a x^2$

50. $f(x) = \log_4 x^3$ **51.** $f(x) = \log_{10} (3x - 4)$ **52.** $f(x) = \log_5 |x|$

53. $f(x) = \log_6 (x^2 - 9)$

Solve using graphing.

54. $2^x > 1$ **55.** $3^x \leq 1$ **56.** $\log_2 x < 0$ **57.** $\log_2 x \geq 4$

58. $\log_2 (x - 3) \leq 5$ **59.** $2^{x+3} > 1$

60. ▦ Approximate each of the following to six decimal places.

$$2^3, \quad 2^{3.1}, \quad 2^{3.14}, \quad 2^{3.141}, \quad 2^{3.1415}, \quad 2^{3.14159}$$

61. ▦ Which is larger, 5^π or π^5?   **62.** ▦ Which is larger, $\sqrt{8^3}$ or $8^{\sqrt{3}}$?   **63.** ▦ Graph $y = (0.745)^x$.

★ ──

Graph.

64. $y = 2^{-x^2}$ **65.** $y = 3^{-(x+1)^2}$ **66.** $y = |2^{x^2} - 8|$

67. Solve $3^{3^x} = 1$.

7.2 PROPERTIES OF LOGARITHMIC FUNCTIONS

[i] Let us now establish some basic properties of logarithmic functions.

THEOREM 3

For any positive numbers x and y,

$$\log_a (x \cdot y) = \log_a x + \log_a y,$$

where a is any positive number different from 1.

Theorem 3 says that the logarithm of a *product* is the *sum* of the logarithms of the factors. Note that the base a must remain constant. The logarithm of a sum is *not* the sum of the logarithms of the summands.

Proof of Theorem 3. Since a is positive and different from 1, it can serve as a logarithm base. Since x and y are assumed positive, they are in the domain of the function $y = \log_a x$. Now let $b = \log_a x$ and $c = \log_a y$.

Writing equivalent exponential equations, we have

$$x = a^b \quad \text{and} \quad y = a^c.$$

Next we multiply, to obtain

$$xy = a^b a^c = a^{b+c}.$$

Now writing an equivalent logarithmic equation, we obtain

$$\log_a (xy) = b + c,$$

or

$$\log_a (xy) = \log_a x + \log_a y,$$

which was to be shown.

Example 1 Express as a sum of logarithms and simplify.

$$\log_2 (4 \cdot 16) = \log_2 4 + \log_2 16$$
$$= 2 + 4 = 6$$

DO EXERCISES 1–4.

THEOREM 4

For any positive number x and any number p,

$$\log_a x^p = p \cdot \log_a x,$$

where a is any logarithm base.

Theorem 4 says that the logarithm of a power of a number is the exponent times the logarithm of the number.

Proof of Theorem 4. Let $b = \log_a x$. Then, writing an equivalent exponential equation, we have $x = a^b$. Next we raise both sides of

Express as a product.

5. $\log_7 4^5$

the latter equation to the pth power. This gives us

$$x^p = (a^b)^p, \quad \text{or} \quad a^{bp}.$$

Now we can write an equivalent logarithmic equation,

$$\log_a x^p = \log_a a^{bp} = bp.$$

But $b = \log_a x$, so we have

$$\log_a x^p = p \cdot \log_a x,$$

which was to be shown.

Examples Express as products.

2. $\log_b 9^{-5} = -5 \cdot \log_b 9$

3. $\log_a \sqrt[4]{5} = \log_a 5^{1/4} = \frac{1}{4} \log_a 5$

DO EXERCISES 5 AND 6.

THEOREM 5

For any positive numbers x and y,

$$\log_a \frac{x}{y} = \log_a x - \log_a y,$$

where a is any logarithm base.

Theorem 5 says that the logarithm of a quotient is the difference of the logarithms (i.e., the logarithm of the dividend minus the logarithm of the divisor).

Proof of Theorem 5. $x/y = x \cdot y^{-1}$, so

$$\log_a \frac{x}{y} = \log_a (xy^{-1}).$$

By Theorem 2,

$$\log_a (xy^{-1}) = \log_a x + \log_a y^{-1},$$

and by Theorem 3,

$$\log_a y^{-1} = -1 \cdot \log_a y,$$

so we have

$$\log_a \frac{x}{y} = \log_a x - \log_a y,$$

which was to be shown.

6. $\log_a \sqrt{5}$

Example 4 Express as sums and differences of logarithms and without exponential notation or radicals.

$$\log_a \frac{\sqrt{17}}{5\pi} = \log_a \sqrt{17} - \log_a 5\pi$$

$$= \log_a 17^{1/2} - (\log_a 5 + \log_a \pi)$$

$$= \frac{1}{2} \log_a 17 - \log_a 5 - \log_a \pi$$

Example 5 Express in terms of logarithms of x, y, and z.

$$\log_a \sqrt[4]{\frac{xy}{z^3}} = \log_a \left(\frac{xy}{z^3}\right)^{1/4}$$

$$= \frac{1}{4} \cdot \log_a \frac{xy}{z^3}$$

$$= \frac{1}{4}[\log_a xy - \log_a z^3]$$

$$= \frac{1}{4}[\log_a x + \log_a y - 3\log_a z]$$

$$= \frac{1}{4}\log_a x + \frac{1}{4}\log_a y - \frac{3}{4}\log_a z$$

Example 6 Express as a single logarithm.

$$\frac{1}{2}\log_a x - 7\log_a y + \log_a z = \log_a \sqrt{x} - \log_a y^7 + \log_a z$$

$$= \log_a \frac{\sqrt{x}}{y^7} + \log_a z$$

$$= \log_a \frac{z\sqrt{x}}{y^7}$$

DO EXERCISES 7–10.

Example 7 Given that $\log_a 2 = 0.301$ and $\log_a 3 = 0.477$, find:

a) $\log_a 6 = \log_a 2 \cdot 3$
   $= \log_a 2 + \log_a 3$
   $= 0.301 + 0.477 = 0.778;$

b) $\log_a \sqrt{3} = \log_a 3^{1/2} = \frac{1}{2} \cdot \log_a 3 = \frac{1}{2} \cdot 0.477 = 0.2385;$

c) $\log_a \frac{2}{3} = \log_a 2 - \log_a 3 = 0.301 - 0.477 = -0.176;$

d) $\log_a 5$; *No way to find, using Theorems 3–5.*
   $(\log_a 5 \neq \log_a 2 + \log_a 3);$

e) $\frac{\log_a 2}{\log_a 3} = \frac{0.301}{0.477} = 0.63.$ Note that we could not use Theorems 3–5; we simply divided.

DO EXERCISE 11.

   For any base a, $\log_a a = 1$. This is easily seen by writing an equivalent exponential equation, $a^1 = a$. Similarly, for any base a, $\log_a 1 = 0$. These facts are important and should be remembered well.

THEOREM 6

   For any base a,

   $$\log_a a = 1 \quad \text{and} \quad \log_a 1 = 0.$$

DO EXERCISES 12–15.

7. Express as a difference.

   a) $\log_a \frac{M}{N}$

   b) $\log_c \frac{1}{4}$

8. Express as sums and differences of logarithms and without exponential notation or radicals.

   $$\log_{10} \frac{4\pi}{\sqrt{23}}$$

9. Express in terms of logarithms of x, y, and z.

   $$\log_a \sqrt{\frac{z^3}{xy}}$$

10. Express as a single logarithm.

   $$5\log_a x - \log_a y + \frac{1}{4}\log_a z$$

11. Given that $\log_a 2 = 0.301$ and $\log_a 3 = 0.477$, find:

   a) $\log_a 9$;
   b) $\log_a \sqrt{2}$;
   c) $\log_a \sqrt[3]{2}$;
   d) $\log_a \frac{3}{2}$;

   e) $\frac{\log_a 3}{\log_a 2}$.

Simplify.

12. $\log_\pi \pi$

13. $\log_9 1$

14. $\log_e 1$

15. $\log_{1/4} \frac{1}{4}$

EXERCISE SET 7.2

[i] Express in terms of logarithms of x, y, and z.

1. $\log_a x^2 y^3 z$ **2.** $\log_a 5xy^4 z^3$ **3.** $\log_b \dfrac{xy^2}{z^3}$ **4.** $\log_c \sqrt[3]{\dfrac{x^4}{y^3 z^2}}$

Express as a single logarithm and simplify if possible.

5. $\dfrac{2}{3} \log_a 64 - \dfrac{1}{2} \log_a 16$ **6.** $\dfrac{1}{2} \log_a x + 3 \log_a y - 2 \log_a x$

7. $\log_a 2x + 3(\log_a x - \log_a y)$ **8.** $\log_a x^2 - 2 \log_a \sqrt{x}$

9. $\log_a \dfrac{a}{\sqrt{x}} - \log_a \sqrt{ax}$ **10.** $\log_a (x^2 - 4) - \log_a (x - 2)$

11. $\log_a (x^3 + y^3) - \log_a (x + y)$ **12.** $\log_a (x - y) + \log_a (x^2 + xy + y^2)$

Express as a sum and/or difference of logarithms.

13. $\log_a \sqrt{1 - x^2}$ **14.** $\log_a \dfrac{x + t}{\sqrt{x^2 - t^2}}$

Given $\log_{10} 2 = 0.301$, $\log_{10} 3 = 0.477$, and $\log_{10} 10 = 1$, find:

15. $\log_{10} 4$.   **16.** $\log_{10} 5$.   **17.** $\log_{10} 50$.   **18.** $\log_{10} 12$.

                        $\left(\text{Hint: } 5 = \dfrac{10}{2}\right)$ $\left(\text{Hint: } 50 = \dfrac{100}{2}\right)$

19. $\log_{10} 60$.   **20.** $\log_{10} \dfrac{1}{3}$.   **21.** $\log_{10} \sqrt{\dfrac{2}{3}}$.   **22.** $\log_{10} \sqrt[5]{12}$.

23. $\log_{10} 90$.   **24.** $\log_{10} \dfrac{9}{8}$.   **25.** $\log_{10} \dfrac{9}{10}$.   **26.** $\log_{10} \dfrac{1}{4}$.

☆ ——————————————————————————————————

Which of the following are false?

27. $\dfrac{\log_a M}{\log_a N} = \log_a M - \log_a N$ **28.** $\dfrac{\log_a M}{\log_a N} = \log_a \dfrac{M}{N}$ **29.** $\dfrac{\log_a M}{c} = \log_a M^{1/c}$

30. $\log_N (M \cdot N)^x = x \log_N M + x$ **31.** $\log_a 2x = 2 \log_a x$ **32.** $\log_a 2x = \log_a 2 + \log_a x$

33. $\log_a (M + N) = \log_a M + \log_a N$ **34.** $\log_a x^3 = 3 \log_a x$

Solve.

35. $\log_\pi \pi^{2x+3} = 4$ **36.** $3^{\log_3 (8x-4)} = 5$ **37.** $4^{2 \log_4 x} = 7$ **38.** $8^{2 \log_8 x + \log_8 x} = 27$

39. $(x + 3) \cdot \log_a a^x = x$ **40.** $\log_a x^2 = 2 \log_a x$

41. $\log_a 5x = \log_a 5 + \log_a x$ **42.** $\log_b \dfrac{5}{x + 2} = \log_b 5 - \log_b (x + 2)$

★ ——————————————————————————————————

43. If $\log_a x = 2$, what is $\log_a \left(\dfrac{1}{x}\right)$?   **44.** If $\log_a x = 2$, what is $\log_{1/a} x$?

Prove the following for any base a and any positive number x.

45. $\log_a \left(\dfrac{1}{x}\right) = -\log_a x$ **46.** $\log_a \left(\dfrac{1}{x}\right) = \log_{1/a} x$

47. Show that $\log_a \left(\dfrac{x + \sqrt{x^2 - 5}}{5}\right) = -\log_a (x - \sqrt{x^2 - 5})$.

48. Graph and compare: $y = \log_2 |x|$ and $y = |\log_2 x|$.

7.3 COMMON LOGARITHMS

Base ten logarithms are know as *common logarithms*. Tables for these logarithms are readily available (Table 3 at the back of this book).

Historical Uses of Logarithms

Before calculators and computers became so readily available, common logarithms were used extensively to do certain kinds of calculations. In fact, that is why logarithms were developed. Today, computations with logarithms are mainly of historical interest; the logarithm *functions* are of modern importance. The study of computations with logarithms can, however, help the student fix the ideas with respect to properties of logarithm functions.

The following is a table of powers of 10, or *logarithms* base 10.

$$
\begin{aligned}
1 &= 10^{0.0000}, && \text{or} && \log_{10} 1 &= 0.0000 \\
2 &= 10^{0.3010}, && \text{or} && \log_{10} 2 &= 0.3010 \\
3 &= 10^{0.4771}, && \text{or} && \log_{10} 3 &= 0.4771 \\
4 &= 10^{0.6021}, && \text{or} && \log_{10} 4 &= 0.6021 \\
5 &= 10^{0.6990}, && \text{or} && \log_{10} 5 &= 0.6990 \\
6 &= 10^{0.7782}, && \text{or} && \log_{10} 6 &= 0.7782 \\
7 &= 10^{0.8451}, && \text{or} && \log_{10} 7 &= 0.8451 \\
8 &= 10^{0.9031}, && \text{or} && \log_{10} 8 &= 0.9031 \\
9 &= 10^{0.9542}, && \text{or} && \log_{10} 9 &= 0.9542 \\
10 &= 10^{1.0000}, && \text{or} && \log_{10} 10 &= 1.0000 \\
11 &= 10^{1.0414}, && \text{or} && \log_{10} 11 &= 1.0414 \\
12 &= 10^{1.0792}, && \text{or} && \log_{10} 12 &= 1.0792 \\
13 &= 10^{1.1139}, && \text{or} && \log_{10} 13 &= 1.1139 \\
14 &= 10^{1.1461}, && \text{or} && \log_{10} 14 &= 1.1461 \\
15 &= 10^{1.1761}, && \text{or} && \log_{10} 15 &= 1.1761 \\
16 &= 10^{1.2041}, && \text{or} && \log_{10} 16 &= 1.2041
\end{aligned}
$$

The exponents are approximate, but accurate to four decimal places. To illustrate how logarithms can be used for computation we will use the above table and do some easy calculations.

Example 1 Find 3×4 using the table of exponents.

$$
\begin{aligned}
3 \times 4 &= 10^{0.4771} \times 10^{0.6021} \\
&= 10^{1.0792} \quad \text{Adding exponents}
\end{aligned}
$$

From the table we see that $10^{1.0792} = 12$, so $3 \times 4 = 12$.

DO EXERCISE 1.

Note from Example 1 that we can find a product by adding the logarithms of the factors and then finding the number having the result as its logarithm. That is, we found the number $10^{1.0792}$. This number is often referred to as the *antilogarithm* of 1.0792. In other words, if

$$f(x) = \log_{10} x,$$

then

$$f^{-1}(x) = \text{antilog}_{10} x = 10^x.$$

John Napier invented logarithms about 1614. The word *logarithm* was derived from two Greek words, *logos*, which means "ratio," and *arithmos*, which means "number."

OBJECTIVES

You should be able to:

[i] Use Table 3 and scientific notation to find common logarithms and antilogarithms.

[ii] (*Optional*). Find products, quotients, powers, and roots, and combinations of these, using logarithms.

1. a) Find 4×2 using the table of exponents.
   b) Find 2×8 using the table of exponents.

Using base 10 logarithms, find the following.

2. 4×2

3. $\dfrac{15}{3}$

4. $\sqrt[3]{8}$

5. 3^2

Use Table 3 to find each logarithm.

6. log 3.14

7. log 9.99

8. log 4.00

Use Table 3 to find each antilogarithm.

9. antilog 0.7589

10. antilog 0.0000

11. antilog 0.5587

In other words, an antilogarithm function is simply an exponential function.

Example 2 Find $\frac{14}{2}$, using base 10 logarithms.

$$\log_{10} \frac{14}{2} = \log_{10} 14 - \log_{10} 2$$
$$= 1.1461 - 0.3010$$
$$\log_{10} \frac{14}{2} = 0.8451$$
$$\frac{14}{2} = \text{antilog}_{10}\, 0.8451 = 7$$

Example 3 Find $\sqrt[4]{16}$, using base 10 logarithms.

$$\log_{10} \sqrt[4]{16} = \log_{10} 16^{1/4} = \frac{1}{4} \cdot \log_{10} 16 = \frac{1}{4} \cdot 1.2041$$
$$\log_{10} \sqrt[4]{16} = 0.3010 \qquad \text{Rounded to four decimal places}$$
$$\sqrt[4]{16} = \text{antilog}_{10}\, 0.3010 = 2$$

Example 4 Find 2^3, using base 10 logarithms.

$$\log_{10} 2^3 = 3 \cdot \log_{10} 2 = 3 \cdot 0.3010 = 0.9030$$
$$\log_{10} 2^3 = 0.9030$$
$$2^3 = \text{antilog}_{10}\, 0.9030 \approx 8$$

Note the rounding error.

DO EXERCISES 2–5.

[i] Finding Common Logarithms Using Tables

We often omit the base, 10, when working with common logarithms.

Table 3 contains logarithms of numbers from 1 to 10. Part of that table is shown below. To illustrate the use of the table, let us find log 5.24. We locate the row headed 5.2, then move across to the column headed 4. We find log 5.24 as the colored entry in the table.

x	0	1	2	3	4	5	6	7	8	9
5.0	0.6990	0.6998	0.7007	0.7016	0.7024	0.7033	0.7042	0.7050	0.7059	0.7067
5.1	0.7076	0.7084	0.7093	0.7101	0.7110	0.7118	0.7126	0.7135	0.7143	0.7152
5.2	0.7160	0.7168	0.7177	0.7185	0.7193	0.7202	0.7210	0.7218	0.7226	0.7235
5.3	0.7234	0.7251	0.7259	0.7267	0.7275	0.7284	0.7292	0.7300	0.7308	0.7316
5.4	0.7324	0.7332	0.7340	0.7348	0.7356	0.7364	0.7372	0.7380	0.7388	0.7396

We can find antilogarithms by reversing this process. For example, antilog $0.7193 = 10^{0.7193} = 5.24$. Similarly, antilog $0.7292 = 5.36$.

DO EXERCISES 6–11.

Using Table 3 and scientific notation we can approximate logarithms of numbers that are not between 1 and 10. First recall that

$$\log_a a^k = k \quad \text{for any number } k. \qquad \text{Theorem 2}$$

Thus

$$\log_{10} 10^k = k \quad \text{for any number } k.$$

Examples

5. $\log 52.4 = \log (5.24 \times 10^1)$ Converting to scientific notation
$= \log 5.24 + \log 10^1$ Theorem 3
$= 0.7193 + 1$

6. $\log 0.524 = \log (5.24 \times 10^{-1})$
$= \log 5.24 + \log 10^{-1}$
$= 0.7193 + (-1)$

7. $\log 52{,}400 = \log (5.24 \times 10^4)$
$= \log 5.24 + \log 10^4$
$= 0.7193 + 4$

8. $\log 0.00524 = \log (5.24 \times 10^{-3})$
$= \log 5.24 + \log 10^{-3}$
$= 0.7193 + (-3)$

DO EXERCISES 12 AND 13.

The preceding examples illustrate the importance of using the base 10 for computation. It allows great economy in the printing of tables. If we know the logarithm of a number from 1 to 10 we can multiply that number by any power of ten and easily determine the logarithm of the resulting number. For any base other than 10 this would not be the case. In each of Examples 5–8, the integer part of the logarithm is the exponent in the scientific notation. This integer is called the *characteristic* of the logarithm. The other part of the logarithm, a number between 0 and 1, is called the *mantissa* of the logarithm. Table 3 contains only mantissas.

Example 9 Find log 0.0538, indicating the characteristic and mantissa.

We first write scientific notation for the number:

$$5.38 \times 10^{-2}.$$

Then we find log 5.38. This is the mantissa:

$$\log 5.38 = 0.7308.$$

The characteristic of the logarithm is the exponent -2. Now $\log 0.0538 = 0.7308 + (-2)$, or -1.2692. When negative characteristics occur, it is often best to name the logarithm so that the characteristic and mantissa are preserved. In the preceding example, we have

$$\log 0.0538 = 0.7308 + (-2) = -1.2692,$$

Use scientific notation and Table 3 to find each logarithm.

12. log 289

13. log 0.000289

Find the following. Use Table 3 and try to write only the answer. Where appropriate, name so that positive mantissas are preserved.

14. log 67,800

15. log 892,000

16. log 45.9

17. log 609,000,000

18. log 0.0782

19. log 0.000111

20. log 0.0079

but the latter notation displays neither the characteristic nor the mantissa. We can rename the characteristic, -2, as $8 - 10$, and then add the mantissa, to obtain

$$8.7308 - 10,$$

thus preserving both mantissa and characteristic.

▤ The characteristic and mantissa are useful when working with logarithm tables, but are not needed on a calculator. For example, on a calculator with a ten-digit readout, we find that

$$\log 0.0538 = -1.269217724,$$

but this shows neither the characteristic nor the mantissa. Check this on your calculator. How can you find the characteristic and mantissa?

Example 10 Find log 0.00687.

We write scientific notation (or at least visualize it):

$$0.00687 = 6.87 \times 10^{-3}.$$

The characteristic is -3, or $7 - 10$. The mantissa, from the table, is 0.8370. Thus $\log 0.00687 = 7.8370 - 10$.

DO EXERCISES 14–20.

Antilogarithms

To find antilogarithms, we reverse the procedure for finding logarithms.

Example 11 Find antilog 2.6085 (or find x such that log x = 2.6085).

$$\text{antilog } 2.6085 = 10^{2.6085}$$
$$= 10^{(2+0.6085)}$$
$$= 10^2 \cdot 10^{0.6085}$$

From the table we can find $10^{0.6085}$, or antilog 0.6085. It is 4.06. Thus we have

$$\text{antilog } 2.6085 = 10^2 \times 4.06, \quad \text{or } 406.$$

Note in this example, we in effect separate the number 2.6085 into an integer and a number between 0 and 1. We use the latter with the table, after which we have scientific notation for our answer.

Example 12 Find antilog 3.7118.

From the table we find that antilog 0.7118 = 5.15. Thus

$$\text{antilog } 3.7118 = 5.15 \times 10^3 \quad \text{Note that 3 is the characteristic.}$$
$$= 5150.$$

Example 13 Find antilog (7.7143 − 10).

The characteristic is −3 and the mantissa is 0.7143.

From the table we find that antilog 0.7143 = 5.18. Thus

$$\text{antilog } (7.7143 - 10) = 5.18 \times 10^{-3}$$
$$= 0.00518.$$

Example 14 Find antilog −2.2857.

We are to find the antilog of a number, but the number is named so that the mantissa is not apparent. To find the mantissa we add 0, naming it 10 − 10:

$$-2.2857 = -2.2857 + (10 - 10)$$
$$= (-2.2857 + 10) - 10 = 7.7143 - 10.$$

Then we proceed as in Example 13. The answer is 0.00518.

DO EXERCISES 21–25.

[ii] Calculations with Logarithms (Optional)

The kinds of calculations in which logarithms were helpful historically are multiplication, division, taking powers, and taking roots.

Example 15 Find $\dfrac{0.0578 \times 32.7}{8460}$.

We write a *plan* for the use of logarithms, then look up all the mantissas at one time. Let

$$N = \frac{0.0578 \times 32.7}{8460}.$$

Then

$$\log N = \log 0.0578 + \log 32.7 - \log 8460.$$

This gives us the plan. We use a straight line to indicate addition and a wavy line to indicate subtraction.

Completion of plan:

$$\log 0.0578 = 8.7619 - 10$$
$$\log 32.7 = \underline{1.5145}$$

$$\log \text{numerator} = 10.2764 - 10$$
$$\log 8460 = \underline{3.9274}$$

$$\log \text{fraction} = 6.3490 - 10$$
$$\text{fraction} = 0.000223 \quad \text{Taking antilog}$$

DO EXERCISE 26.

Find the following. Use Table 3. Try to write only the answer.

21. If log x = 4.8069, find x.

22. $10^{4.8069}$

23. $10^{3.9325}$

24. antilog 6.6284 − 10

25. antilog −1.9788

26. Use logarithms to find
$$\frac{88.4 \times 0.00641}{9.43}.$$

27. $(92.8)^3 \times \sqrt[5]{0.986}$

Example 16 Use logarithms to find $\sqrt[4]{0.325} \times (4.23)^2$.

If

$$N = \sqrt[4]{0.325} \times (4.23)^2,$$

then

$$\log N = \frac{1}{4} \log 0.325 + 2 \log 4.23;$$

this yields the plan. Note that powers and roots are involved. In such cases it is easier to first find the following logs:

$$\log 0.325 = 9.5119 - 10$$
$$\log 4.23 = 0.6263.$$

Consider $9.5119 - 10$. Since we will be dividing this number by 4, but -10 is not divisible by 4, we rename this number:

$$\log 0.325 = 39.5119 - 40.$$

Now the negative term is divisible by 4. We divide log 0.325 by 4 and multiply log 4.23 by 2. Then we have

$$\log \sqrt[4]{0.325} = \quad 9.8780 - 10$$
$$\log (4.23)^2 = \quad \underline{1.2526}$$

$$\log \text{product} = 11.1306 - 10$$
$$= \quad 1.1306$$
$$\text{product} = 13.5.$$

DO EXERCISE 27.

EXERCISE SET 7.3

[i] Use Table 3 to find each of the following. Where appropriate, write so that (positive) mantissas are preserved.

1. log 2.46　　　　**2.** log 7.5　　　　**3.** log 347　　　　**4.** log 8720

5. log 52.5　　　　**6.** log 20.8　　　　**7.** log 624,000　　　　**8.** log 13,400

9. log 0.0702　　　　**10.** log 0.640　　　　**11.** log 0.000216　　　　**12.** log 0.173

13. antilog 2.3674　　　　**14.** antilog 1.9222　　　　**15.** antilog $8.2553 - 10$　　　　**16.** antilog $6.6294 - 10$

17. antilog -5.9788　　　　**18.** antilog -2.2628　　　　**19.** $10^{1.4014}$　　　　**20.** $10^{3.6590}$

21. $10^{8.9881-10}$　　　　**22.** $10^{7.5391-10}$

Solve for x.

23. $\log x = 0.6522$　　　　　　　　　　　　**24.** $\log x = 4.8156$

25. $\log x = 8.1239 - 10$　　　　　　　　　**26.** $\log x = -1.0218$

[ii] (*Optional*). Use logarithms to compute. When exact values do not occur in the table, use the nearest values.

27. 3.14×60.4　　　　**28.** 541×0.0152　　　　**29.** $286 \div 1.05$　　　　**30.** $12.8 \div 81.6$

31. $\sqrt{76.9}$　　　　**32.** $\sqrt[3]{56.9}$　　　　**33.** $(1.36)^{4.2}$　　　　**34.** $(0.727)^{3.6}$

35. $\dfrac{70.7 \times (10.6)^2}{18.6 \times \sqrt{276}}$　　　　**36.** $\sqrt[3]{\dfrac{3.24 \times (3.16)^2}{78.4 \times 24.6}}$

Find each of the following. Round to six decimal places.

37. ▦ log 56,789   **38.** ▦ log 0.0111347   **39.** ▦ log (log 3)

Find each antilogarithm. Since antilog$_{10}$ x = 10^x, you can find the antilogarithm, base 10, of a number x by raising 10 to the power x.

40. ▦ $10^{0.4356}$ **41.** ▦ antilog$_{10}$ 7.8943   **42.** ▦ antilog$_{10}$ (−7.5689)   **43.** ▦ $10^{-3.23445678}$

7.4 INTERPOLATION

Example 1 Find log 34,870.

a) Find the characteristic. Since 34,870 = 3.487 × 10^4, the characteristic is 4.

b) Find the mantissa. From Table 3 we have:

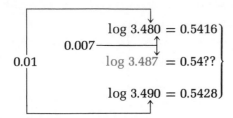

The tabular difference is 0.0012.

The tabular difference (difference between consecutive values in the table) is 0.0012. Now 3.487 is $\frac{7}{10}$ of the way from 3.480 to 3.490. So we take 0.7 of 0.0012, which is 0.00084, and round it to 0.0008. We add this to 0.5416. The mantissa that results is 0.5424.

c) Add the characteristic and mantissa:

$$\log 34,870 = 4.5424.$$

With practice you will take 0.7 of 12, forgetting the zeros but adding in the same way.

DO EXERCISE 1.

Example 2 Find log 0.009543.

a) Find the characteristic. Since 0.009543 = 9.543 × 10^{-3}, the characteristic is −3, or 7 − 10.

b) Find the mantissa. From Table 3 we have:

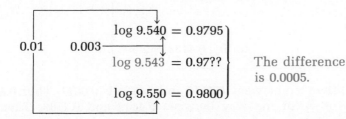

The difference is 0.0005.

OBJECTIVE

You should be able to:

Use linear interpolation to find logarithms and antilogarithms from a table.

1. Find log 4562.

2. Find log 0.02387.

Now 9.543 is $\frac{3}{10}$ of the way from 9.540 to 9.550, so we take 0.3 of 0.0005, which is 0.00015, and round it to 0.0002. We add this to 0.9795. The mantissa that results is 0.9797.

c) Add the characteristic and the mantissa:

$$\log 0.009543 = 7.9797 - 10.$$

DO EXERCISE 2.

Antilogarithms

We interpolate when finding antilogarithms, using the table in reverse.

Example 3 Find antilog 4.9164.

a) The characteristic is 4. The mantissa is 0.9164.

b) Find the antilog of the mantissa, 0.9164. From Table 3, we have:

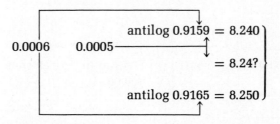

3. Find $10^{3.4557}$.

The difference between 0.9159 and 0.9165 is 0.0006. Thus 0.9164 is $\frac{0.0005}{0.0006}$, or $\frac{5}{6}$, of the way between 0.9159 and 0.9165. Then antilog 0.9164 is $\frac{5}{6}$ of the way between 8.240 and 8.250, or $\frac{5}{6}(0.010)$, which is $0.00833\ldots$, and round it to 0.008. Thus the antilog of the mantissa is 8.248.

Thus antilog $4.9164 = 8.248 \times 10^4 = 82,480.$

DO EXERCISE 3.

Example 4 Find antilog $(7.4122 - 10)$.

a) The characteristic is -3. The mantissa is 0.4122.

b) Find the antilog of the mantissa, 0.4122. From Table 3 we have:

$$
\begin{array}{c}
0.0017 \quad 0.0006
\begin{cases}
\text{antilog } 0.4116 = 2.580 \\
\qquad\qquad\qquad = 2.58? \\
\text{antilog } 0.4133 = 2.590
\end{cases}
\end{array}
$$

The difference between 0.4116 and 0.4133 is 0.0017. Thus 0.4122 is $\frac{0.0006}{0.0017}$, or $\frac{6}{17}$, of the way between 0.4116 and 0.4133. Then antilog 0.4122 is $\frac{6}{17}$ of the way between 2.580 and 2.590, or $\frac{6}{17}(0.010)$, which is 0.0035, to four places. We round it to 0.004. Thus the antilog of the mantissa is 2.584.

So antilog $(7.4122 - 10) = 2.584 \times 10^{-3} = 0.002584$.

DO EXERCISE 4.

4. Find antilog $(6.7749 - 10)$.

Calculations

Calculations with logarithms, using Table 3, can be done with four-digit precision when interpolation is used in finding both logarithms and antilogarithms.

EXERCISE SET 7.4

[i] Find each of the following logarithms using interpolation and Table 3.

1. log 41.63

2. log 472.1

3. log 2.944

4. log 21.76

5. log 650.2

6. log 37.37

7. log 0.1425

8. log 0.0904

9. log 0.004257

10. log 4518

11. log 0.1776

12. log 0.08356

13. log 600.6

14. log (log 3)

15. log (log 5)

Find each of the following antilogarithms using interpolation and Table 3.

16. antilog 1.6350

17. antilog 2.3512

18. antilog 0.6478

19. antilog 1.1624

20. antilog 0.0342

21. antilog 4.8453

22. antilog 9.8561 − 10

23. antilog 8.9659 − 10

24. antilog 7.4128 − 10

25. antilog 9.7278 − 10

26. antilog 8.2010 − 10

27. antilog 7.8630 − 10

☆

Use logarithms and interpolation to do the following calculations. Use four-digit precision. Answers may be checked using a calculator.

28. $\dfrac{35.24 \times (16.77)^3}{12.93 \times \sqrt{276.2}}$

29. $\sqrt[5]{\dfrac{16.79 \times (4.234)^3}{18.81 \times 175.3}}$

7.5 EXPONENTIAL AND LOGARITHMIC EQUATIONS

[i] Exponential Equations

An equation with variables in exponents, such as $3^{2x-1} = 4$, is called an *exponential equation*. We can solve such equations by taking logarithms on both sides and then using Theorem 4.*

---

*The use of a calculator is recommended for the rest of this chapter.

OBJECTIVES

You should be able to:

[i] Solve exponential and logarithmic equations.

[ii] Solve applied problems involving exponential and logarithmic equations.

1. Solve $2^x = 7$.

Example 1 Solve $3^x = 8$.

$$\log 3^x = \log 8 \qquad \text{Taking log on both sides.}$$

Remember $\log m = \log_{10} m$.

$$x \log 3 = \log 8 \qquad \text{Using Theorem 4}$$

$$x = \frac{\log 8}{\log 3} \qquad \text{Solving for x}$$

$$x \approx \frac{0.9031}{0.4771} \approx 1.8929 \qquad \text{We look up the logs, or find}$$

them on a calculator, and divide.

DO EXERCISE 1.

Example 2 Solve $2^{3x-5} = 16$.

$$\log 2^{3x-5} = \log 16 \qquad \text{Taking log on both sides}$$
$$(3x - 5) \log 2 = \log 16 \qquad \text{Using Theorem 4}$$

$$3x - 5 = \frac{\log 16}{\log 2}$$

$$x = \frac{\frac{\log 16}{\log 2} + 5}{3} \approx \frac{\frac{1.2041}{0.3010} + 5}{3} \qquad \begin{array}{l}\text{Solving for x and} \\ \text{evaluating logarithms}\end{array}$$

$$x \approx 3.0001 \qquad \text{Calculating}$$

The answer is approximate because the logarithms are approximate.

DO EXERCISE 2.

2. Solve $4^{2x-3} = 64$, using Table 3.

Example 3 Solve $\dfrac{e^x + e^{-x}}{2} = t$, for x.

Note that we are to solve for x. However, we have more than one term with x in the exponent. To get a single expression with x in the exponent, we do the following:

$$e^x + e^{-x} = 2t \qquad \text{Multiplying by 2}$$

$$e^x + \frac{1}{e^x} = 2t \qquad \begin{array}{l}\text{Rewriting to eliminate the minus} \\ \text{sign in an exponent}\end{array}$$

$$e^{2x} + 1 = 2t \cdot e^x \qquad \text{Multiplying on both sides by } e^x$$
$$(e^x)^2 - 2t \cdot e^x + 1 = 0$$

This equation is reducible to quadratic, with $u = e^x$. Using the quadratic formula, we obtain

$$e^x = \frac{2t \pm \sqrt{4t^2 - 4}}{2} = t \pm \sqrt{t^2 - 1}$$

$$\log e^x = \log (t \pm \sqrt{t^2 - 1}) \qquad \text{Taking log on both sides}$$
$$x \log e = \log (t \pm \sqrt{t^2 - 1}) \qquad \text{Using Theorem 4}$$

$$x = \frac{\log (t \pm \sqrt{t^2 - 1})}{\log e}.$$

If e should be a number suitable as a logarithm base, we might also take base e logarithms on both sides. The denominator would then be $\log_e e$, or 1, and we would have

$$x = \log_e (t \pm \sqrt{t^2 - 1}).$$

DO EXERCISE 3.

Logarithmic Equations

Equations that contain logarithmic expressions are called *logarithmic equations*. We solved some equations of this type in Section 7.1 by converting to an equivalent exponential equation. For example, to solve $\log_2 x = -3$, we convert to $x = 2^{-3}$ and find that $x = \frac{1}{8}$.

To solve logarithmic equations we first try to obtain a single logarithmic expression on one side of the equation and then write an equivalent exponential equation.

Example 4 Solve $\log_3 (5x + 7) = 2$.

We already have a single logarithmic expression, so we write an equivalent exponential equation:

$$5x + 7 = 3^2 \qquad \text{Writing an equivalent exponential equation}$$
$$5x + 7 = 9$$
$$5x = 2$$
$$x = \frac{2}{5}$$

Check:
$$\frac{\log_3 (5x + 7) \;=\; 2}{}$$

$$
\begin{array}{c|c}
\log_3 \left(5 \cdot \dfrac{2}{5} + 7\right) & 2 \\
\log_3 (2 + 7) & \\
\log_3 9 & \\
2 & \\
\end{array}
$$

DO EXERCISES 4 AND 5.

Example 5 Solve $\log x + \log (x - 3) = 1$.

Here we must first obtain a single logarithmic equation:

$$\log x + \log (x - 3) = 1$$
$$\log x (x - 3) = 1 \qquad \text{Using Theorem 3 to obtain a single logarithm}$$
$$\log_{10} x(x - 3) = 1$$
$$x(x - 3) = 10^1 \qquad \text{Converting to an equivalent exponential equation}$$
$$x^2 - 3x = 10$$
$$x^2 - 3x - 10 = 0$$
$$(x + 2)(x - 5) = 0 \qquad \text{Factoring and principle of zero products}$$
$$x = -2 \quad \text{or} \quad x = 5$$

3. Solve $\dfrac{e^x - e^{-x}}{2} = t$, for x.

4. Solve.
$$\log_5 x = 3$$

5. Solve.
$$\log_4 (8x - 6) = 3$$

6. Solve $\log x + \log (x + 3) = 1$.

Possible solutions to logarithmic equations must be checked because domains of logarithmic functions consist only of positive numbers.

Check:

$\log x + \log (x - 3) = 1$		$\log x + \log (x - 3) = 1$	
$\log (-2) + \log (-2 - 3)$	1	$\log 5 + \log (5 - 3)$	1
		$\log 5 + \log 2$	
		$\log 10$	
		1	

The number -2 is not a solution because negative numbers do not have logarithms. The solution is 5.

DO EXERCISE 6.

[ii] Applications

Exponential and logarithmic functions and equations have many applications. We shall consider a few of them.

Example 6 (*Compound interest*). The amount A that principal P will be worth after t years at interest rate i, compounded annually, is given by the formula $A = P(1 + i)^t$. Suppose \$4000 principal is invested at 6% interest and yields \$5353. For how many years was it invested?

7. \$5000 was invested at 14%, compounded annually, and it yielded \$18,540. For how long was it invested?

Using the formula $A = P(1 + i)^t$, we have

$$5353 = 4000(1 + 0.06)^t, \quad \text{or} \quad 5353 = 4000(1.06)^t.$$

Solving for t, we have

$$\log 5353 = \log 4000(1.06)^t \qquad \text{Taking log on both sides}$$
$$\log 5353 = \log 4000 + t \log 1.06 \qquad \text{Theorems 3 and 4}$$
$$\frac{\log 5353 - \log 4000}{\log 1.06} = t \qquad \text{Solving for } t$$
$$\frac{3.7286 - 3.6021}{0.0253} = t \qquad \text{Evaluating logarithms}$$
$$5 = t.$$

The money was invested for 5 years.

▦ We can use a calculator for an approximate *check:*

$5353 = 4000(1.06)^t$	
5353	$4000(1.06)^5$
	$4000(1.338226)$
	5352.904

DO EXERCISE 7.

Example 7 (*Loudness of sound*). The sensation of loudness of sound is not proportional to the energy intensity, but rather is a logarithmic function. *Loudness*, in Bels (after Alexander Graham Bell), of a sound of intensity I is defined to be

$$L = \log \frac{I}{I_0},$$

where I_0 is the minimum intensity detectable by the human ear (such as the tick of a watch at 20 ft under quiet conditions). When a sound is 10 times as intense as another, its loudness is 1 Bel greater. If a sound is 100 times as intense as another, it is louder by 2 Bels, and so on. The Bel is a large unit, so a subunit, a *decibel,* is generally used. For L in decibels, the formula is as follows:

$$L = 10 \log \frac{I}{I_0}.$$

a) Find the loudness, in decibels, of the sound in a radio studio, for which the intensity I is 199 times I_0.

We substitute into the formula and calculate, using Table 3:

$$L = 10 \log \frac{199 \cdot I_0}{I_0} = 10 \log 199$$

$$= 10(2.2989)$$
$$= 23 \text{ decibels.}$$

b) Find the loudness of the sound of a heavy truck, for which the intensity is 10^9 times I_0.

$$L = 10 \log \frac{10^9 \cdot I_0}{I_0} = 10 \log 10^9$$

$$= 10 \cdot 9$$
$$= 90 \text{ decibels.}$$

8. Find the loudness, in decibels, of the sound in a library that is 2510 times as intense as the minimum intensity I_0.

9. Find the loudness, in decibels, of conversational speech, having an intensity that is 10^6 times as intense as the minimum, I_0.

DO EXERCISES 8 AND 9.

Example 8 (*Earthquake magnitude*). The magnitude R (on the Richter scale) of an earthquake of intensity I is defined as follows:

$$R = \log \frac{I}{I_0},$$

where I_0 is a minimum intensity used for comparison. The Mexico City earthquake of 1978 had an intensity $10^{7.85}$ times I_0. What is its magnitude on the Richter scale?

We substitute into the formula:

$$R = \log \frac{10^{7.85} \cdot I_0}{I_0} = \log 10^{7.85} = 7.85.$$

10. An earthquake has an intensity that is $10^{7.8}$ times I_0. What is its magnitude on the Richter scale?

DO EXERCISE 10.

Example 9 (*Forgetting*). Here is a mathematical model from psychology. A group of people take a test and make an average score of S. After a time t they take an equivalent form of the same test. At that time the

11. Students in an accounting class took a final exam and then equivalent forms of the same test at monthly intervals. The average score $S(t)$, after t months, was found to be given by

$$S(t) = 68 - 20 \log (t + 1).$$

a) What was the initial average score?

b) What was the average score after 4 months?

c) What was the average score after 24 months?

average score is $S(t)$. According to this model, $S(t)$ is given by the following function:

$$S(t) = A - B \log (t + 1),$$

where t is in months and the constants A and B are determined by experiment in various kinds of learning situations. The model is appropriate only over the interval $[0, 10^{A/B} - 1]$. Students in a zoology class took a final exam, and took equivalent forms of the test at monthly intervals thereafter. The average scores were found to be given by the function

$$S(t) = 78 - 15 \log (t + 1).$$

What was the average score (a) when they took the test originally? (b) after 4 months?

We substitute into the equation defining the function:

a) $S(0) = 78 - 15 \log (0 + 1)$
$\qquad = 78 - 15 \log 1$
$\qquad = 78 - 0 = 78.$ Original score

b) $S(4) = 78 - 15 \log (4 + 1)$
$\qquad = 78 - 15 \log 5$
$\qquad = 78 - 15 \cdot 0.6990$
$\qquad = 78 - 10.49 = 67.51.$ Score after 4 months

DO EXERCISE 11.

EXERCISE SET 7.5

[i] Solve.

1. $2^x = 32$

2. $3^{x-1} + 3 = 30$

3. $4^{2x} = 8^{3x-4}$

4. $3^{x^2+4x} = \dfrac{1}{27}$

5. $3^{5x} \cdot 9^{x^2} = 27$

6. $4^x = 7$

7. $2^x = 3^{x-1}$

8. $3^{x+2} = 5^{x-1}$

9. $\log x + \log (x - 9) = 1$

10. $\log x - \log (x + 3) = -1$

11. $\log (x + 9) - \log x = 1$

12. $\log (2x + 1) - \log (x - 2) = 1$

13. $\log_4 (x + 3) + \log_4 (x - 3) = 2$

14. $\log_8 (x + 1) - \log_8 x = \log_8 4$

15. $\log x^2 = (\log x)^2$

16. $(\log_3 x)^2 - \log_3 x^2 = 3$

17. $\log_3 (\log_4 x) = 0$

18. $\log (\log x) = 2$

19. Solve for x. (*Hint:* Use $\log_e$ in the last step.)

$$\frac{e^x - e^{-x}}{2} = t$$

20. Solve for x.

$$3^x + 3^{-x} = t$$

21. Solve for x.

$$\frac{5^x - 5^{-x}}{5^x + 5^{-x}} = t$$

22. Solve for x.

$$\frac{e^x + e^{-x}}{e^x - e^{-x}} = t$$

[ii] Solve.

23. (*Doubling time*). How many years will it take an investment of $1000 to double itself when interest is compounded annually at 6%?

24. (*Tripling time*). How many years will it take an investment of $1000 to triple itself when interest is compounded annually at 5%?

25. Find the loudness of the sound of an automobile, having an intensity 3,100,000 times I_0.

26. Find the loudness of the sound of a dishwasher, having an intensity 2,500,000 times I_0.

27. Find the loudness of the threshold of sound pain, for which the intensity is 10^{14} times I_0.

28. Find the loudness of a jet aircraft, having an intensity 10^{12} times I_0.

29. The Los Angeles earthquake of 1971 had an intensity $10^{6.7}$ times I_0. What was its magnitude on the Richter scale?

30. The San Francisco earthquake of 1906 had an intensity $10^{8.25}$ times I_0. What was its magnitude on the Richter scale?

31. An earthquake has a magnitude of 5 on the Richter scale. What is its intensity?

32. An earthquake has a magnitude of 7 on the Richter scale. What is its intensity?

33. Students in an industrial mathematics course take a final exam and are then retested at monthly intervals. The forgetting function is given by the equation $S(t) = 82 - 18 \log (t + 1)$.

    a) What was the average score originally on the final exam?

    b) What was the average score after 5 months had elapsed?

34. Students graduating from a cosmetology curriculum take a final exam and are then retested at monthly intervals. The forgetting function is given by the equation $S(t) = 75 - 20 \log (t + 1)$.

    a) What was the average score originally on the final exam?

    b) What was the average score after 6 months had elapsed?

35. Refer to Exercise 33. How much time will elapse before the average score has decreased to 64?

36. Refer to Exercise 34. How much time will elapse before the average score has decreased to 61?

37. In chemistry, pH is defined as follows:

$$pH = -\log [H^+],$$

where $[H^+]$ is the hydrogen ion concentration in moles per liter. For example, the hydrogen ion concentration in milk is 4×10^{-7} moles per liter, so

$$pH = -\log (4 \times 10^{-7}) = -[\log 4 + (-7)] \approx 6.4.$$

For tomatoes, $[H^+]$ is about 6.3×10^{-5}. Find the pH.

38. For eggs, $[H^+]$ is about 1.6×10^{-8}. Find the pH.

☆

Solve.

39. $\log \sqrt{x} = \sqrt{\log x}$

40. $\log_5 \sqrt{x^2 + 1} = 1$

41. $(\log_a x)^{-1} = \log_a x^{-1}$

42. $|\log_5 x| = 2$

43. $\log_3 |x| = 2$

44. $\dfrac{(e^{3x+1})^2}{e^4} = e^{10x}$

45. $\dfrac{\sqrt{(e^{2x} \cdot e^{-5x})^{-4}}}{e^x \div e^{-x}} = e^7$

46. $\log x^{\log x} = 4$

47. Solve $y = ax^n$, for n. Use $\log_x$.

48. Solve $y = ke^{at}$, for t. Use $\log_e$.

49. Solve for t. Use $\log_e$.

$$P = P_0 e^{rt/100}$$

50. Solve for t. Use $\log_e$.

$$I = \frac{E}{R}(1 - e^{-(Rt/L)})$$

51. Solve for t.

$$T = T_0 + (T_1 - T_0)10^{-kt}$$

52. Solve for n. Use $\log_V$.

$$PV^n = c$$

53. Solve for Q.

$$\log_a Q = \frac{1}{3} \log_a y + b$$

54. Solve for y.

$$\log_a y = 2x + \log_a x$$

Solve for x.

55. $x^{\log x} = \dfrac{x^3}{100}$

56. $x^{\log x} = 100x$

★ ───

Solve.

57. $|\log_5 x| + 3 \log_5 |x| = 4$ **58.** $|\log_a x| = \log_a |x|$ **59.** $(0.5)^z < \dfrac{4}{5}$ **60.** $8x^{0.3} - 8x^{-0.3} = 63$

61. Solve the system of equations.
$$5^{x+y} = 100,$$
$$3^{2x-y} = 1000$$

62. Given that
$$\log_2 [\log_3 (\log_4 x)] =$$
$$\log_3 [\log_2 (\log_4 y)] =$$
$$\log_4 [\log_3 (\log_2 z)] = 0,$$
find $x + y + z$.

63. If $2 \log_3 (x - 2y) = \log_3 x + \log_3 y$, find $\dfrac{x}{y}$.

64. Find the ordered pair (x, y) for which $4^{\log_{16} 27} = 2^x 3^y$.

───

OBJECTIVES

You should be able to:

[i] Graph exponential functions, base e.

[ii] Use a calculator or Table 5 to find natural logarithms.

[iii] Solve certain exponential equations, base e.

[iv] Solve applied problems involving exponential functions.

[v] Find the logarithm, base a, for any a, of a number using common or natural logarithms.

1. ▦ Find $\left(1 + \dfrac{1}{n}\right)^n$ when $n = 5$.

2. ▦ Find $\left(1 + \dfrac{1}{n}\right)^n$ when $n = 10$.

7.6 THE NUMBER e; NATURAL LOGARITHMS; CHANGE OF BASE

The Number e

One of the most important numbers is a certain irrational number, with a nonrepeating decimal representation. This number is usually named e, and it arises in a number of different ways. One of these is the following.

Consider the compound interest formula

$$A = p(1 + r)^t,$$

where A is the amount that the investment is worth, p is the principal, r is the rate of interest, and t is the time in years. Interest is usually compounded more often than once a year, say n times per year. In this event we alter the formula. The rate of interest for an *interest period* is r/n and the number of interest periods is $n \cdot t$, for t years. Thus we have

$$A = p\left(1 + \frac{r}{n}\right)^{nt}.$$

Suppose you invest one dollar for one year at 100%. Since $r = 100\%$, or 1, this makes the formula simple:

$$A = \left(1 + \frac{1}{n}\right)^n.$$

The amount A is now a function of n. The more often interest is compounded, the greater A becomes. Let us look at some values of this function.

COMPOUNDING	n	A
Annually	1	$(1 + \frac{1}{1})^1 = 2$
Semiannually	2	$(1 + \frac{1}{2})^2 = 2.25$
Quarterly	4	$(1 + \frac{1}{4})^4 = 2.44 \ldots$

According to this table, for semiannual compounding the dollar grows to $2.25 in one year. For quarterly compounding it grows to $2.44. What would you expect to happen if interest were compounded daily, or once each minute?

DO EXERCISES 1 AND 2.

Let us look at some more function values.

COMPOUNDING	n	A
Daily	365	2.71456...
Once per hour	8,760	2.71812...
Once per minute	525,600	2.71828...

This result may be surprising. The function values approach a limit as n increases. No matter how often interest is compounded your investment will not amount to more than $2.72 for the year. The limit that these function values approach is the number called e:

$$e = 2.718281828459\ldots.$$

Natural Logarithms and Exponential Functions

[i] Exponential Functions, Base e

The exponential function, base e, is important in many applications, and its inverse, the logarithm function base e, is also important in mathematical theory and in applications as well. Logarithms to the base e are called *natural logarithms*. Most scientific calculators have an $\boxed{e^x}$ key. Table 4 at the back of the book gives function values for e^x and e^{-x}. Using a calculator or these tables we can construct graphs of $y = e^x$ and $y = \log_e x$.

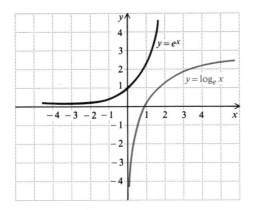

DO EXERCISES 3 AND 4.

[ii] Natural Logarithms

The number $\log_e x$ is abbreviated $\ln x$; that is, $\ln x = \log_e x$. The following is a restatement of the basic theorems of logarithms in terms of natural logarithms.

$$\ln xy = \ln x + \ln y \qquad \textbf{(Theorem 3)}$$

$$\ln x^p = p \ln x \qquad \textbf{(Theorem 4)}$$

$$\ln \frac{x}{y} = \ln x - \ln y \qquad \textbf{(Theorem 5)}$$

$$\ln e^k = k \qquad \textbf{(Theorem 2)}$$

3. Using the same axes, graph

a) $y = 2^x$,

b) $y = 4^x$,

c) $y = e^x$.

Use Table 4 to obtain values of e^x.

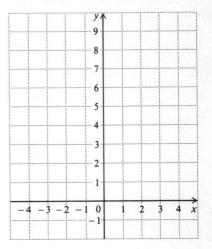

4. Graph $y = e^{-x}$.

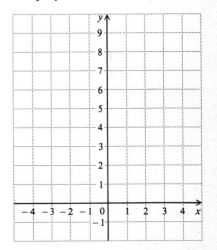

▦ Find each logarithm. Round to six decimal places.

5. ln 2

6. ln 100

7. ln 0.07432

8. ln 0.9999

Find each logarithm. Use Table 5.

9. ln 8.13

10. ln 81,300

11. ln 0.0813

12. ln 2000

13. ln 0.0001

If you have an ⌐ln⌐ key on your calculator, you can find such logarithms directly; for example, to six decimal places,

$$\ln 5.24 = 1.656321, \quad \ln 52.4 = 3.958907, \quad \text{and} \quad \ln 0.001277 = -6.663242.$$

From ln 5.24 and ln 52.4 we note that natural logarithms do not have characteristics and mantissas. Common logarithms have characteristics and mantissas because our numeration system is based on 10. For any base other than 10, logarithms have neither characteristics nor mantissas.

DO EXERCISES 5–8.

If you do not have a calculator with a natural logarithm key, you can use Table 5. Part of Table 5 is shown below. It shows some values of ln x.

x	\multicolumn{10}{c}{ln x}
	0.00	0.01	0.02	0.03	0.04	0.05	0.06	0.07	0.08	0.09
5.0	1.6094	1.6114	1.6134	1.6154	1.6174	1.6194	1.6214	1.6233	1.6253	1.6273
5.1	1.6292	1.6312	1.6332	1.6351	1.6371	1.6390	1.6409	1.6429	1.6448	1.6467
5.2	1.6487	1.6506	1.6525	1.6544	1.6563	1.6582	1.6601	1.6620	1.6639	1.6658
5.3	1.6677	1.6696	1.6715	1.6734	1.6752	1.6771	1.6790	1.6808	1.6827	1.6845
5.4	1.6864	1.6882	1.6901	1.6919	1.6938	1.6956	1.6974	1.6993	1.7011	1.7029

Example 1 Find ln 5.24. Use Table 5.

To find ln 5.24, locate the row headed 5.2, then move across to the column headed 0.04. Note the colored number in the table.

$$\ln 5.24 = 1.6563$$

We can find natural logarithms of numbers not in the table as follows. We first write scientific notation for the number.

Example 2 Find ln 5240. Use Table 5.

$$
\begin{aligned}
\ln 5240 &= \ln (5.24 \times 10^3) \\
&= \ln 5.24 + \ln 10^3 && \text{Theorem 3} \\
&= \ln 5.24 + 3 \ln 10 && \text{Theorem 4} \\
&= 1.6563 + 6.9078 && \text{Find ln 5.24 in the body of Table 5} \\
& && \text{and 3 ln 10 at the bottom.} \\
&= 8.5641
\end{aligned}
$$

Example 3 Find ln 0.000524. Use Table 5.

$$
\begin{aligned}
\ln 0.000524 &= \ln (5.24 \times 10^{-4}) \\
&= \ln 5.24 + \ln 10^{-4} && \text{Theorem 3} \\
&= \ln 5.24 - 4 \ln 10 && \text{Theorem 4} \\
&= 1.6563 - 9.2103 && \text{Find ln 5.24 in the body of Table} \\
& && \text{5 and 4 ln 10 at the bottom.} \\
&= -7.5540
\end{aligned}
$$

DO EXERCISES 9–13.

[iii] Exponential Equations

Natural logarithms can be used to solve exponential equations involving base e.

Example 4 Solve $e^t = 40$.

We could take the common logarithm on both sides as we did in Section 7.5, but taking the natural logarithm makes the simplification on the left easier. We also know how to find $\ln 40$.

$$e^t = 40$$
$$\ln e^t = \ln 40 \qquad \text{Taking the natural log on both sides}$$
$$t = \ln 40 \qquad \text{Theorem 2}$$
$$t \approx 3.7 \qquad \text{Using a calculator or Table 5}$$

DO EXERCISES 14 AND 15.

[iv] Applications

There are many applications of exponential functions, base e. We consider a few.

Example 5 (*Population growth*). One mathematical model for describing population growth is the formula

$$P = P_0 e^{kt},$$

where P_0 is the number of people at time 0, P is the number of people at time t, and k is a positive constant depending on the situation. The population of the United States in 1970 was 208 million. In 1980 it was 225 million. Use these data to find the value of k and then use the model to predict the population in 2000.

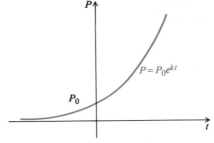

Time t begins with 1970. That is, $t = 0$ in 1970, $t = 10$ in 1980, and so on. Substituting the data into the formula, we get

$$225 = 208e^{k \cdot 10}.$$

We solve for k:

$$\ln 225 = \ln 208e^{10k} \qquad \text{Taking the natural logarithm on both sides}$$
$$\ln 225 = \ln 208 + \ln e^{10k} \qquad \text{Theorem 3}$$
$$\ln 225 = \ln 208 + 10k \qquad \text{Theorem 2}$$
$$k = \frac{\ln 225 - \ln 208}{10} \qquad \text{Solving for } k$$
$$k = \frac{5.4161 - 5.3375}{10} \qquad \text{Using a calculator or Table 5}$$
$$k = 0.008 \qquad \text{Calculating}$$

Solve for t.

14. $e^t = 80$

15. $e^{-0.06t} = 0.07$

16. The population of Tempe, Arizona, was 25,000 in 1960. In 1969 it was 52,000.

a) Use these data to determine k in the growth model.

b) Use these data to predict the population of Tempe in 1984.

To find the population in 2000, we will use $P_0 = 208$ (population in the year 1970). We then have

$$P = 208e^{0.008t}.$$

In 2000, t will be 30, so we have

$$P = 208e^{0.008(30)}$$

$$= 208e^{0.24}.$$

Using a calculator or Table 4, we find that $e^{0.24} = 1.2712$. Multiplying by 208 gives us about 264 million. This is our prediction for the population of the United States in the year 2000.

DO EXERCISE 16.

Example 6 (*Radioactive decay*). In a radioactive substance such as radium, some of the atoms are always ceasing to become radioactive. Thus the amount of a radioactive substance decreases. This is called radioactive *decay*. A model for radioactive decay is as follows:

$$N = N_0 e^{-kt},$$

where N_0 is the amount of a radioactive substance at time 0, N is the amount at time t, and k is a positive constant depending on the situation. Strontium 90 has a *half-life* of 25 years. This means that half of a sample of the substance will cease to become radioactive in 25 yr. Find k in the formula and then use the formula to find how much of a 36-gram sample will remain after 100 years.

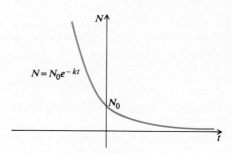

When $t = 25$ (half-life), N will be half of N_0, so we have

$$\frac{1}{2}N_0 = N_0 e^{-25k} \quad \text{or} \quad \frac{1}{2} = e^{-25k}.$$

We take the natural log on both sides:

$$\ln \frac{1}{2} = \ln e^{-25k} = -25k.$$

Thus

$$k = -\frac{\ln 0.5}{25} \approx 0.0277.$$

Now to find the amount remaining after 100 years, we use the formula

$$N = 36e^{-0.0277 \cdot 100}$$
$$= 36e^{-2.77}.$$

From a calculator or Table 4 we find that $e^{-2.8} = 0.0608$, and thus

$$N \approx 2.2 \text{ grams.}$$

DO EXERCISE 17.

Example 7 (*Atmospheric pressure*). Under standard conditions of temperature, the atmospheric pressure at height h is given by

$$P = P_0 e^{-kh},$$

where P is the pressure, P_0 is the pressure where $h = 0$, and k is a positive constant. Standard sea level pressure is 1013 millibars. Suppose that the pressure at 18,000 ft is half that at sea level. Then find k and find the pressure at 1000 ft.

At 18,000 ft P is half of P_0, so we have

$$\frac{P_0}{2} = P_0 e^{-18,000k}.$$

Taking natural logarithms on both sides, we obtain

$$\ln \frac{1}{2} = -18,000k, \quad \text{or} \quad -\ln 2 = -18,000k.$$

Then

$$k = \frac{\ln 2}{18,000} = 3.85 \times 10^{-5}.$$

To find the pressure at 1000 ft we use the fact that $P_0 = 1013$ and we let $h = 1000$:

$$P = 1013e^{-3.85 \times 10^{-5} \times 1000} = 1013e^{-0.0385}.$$

Rounding the exponent to -0.04 and using a calculator or Table 4, we calculate the pressure to be 973 millibars at 1000 ft.

DO EXERCISE 18.

[v] Logarithm Tables and Change of Base

The following theorem shows how we can change logarithm bases. The theorem can be applied to find the logarithm of a number to any base, using a table of common or natural logarithms.

THEOREM 7

For any bases a and b, and any positive number M,

$$\log_b M = \frac{\log_a M}{\log_a b}.$$

17. Radioactive bismuth has a half-life of 5 days. A scientist buys 224 grams of it. How much of it will remain radioactive in 30 days?

18. Calculate the atmospheric pressure at 10,000 ft.

19. Find $\log_5 125$.

20. Find $\log_6 4870$.

21. Use 2.718 for e. Find $\log_{10} e$.

22. Find $\log_e 1030$.

23. Find $\log_e 0.457$.

Proof. Let $x = \log_b M$. Then $b^x = M$, so that

$$\log_a M = \log_a b^x, \quad \text{or} \quad x \log_a b.$$

We now solve for x:

$$x = \log_b M = \frac{\log_a M}{\log_a b}.$$

This is the desired formula.

Example 8 Find $\log_3 81$.

$$\log_3 81 = \frac{\log_{10} 81}{\log_{10} 3}$$

$$= \frac{1.9085}{0.4771} \approx 4.0$$

Example 9 Find $\log_5 346$.

$$\log_5 346 = \frac{\log_{10} 346}{\log_{10} 5}$$

$$= \frac{2.5391}{0.6990} \approx 3.6325$$

DO EXERCISES 19–21.

If in Theorem 7 we let $a = 10$ and $b = e$, we obtain a formula for changing from common to natural logarithms:

$$\log_e M = \frac{\log_{10} M}{\log_{10} e}.$$

Now $\log_{10} e \approx 0.4343$, so we have the following.

THEOREM 8

$$\log_e M = \frac{\log_{10} M}{0.4343}$$

Example 10 Find $\log_e 257$.

$$\log_e 257 = \frac{\log_{10} 257}{0.4343}$$

$$= \frac{2.4099}{0.4343}$$

$$= 5.5489$$

DO EXERCISES 22 AND 23.

We can make use of this change-of-base procedure to solve certain exponential equations.

Example 11 Solve $3^x = 8$.

We take logarithms, base 3, on both sides. Then we have

$$\log_3 3^x = \log_3 8$$
$$x = \log_3 8 \qquad \text{Theorem 2}$$

To get an approximation we change to base 10 and divide:

$$x = \log_3 8$$
$$= \frac{\log_{10} 8}{\log_{10} 3}$$
$$= \frac{0.9031}{0.4771}$$
$$= 1.8929.$$

Compare this with Examples 1 and 10 in Section 7.5.

DO EXERCISE 24.

24. Solve $2^x = 7$.

EXERCISE SET 7.6

[i] Graph.

1. $y = e^{2x}$

2. $y = e^{0.5x}$

3. $y = e^{-2x}$

4. $y = e^{-0.5x}$

5. $y = 1 - e^{-x}$, for $x \geq 0$

6. $y = 2(1 - e^{-x})$, for $x \geq 0$

[ii] Find each natural logarithm to four decimal places. Use a calculator or Table 7. Answers in the back of the book have been found using Table 7. If you use a calculator, there may be some variance in the last decimal place.

7. ln 1.88

8. ln 18.8

9. ln 0.0188

10. ln 0.188

11. ln 2.13

12. ln 213

13. ln 0.213

14. ln 0.00213

15. ln 4500

16. ln 81,000

17. ln 0.00056

18. ln 0.999

19. ln 0.08

20. ln 0.0471

21. ln 980,000

22. ln 765,000,000

[iii] Solve.

23. $e^t = 100$

24. $e^t = 1000$

25. $e^x = 60$

26. $e^k = 90$

27. $e^{-t} = 0.1$

28. $e^{-t} = 0.01$

29. $e^{-0.02k} = 0.06$

30. $e^{0.07t} = 2$

[iv] Solve.

31. (*Population growth*). The population of Dallas was 680,000 in 1960. In 1969 it was 815,000. Find k in the growth formula and estimate the population in 1990.

32. (*Population growth*). The population of Kansas City was 475,000 in 1960. In 1970 it was 507,000. Find k in the growth formula and estimate the population in 2000.

33. (*Radioactive decay*). The half-life of polonium is 3 minutes. After 30 minutes, how much of a 410-gram sample will remain radioactive?

34. (*Radioactive decay*). The half-life of a lead isotope is 22 yr. After 66 yr, how much of a 1000-gram sample will remain radioactive?

35. (*Radioactive decay*). A certain radioactive substance decays from 66,560 grams to 6.5 grams in 16 days. What is its half-life?

36. Ten grams of uranium will decay to 2.5 grams in 496,000 years. What is its half-life?

37. (*Radiocarbon dating*). Carbon-14, an isotope of carbon, has a half-life of 5750 years. Organic objects contain carbon-14 as well as nonradioactive carbon, in known proportions. When a living organism dies, it takes in no more carbon. The carbon-14 decays, thus changing the proportions of the kinds of carbon in the organism. By determining the amount of carbon-14 it is possible to determine how long the organism has been dead, hence how old it is.

a) How old is an animal bone that has lost 30% of its carbon-14?

b) A mummy discovered in the pyramid Khufu in Egypt had lost 46% of its carbon-14. Determine its age.

38. (*Radiocarbon dating*).

a) How old is an animal bone that has lost 20% of its carbon-14?

b) The Statue of Zeus at Olympia in Greece is one of the seven wonders of the world. It is made of gold and ivory. The ivory was found to have lost 35% of its carbon-14. Determine the age of the statue.

39. (*Atmospheric pressure*). What is the pressure at the top of Mt. Shasta in California, 14,162 ft high?

40. (*Atmospheric pressure*). Blood will boil when atmospheric pressure goes below about 62 millibars. At what altitude, in an unpressurized vehicle, will a pilot's blood boil?

41. (*Consumer price index*). The *consumer price index* compares the costs of goods and services over various years. The base year is 1967. The same goods and services that cost $100 ($P_0$) in 1967 cost $184.50 in 1977. Assuming the exponential model:

a) Find the value k and write the equation.

b) Estimate what the same goods and services will cost in 1987.

c) When did the same goods and services cost double that of 1967?

42. (*Cost of a double-dip ice cream cone*). In 1970 the cost of a double-dip ice cream cone was 52¢. In 1978 it was 66¢. Assuming the exponential model:

a) Find the value k and write the equation.

b) Estimate the cost of a cone in 1986.

c) When will the cost of a cone be twice that of 1978?

[v] Find the following.

43. $\log_4 20$ **44.** $\log_8 0.99$ **45.** $\log_5 0.78$ **46.** $\log_{12} 15{,}000$

47. $\log_e 12$ (Do not use Table 5.) **48.** $\log_e 0.77$ (Do not use Table 5.)

☆ ─────────────────────────────────

Solve for t.

49. $P = P_0 e^{kt}$ **50.** $P = P_0 e^{-kt}$

Verify each of the following.

51. $\ln x = \dfrac{\log x}{\log e} \approx 2.3026 \log x$ **52.** $\log x = \dfrac{\ln x}{\ln 10} \approx 0.4343 \ln x$

53. ▦ Which is larger, e^π or π^e? **54.** ▦ Which is larger, $e^{\sqrt{\pi}}$ or $\sqrt{e^\pi}$?

55. ▦ Given $f(x) = (1 + x)^{1/x}$, find $f(1)$, $f(0.5)$, $f(0.2)$, $f(0.1)$, $f(0.01)$, and $f(0.001)$ to six decimal places. This sequence of numbers approaches the number e.

56. ▦ Given $f(t) = t^{1/(t-1)}$, find $f(0.5)$, $f(0.9)$, $f(0.99)$, $f(0.999)$, and $f(0.9999)$ to six decimal places. This sequence of numbers approaches the number e.

57. Find a (simple) formula for radioactive decay involving H, the half-life.

58. The time required for a population to double is called the *doubling time*.

a) Find an expression relating T, the doubling time, to k, the constant in the growth equation.

b) Find a (simple) formula for population growth involving T.

Prove the following for any logarithm bases a and b.

59. $\log_a b = \dfrac{1}{\log_b a}$

60. $a^{(\log_b M) \div (\log_b a)} = M$

61. $a^{(\log_b M)(\log_b a)} = M^{(\log_b a)^2}$

62. $\log_a (\log_a x) = \log_a (\log_b x) - \log_a (\log_b a)$

CHAPTER 7 REVIEW

[7.1, ì] Graph.

[7.6, ì] **1.** $y = \log_2 (x - 1)$ **2.** $y = e^{0.4x}$ **3.** $y = e^{-x} - 3$

[7.1, iì] **4.** Write an exponential equation equivalent to $\log_8 \dfrac{1}{4} = -\dfrac{2}{3}$.

5. Write a logarithmic equation equivalent to $7^{2.3} = x$.

[7.1, iiì] Solve.

6. $\log_x 64 = 3$ **7.** $\log_{16} 4 = x$ **8.** $\log_5 125 = x$

[7.2, ì] **9.** Write an equivalent expression containing a single logarithm.

$$\frac{1}{2} \log_b a + \frac{3}{2} \log_b c - 4 \log_b d$$

[7.2, ì] Given that $\log_a 2 = 0.301$, $\log_a 3 = 0.477$, and $\log_a 7 = 0.845$, find:

10. $\log_a 18$. **11.** $\log_a \dfrac{7}{2}$. **12.** $\log_a \dfrac{1}{4}$. **13.** $\log_a \sqrt{3}$.

14. Express in terms of logarithms of M and N: $\log \sqrt[3]{M^2/N}$.

[7.3, ì] Using Table 2, find:

15. $\log 26.3$. **16.** $\log 0.00806$. **17.** $10^{1.8686}$. **18.** antilog $(8.4409 - 10)$.

[7.6, v] **19.** $\log_5 290$ (round to the nearest tenth).

[7.3, iì] **20.** Use logarithms to compute $(0.0524)^2 \cdot \sqrt{0.0638}$.

[7.6, iì] Using Table 7, find:

21. $\ln 8.9$ **22.** $\ln 560$. **23.** $\ln 0.0462$.

[7.1, iv] **24.** Simplify $\log_{12} 12^{x^2+1}$.

[7.5, ì] Solve.

25. $3^{1-x} = 9^{2x}$ **26.** $\log (x^2 - 1) - \log (x - 1) = 1$

[7.5, iì] **27.** How many years will it take an investment of $1000 to double if interest is compounded annually at 13%?

[7.5, iì] **28.** What is the loudness, in decibels, of a sound whose intensity is 1000 times I_0?

[7.6, iv] **29.** The half-life of a radioactive substance is 15 days. How much of a 25-gram sample will remain radioactive after 30 days?

[7.5,]   Solve.

30. $\log x^2 = \log x$ **31.** $\log_2 (x - 1) + \log_2 (x + 1) = 3$
32. $\log 2 + 2 \log x = \log (5x + 3)$ **33.** $\log x = \ln x$

34. $|\log_4 x| = 3$
35. Graph $y = |\log_3 x|$.   **36.** Graph $y = |e^x - 4|$.

Find the domain.

37. $f(x) = \dfrac{17}{\sqrt{5 \ln x - 6}}$ **38.** $f(x) = \dfrac{8}{e^{4x} - 10}$

Appendix 1

HANDLING DIMENSION SYMBOLS

SPEED

Speed is often measured by measuring a distance and a time and then dividing the distance by the time (this is *average* speed):

$$\text{Speed} = \frac{\text{distance}}{\text{time}}.$$

If a distance is measured in kilometers and the time required to travel that distance is measured in hours, the speed will be computed in *kilometers per hour* (km/h). For example, if a car travels 100 km in 2 h, the average speed is

$$\frac{100\ \text{km}}{2\ \text{h}}, \quad \text{or} \quad 50\frac{\text{km}}{\text{h}}.$$

DIMENSION SYMBOLS

The symbol 100 km/2 h looks as if we are dividing 100 km by 2 h. It may be argued that we cannot divide 100 km by 2 h (we can only divide 100 by 2). Nevertheless, it is convenient to treat dimension symbols such as *kilometers, hours, feet, seconds,* and *grams* much like numerals and variables, for the reason that correct results can thus be obtained mechanically. Compare, for example,

$$\frac{100x}{2y} = \frac{100}{2} \cdot \frac{x}{y} = 50\frac{x}{y}$$

with

$$\frac{100\ \text{km}}{2\ \text{h}} = \frac{100}{2} \cdot \frac{\text{km}}{\text{h}} = 50\frac{\text{km}}{\text{h}}.$$

This comparison holds in other situations as shown in the following examples.

Example 1 Compare

$$3\ \text{cm} + 2\ \text{cm} = (3 + 2)\text{cm} = 5\ \text{cm}$$

with

$$3x + 2x = (3 + 2)x = 5x.$$

This is similar to using the distributive law.

Example 2 Compare

$$4\ \text{cm} \cdot 3\ \text{cm} = 4 \cdot 3 \cdot \text{cm} \cdot \text{cm} = 12\ \text{cm}^2\ (\text{sq cm})$$

with

$$4x \cdot 3x = 4 \cdot 3 \cdot x \cdot x = 12x^2.$$

Example 3 Compare

$$5\ \text{men} \cdot 8\ \text{h} = 5 \cdot 8 \cdot \text{man-h} = 40\ \text{man-h}$$

with

$$5x \cdot 8y = 5 \cdot 8 \cdot x \cdot y = 40xy.$$

Example 4 Compare

$$3\frac{\text{dollars}}{\text{yd}} \cdot 5\,\text{yd} = 3 \cdot 5 \cdot \frac{\text{dollars}}{\text{yd}} \cdot \text{yd} = 15\,\text{dollars}$$

with

$$3\frac{y}{x} \cdot 5x = 3 \cdot 5 \cdot \frac{y}{x} \cdot x = 15y.$$

Example 5 Compare

$$48\,\text{in.} \cdot \frac{1\,\text{ft}}{12\,\text{in.}} = \frac{48}{12} \cdot \text{in.} \cdot \frac{\text{ft}}{\text{in.}} = 4\frac{\text{in.}}{\text{in.}} \cdot \text{ft} = 4\,\text{ft}$$

with

$$48x \cdot \frac{y}{12x} = \frac{48}{12} \cdot x \cdot \frac{y}{x} = 4 \cdot \frac{x}{x} \cdot y = 4y.$$

In each of the examples given, dimension symbols are treated as though they were variables or numerals, and as though a symbol like "3 cm" represented a product of "3" by "cm". A symbol like km/h is treated as if it represents a division of kilometers by hours.

Any two measures can be "multiplied" or "divided." For example,

$$6\,\text{gm} \cdot 4\,\text{cm} = 24\,\text{gm-cm},$$

$$7\,\text{in.} \cdot 3\,\text{ft} = 21\,\text{in.-ft},$$

$$3\,\text{m} \cdot 4\,\text{sec} = 12\,\text{m-sec},$$

$$\frac{8\,\text{lb}}{2\,\text{ft}} = 4\frac{\text{lb}}{\text{ft}},$$

$$\frac{3\,\text{cm} \cdot 8\,\text{days}}{6\,\text{gm}} = 4\frac{\text{cm-day}}{\text{gm}}.$$

These "multiplications" and "divisions" may not have sensible interpretations; for example,

$$2\,\text{barns} \cdot 4\,\text{dances} = 8\,\text{barn-dances}.$$

However, the fact that such amusing examples exist causes us no trouble since they do not come up in practice.

CHANGES OF UNIT

Sometimes a change of unit can be achieved by successive substitutions.

Example 1 Change 25 m to cm.

25 m = 25 · 1 m

    = 25 · 100 cm   Substituting **100 cm** for **1 m**

    = 25 · 100 · 1 cm

    = 2500 cm

The notion of "multiplying by one" can also be used to change units in other situations.

Example 2 Change 7.2 in. to yd.

7.2 in. = 7.2 in. · 1 · 1

$$= 7.2 \text{ in.} \cdot \boxed{\frac{1 \text{ ft}}{12 \text{ in.}}} \cdot \boxed{\frac{1 \text{ yd}}{3 \text{ ft}}}$$

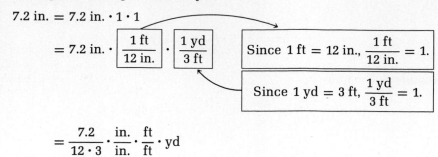

Since 1 ft = 12 in., $\dfrac{1 \text{ ft}}{12 \text{ in.}} = 1$.

Since 1 yd = 3 ft, $\dfrac{1 \text{ yd}}{3 \text{ ft}} = 1$.

$$= \frac{7.2}{12 \cdot 3} \cdot \frac{\text{in.}}{\text{in.}} \cdot \frac{\text{ft}}{\text{ft}} \cdot \text{yd}$$

$$= 0.2 \text{ yd.}$$

Example 3 Change $50 \dfrac{\text{km}}{\text{h}}$ to $\dfrac{\text{m}}{\text{sec}}$.

$$50 \frac{\text{km}}{\text{h}} = 50 \frac{\text{km}}{\text{h}} \cdot \frac{1000 \text{ m}}{1 \text{ km}} \cdot \frac{1 \text{ h}}{60 \text{ min}} \cdot \frac{1 \text{ min}}{60 \text{ sec}}$$

$$= \frac{50 \cdot 1000}{60 \cdot 60} \cdot \frac{\text{km}}{\text{km}} \cdot \frac{\text{h}}{\text{h}} \cdot \frac{\text{min}}{\text{min}} \cdot \frac{\text{m}}{\text{sec}}$$

$$= 13.89 \frac{\text{m}}{\text{sec}}$$

Since 1000 m = 1 k,
1 h = 60 min,
1 min = 60 sec)

Appendix 2
ALGEBRA REVIEW EXERCISES

EXERCISE SET R-1. EXPONENTS AND SCIENTIFIC NOTATION

Simplify.

1. $2^3 \cdot 2^{-4}$

2. $b^2 \cdot b^{-2}$

3. $4^2 \cdot 4^{-5} \cdot 4^6$

4. $2x^3 \cdot 3x^2$

5. $(5a^2b)(3a^{-3}b^4)$

6. $(2x)^3(3x)^2$

7. $(6x^5y^{-2}z^3)(-3x^2y^3z^{-2})$

8. $\dfrac{b^{40}}{b^{37}}$

9. $\dfrac{x^2y^{-2}}{x^{-1}y}$

10. $\dfrac{9a^2}{(-3a)^2}$

11. $\dfrac{24a^5b^3}{8a^4b}$

12. $\dfrac{12x^2y^3z^{-2}}{21xy^2z^3}$

13. $(2ab^2)^3$

14. $(-2x^3)^4$

15. $-(2x^3)^4$

16. $(6a^2b^3c)^2$

17. $(-5c^{-1}d^{-2})^{-2}$

18. $\dfrac{4^{-2} + 2^{-4}}{8^{-1}}$

19. $\dfrac{(-2)^4 + (-4)^2}{(-1)^8}$

20. $\dfrac{(3a^2b^{-2}c^4)^3}{(2a^{-1}b^2c^{-3})^2}$

21. $\dfrac{6^{-2}x^{-3}y^2}{3^{-3}x^{-4}y}$

Convert to scientific notation.

22. 58,000,000

23. 365,000

24. .0000027

25. 0.027

26. 0.05

Convert to decimal notation.

27. 5×10^{-4}

28. 7.8×10^6

29. 8.54×10^{-7}

EXERCISE SET R-2. ADDITION AND SUBTRACTION

Add.

1. $5x^2y - 2xy^2 + 3xy - 5$ and
    $-2x^2y - 3xy^2 + 4xy + 7$

2. $-3pq^2 - 5p^2q + 4pq + 3$ and
    $-7pq^2 + 3pq - 4p + 2q$

3. $2x + 3y + z - 7$ and
    $4x - 2y - z + 8$ and
    $-3x + y - 2z - 4$

4. $7x\sqrt{y} - 3y\sqrt{x} + \dfrac{1}{5}$ and

    $-2x\sqrt{y} - y\sqrt{x} - \dfrac{3}{5}$

5. $0.5p^2\sqrt{q} + 1.2\sqrt[3]{pq} + 0.75\sqrt{p}$ and
    $-1.3p^2\sqrt{q} - 1.1\sqrt[3]{pq} + 0.13\sqrt{p}$

6. $3xy^{-2} - 4xy + 2\sqrt{xy} - 1$ and
    $-5xy^{-2} + 2xy - 5\sqrt{xy}$

Subtract.

7. $(3x^2 - 2x - x^3 + 2) - (5x^2 - 8x - x^3 + 4)$

8. $(4a - 2b - c + 3d) - (-2a + 3b + c - d)$

9. $(x^4 - 3x^2 + 4x) - (3x^3 + x^2 - 5x + 3)$

10. $(7x\sqrt{y} - 4y\sqrt{x} + 7.5) - (-2x\sqrt{y} - y\sqrt{x} - 1.6)$

11. $(3xy^{-2} + 2\sqrt{xy} - 1) - (-5xy^{-2} + 2xy + 5\sqrt{xy})$

EXERCISE SET R-3. MULTIPLICATION

Multiply.

1. $2x^2 + 4x + 16$ and $3x - 4$

2. $4a^2b - 2ab + 3b^2$ and $ab - 2b + 1$

3. $(2y + 3)(3y - 2)$

4. $(2x + 3y)(2x + y)$

5. $\left(4x^2 - \dfrac{1}{2}y\right)\left(3x + \dfrac{1}{4}y\right)$

6. $(\sqrt{2}x^2 - y^2)(\sqrt{2}x - 2y)$

7. $(2x + 3y)^2$

8. $(2x^2 - 3y)^2$

9. $(2x^3 + 3y^2)^2$

10. $\left(\dfrac{1}{2}x^2 - \dfrac{3}{5}y\right)^2$

11. $(0.5x + 0.7y^2)^2$

12. $(3x - 2y)(3x + 2y)$

13. $(x^2 + yz)(x^2 - yz)$

14. $(3x^2 - \sqrt{2})(3x^2 + \sqrt{2})$

15. $(2x + 3y + 4)(2x + 3y - 4)$

16. $(x^2 + 3y + y^2)(x^2 + 3y - y^2)$

17. $(x + 1)(x - 1)(x^2 + 1)$

18. $(2x + y)(2x - y)(4x^2 + y^2)$

EXERCISE SET R-4. FACTORING

Factor.

1. $18a^2b - 15ab^2$

2. $a(b - 2) + c(b - 2)$

3. $x^2 + 9x + 18$

4. $9x^2 - 25$

5. $4xy^4 - 4xz^2$

6. $y^2 - 6y + 9$

7. $1 - 8x + 16x^2$

8. $4x^2 - 5$

9. $x^2y^2 - 14xy + 49$

10. $4ax^2 + 20ax - 56a$

11. $a^2 + 2ab + b^2 - c^2$

12. $x^2 + 2xy + y^2 - a^2 - 2ab - b^2$

13. $5y^4 - 80x^4$

14. $x^3 + 8$

15. $3x^3 - \dfrac{3}{8}$

16. $x^3 + 0.001$

17. $3z^3 - 24$

EXERCISE SET R-5. EQUATIONS AND INEQUALITIES

Solve.

1. $4x + 12 = 60$

2. $4 + \dfrac{1}{2}x = 1$

3. $y + 1 = 2y - 7$

4. $5x - 2 + 3x = 2x + 6 - 4x$

5. $1.9x - 7.8 + 5.3x = 3.0 + 1.8x$

6. $7(3x + 6) = 11 - (x + 2)$

7. $2x - (5 + 7x) = 4 - [x - (2x + 3)]$

8. $(2x - 3)(3x - 2) = 0$

9. $x(x - 1)(x + 2) = 0$

10. $3x^2 + x - 2 = 0$

11. $(x - 1)(x + 1) = 5(x - 1)$

12. $x + 6 < 5x - 6$

13. $3x - 3 + 2x \geq 1 - 7x - 9$

EXERCISE SET R-6. QUADRATIC EQUATIONS

Solve for x.

1. $2x^2 = 14$

2. $9x^2 = 5$

3. $ax^2 = b$

4. $(x - 7)^2 = 5$

5. $(x - h)^2 = a$

Solve by completing the square.

6. $x^2 + 6x + 4 = 0$

7. $y^2 + 7y - 30 = 0$

8. $5x^2 - 4x - 2 = 0$

9. $2x^2 + 7x - 15 = 0$

Solve using the quadratic formula.

10. $x^2 + 4x - 5 = 0$

11. $2y^2 - 3y - 2 = 0$

12. $3t^2 + 8t + 5 = 0$

13. $u^2 + 12u - 3 = 0$

Solve using any method.

14. $2x^2 - x - 6 = 0$

15. $x^2 + 3x = 8$

16. $x + \dfrac{1}{x} = \dfrac{13}{6}$

17. $t^2 + 0.2t - 0.3 = 0$

EXERCISE SET R-7. FRACTIONAL EXPRESSIONS

In Exercises 1-3, simplify.

1. $\dfrac{25x^2y^2}{10xy^2}$

2. $\dfrac{x^2 - 4}{x^2 + 5x + 6}$

3. $\dfrac{x^2 - 3x + 2}{x^2 + x - 2}$

Multiply or divide and simplify.

4. $\dfrac{x^2 - y^2}{(x - y)^2} \cdot \dfrac{1}{x + y}$

5. $\dfrac{x^2 - 2x - 35}{2x^3 - 3x^2} \cdot \dfrac{4x^3 - 9x}{7x - 49}$

6. $\dfrac{a^2 - a - 6}{a^2 - 7a + 12} \cdot \dfrac{a^2 - 2a - 8}{a^2 - 3a - 10}$

7. $\dfrac{m^2 - n^2}{r + s} \div \dfrac{m - n}{r + s}$

8. $\dfrac{3x + 12}{2x - 8} \div \dfrac{(x + 4)^2}{(x - 4)^2}$

9. $\dfrac{x^2 - y^2}{x^3 - y^3} \cdot \dfrac{x^2 + xy + y^2}{x^2 + 2xy + y^2}$

10. $\dfrac{(x - y)^2 - z^2}{(x + y)^2 - z^2} \div \dfrac{x - y + z}{x + y - z}$

EXERCISE SET R-8. FRACTIONAL EXPRESSIONS

Add or subtract and simplify.

1. $\dfrac{3}{2a + 3} + \dfrac{2a}{2a + 3}$

2. $\dfrac{y}{y - 1} + \dfrac{2}{1 - y}$

3. $\dfrac{x}{2x - 3y} - \dfrac{y}{3y - 2x}$

4. $\dfrac{3}{x + 2} + \dfrac{2}{x^2 - 4}$

5. $\dfrac{y}{y^2 - y - 20} + \dfrac{2}{y + 4}$

6. $\dfrac{3}{x + y} + \dfrac{x - 5y}{x^2 - y^2}$

7. $\dfrac{9x + 2}{3x^2 - 2x - 8} + \dfrac{7}{3x^2 + x - 4}$

8. $\dfrac{5a}{a - b} + \dfrac{ab}{a^2 - b^2} + \dfrac{4b}{a + b}$

9. $\dfrac{7}{x + 2} - \dfrac{x + 8}{4 - x^2} + \dfrac{3x - 2}{4 - 4x + x^2}$

10. $\dfrac{1}{x + 1} - \dfrac{x}{x - 2} + \dfrac{x^2 + 2}{x^2 - x - 2}$

EXERCISE SET R-9. COMPLEX FRACTIONAL EXPRESSIONS

Simplify.

1. $\dfrac{\dfrac{x^2 - y^2}{xy}}{\dfrac{x - y}{y}}$

2. $\dfrac{x - \dfrac{1}{x}}{x + \dfrac{1}{x}}$

3. $\dfrac{c + \dfrac{8}{c^2}}{1 + \dfrac{2}{c}}$

4. $\dfrac{x^2 + xy + y^2}{\dfrac{x^2}{y} - \dfrac{y^2}{x}}$

5. $\dfrac{\dfrac{x}{y} - \dfrac{y}{x}}{\dfrac{1}{y} + \dfrac{1}{x}}$

6. $\dfrac{\dfrac{x^2}{y^2} - \dfrac{y^2}{x^2}}{\dfrac{x}{y} + \dfrac{y}{x}}$

7. $\dfrac{\dfrac{a}{1-a} + \dfrac{1+a}{a}}{\dfrac{1-a}{a} + \dfrac{a}{1+a}}$

8. $\dfrac{\dfrac{1}{a^2} + \dfrac{2}{ab} + \dfrac{1}{b^2}}{\dfrac{1}{a^2} - \dfrac{1}{b^2}}$

EXERCISE SET R-10. RADICAL NOTATION

Simplify.

1. $\sqrt{(-11)^2}$

2. $\sqrt{16x^2}$

3. $\sqrt{(b+1)^2}$

4. $\sqrt[3]{-27x^3}$

5. $\sqrt{x^2 - 4x + 4}$

6. $\sqrt[5]{32}$

7. $\sqrt[5]{-32}$

8. $\sqrt{48}$

9. $\sqrt[3]{135}$

10. $\sqrt{128c^2d^{-4}}$

11. $\sqrt{3}\,\sqrt{6}$

12. $\sqrt[3]{2x^3y}\,\sqrt{12xy}$

13. $\sqrt[3]{-3x^2y}\,\sqrt{36x}$

14. $\sqrt[3]{2(x+4)}\,\sqrt[3]{4(x+4)^4}$

15. $\sqrt[3]{4(x+1)^2}\,\sqrt[3]{18(x+1)^2}$

16. $\dfrac{\sqrt{21ab^2}}{\sqrt{3ab}}$

17. $\dfrac{\sqrt[3]{40m}}{\sqrt[3]{5m}}$

18. $\dfrac{\sqrt[3]{3x^2}}{\sqrt[3]{24x^5}}$

19. $\dfrac{\sqrt{a^2 - b^2}}{\sqrt{a-b}}$

20. $\sqrt{\dfrac{9a^2}{8b}}$

EXERCISE SET R-11. RADICAL NOTATION

Simplify.

1. $8\sqrt{2} - 6\sqrt{20} - 5\sqrt{8}$

2. $2\sqrt[3]{-8x^2} + 5\sqrt[3]{27x^2} - 3\sqrt[3]{-x^3}$

3. $3\sqrt{3y^2} - \dfrac{y\sqrt{48}}{\sqrt{2}} + \sqrt{\dfrac{12}{4y^{-2}}}$

4. $(\sqrt{3} - \sqrt{2})(\sqrt{3} + \sqrt{2})$

5. $(\sqrt{m^2y} - 2\sqrt{x})(3m\sqrt{y} + \sqrt{x})$

6. $(\sqrt{x+3} - \sqrt{3})(\sqrt{x+3} + \sqrt{3})$

Rationalize the denominator.

7. $\dfrac{6}{3 + \sqrt{5}}$

8. $\sqrt[3]{\dfrac{16}{9}}$

9. $\dfrac{4\sqrt{x} - 3\sqrt{y}}{2\sqrt{x} + 5\sqrt{y}}$

Rationalize the numerator.

10. $\dfrac{\sqrt{2} + \sqrt{5a}}{6}$

11. $\dfrac{\sqrt{x+1} + 1}{\sqrt{x+1} - 1}$

12. $\dfrac{\sqrt{a+3} - \sqrt{3}}{3}$

EXERCISE SET R-12. RADICALS AND EXPONENTS

Convert to radical notation and simplify.

1. $x^{3/4}$

2. $16^{3/4}$

3. $x^{5/4}y^{-(3/4)}$

4. $a^{3/2}b^{-(1/2)}$

Convert to exponential notation and simplify.

5. $\sqrt[3]{20^2}$

6. $(\sqrt[4]{13})^5$

7. $\sqrt[3]{\sqrt{11}}$

8. $\sqrt{5} \sqrt[3]{5}$ **9.** $\sqrt[5]{32^2}$ **10.** $\sqrt[3]{8y^6}$

11. $\sqrt[3]{a^2 + b^2}$ **12.** $\sqrt[3]{27a^3b^9}$ **13.** $\sqrt[6]{\dfrac{m^{12}n^{24}}{64}}$

Simplify and then write radical notation.

14. $(2a^{3/2})(4a^{1/2})$ **15.** $\left(\dfrac{x^6}{9b^{-4}}\right)^{-(1/2)}$ **16.** $\dfrac{x^{2/3}y^{5/6}}{x^{-(1/3)}y^{1/2}}$

Write an expression containing a single radical.

17. $\sqrt[3]{6} \sqrt{2}$ **18.** $\sqrt[4]{xy} \sqrt[3]{x^2y}$ **19.** $\sqrt[3]{a^4} \sqrt{a^3}$

20. $\dfrac{\sqrt{(a + x)^3} \sqrt[3]{(a + x)^2}}{\sqrt[4]{a + x}}$

TABLES

TABLES

Table 1 Powers, Roots, and Reciprocals

n	n^2	n^3	$\sqrt{n}$	$\sqrt[3]{n}$	$\sqrt{10n}$	$\frac{1}{n}$	n	n^2	n^3	$\sqrt{n}$	$\sqrt[3]{n}$	$\sqrt{10n}$	$\frac{1}{n}$
1	1	1	1.000	1.000	3.162	1.0000	51	2,601	132,651	7.141	3.708	22.583	.0196
2	4	8	1.414	1.260	4.472	.5000	52	2,704	140,608	7.211	3.733	22.804	.0192
3	9	27	1.732	1.442	5.477	.3333	53	2,809	148,877	7.280	3.756	23.022	.0189
4	16	64	2.000	1.587	6.325	.2500	54	2,916	157,464	7.348	3.780	23.238	.0185
5	25	125	2.236	1.710	7.071	.2000	55	3,025	166,375	7.416	3.803	23.452	.0182
6	36	216	2.449	1.817	7.746	.1667	56	3,136	175,616	7.483	3.826	23.664	.0179
7	49	343	2.646	1.913	8.367	.1429	57	3,249	185,193	7.550	3.849	23.875	.0175
8	64	512	2.828	2.000	8.944	.1250	58	3,364	195,112	7.616	3.871	24.083	.0172
9	81	729	3.000	2.080	9.487	.1111	59	3,481	205,379	7.681	3.893	24.290	.0169
10	100	1,000	3.162	2.154	10.000	.1000	60	3,600	216,000	7.746	3.915	24.495	.0167
11	121	1,331	3.317	2.224	10.488	.0909	61	3,721	226,981	7.810	3.936	24.698	.0164
12	144	1,728	3.464	2.289	10.954	.0833	62	3,844	238,328	7.874	3.958	24.900	.0161
13	169	2,197	3.606	2.351	11.402	.0769	63	3,969	250,047	7.937	3.979	25.100	.0159
14	196	2,744	3.742	2.410	11.832	.0714	64	4,096	262,144	8.000	4.000	25.298	.0156
15	225	3,375	3.873	2.466	12.247	.0667	65	4,225	274,625	8.062	4.021	25.495	.0154
16	256	4,096	4.000	2.520	12.648	.0625	66	4,356	287,496	8.124	4.041	25.690	.0152
17	289	4,913	4.123	2.571	13.038	.0588	67	4,489	300,763	8.185	4.062	25.884	.0149
18	324	5,832	4.243	2.621	13.416	.0556	68	4,624	314,432	8.246	4.082	26.077	.0147
19	361	6,859	4.359	2.668	13.784	.0526	69	4,761	328,509	8.307	4.102	26.268	.0145
20	400	8,000	4.472	2.714	14.142	.0500	70	4,900	343,000	8.367	4.121	26.458	.0143
21	441	9,261	4.583	2.759	14.491	.0476	71	5,041	357,911	8.426	4.141	26.646	.0141
22	484	10,648	4.690	2.802	14.832	.0455	72	5,184	373,248	8.485	4.160	26.833	.0139
23	529	12,167	4.796	2.844	15.166	.0435	73	5,329	389,017	8.544	4.179	27.019	.0137
24	576	13,824	4.899	2.884	15.492	.0417	74	5,476	405,224	8.602	4.198	27.203	.0135
25	625	15,625	5.000	2.924	15.811	.0400	75	5,625	421,875	8.660	4.217	27.386	.0133
26	676	17,576	5.099	2.962	16.125	.0385	76	5,776	438,976	8.718	4.236	27.568	.0132
27	729	19,683	5.196	3.000	16.432	.0370	77	5,929	456,533	8.775	4.254	27.749	.0130
28	784	21,952	5.292	3.037	16.733	.0357	78	6,084	474,552	8.832	4.273	27.928	.0128
29	841	24,389	5.385	3.072	17.029	.0345	79	6,241	493,039	8.888	4.291	28.107	.0127
30	900	27,000	5.477	3.107	17.321	.0333	80	6,400	512,000	8.944	4.309	28.284	.0125
31	961	29,791	5.568	3.141	17.607	.0323	81	6,561	531,441	9.000	4.327	28.460	.0123
32	1,024	32,768	5.657	3.175	17.889	.0312	82	6,724	551,368	9.055	4.344	28.636	.0122
33	1,089	35,937	5.745	3.208	18.166	.0303	83	6,889	571,787	9.110	4.362	28.810	.0120
34	1,156	39,304	5.831	3.240	18.439	.0294	84	7.056	592,704	9.165	4.380	28.983	.0119
35	1,225	42,875	5.916	3.271	18.708	.0286	85	7,225	614,125	9.220	4.397	29.155	.0118
36	1,296	46,656	6.000	3.302	18.974	.0278	86	7,396	636,056	9.274	4.414	29.326	.0116
37	1,369	50,653	6.083	3.332	19.235	.0270	87	7,569	658,503	9.327	4.431	29.496	.0115
38	1,444	54,872	6.164	3.362	19.494	.0263	88	7,744	681,472	9.381	4.448	29.665	.0114
39	1,521	59,319	6.245	3.391	19.748	.0256	89	7,921	704,969	9.434	4.465	29.833	.0112
40	1,600	64,000	6.325	3.420	20.000	.0250	90	8,100	729,000	9.487	4.481	30.000	.0111
41	1,681	68,921	6.403	3.448	20.248	.0244	91	8,281	753,571	9.539	4.498	30.166	.0110
42	1,764	74,088	6.481	3.476	20.494	.0238	92	8,464	778,688	9.592	4.514	30.332	.0109
43	1,849	79,507	6.557	3.503	20.736	.0233	93	8,649	804,357	9.644	4.531	30.496	.0108
44	1,936	85,184	6.633	3.530	20.976	.0227	94	8,836	830,584	9.695	4.547	30.659	.0106
45	2,025	91,125	6.708	3.557	21.213	.0222	95	9,025	857,375	9.747	4.563	30.822	.0105
46	2,116	97,336	6.782	3.583	21.448	.0217	96	9,216	884,736	9.798	4.579	30.984	.0104
47	2,209	103,823	6.856	3.609	21.679	.0213	97	9,409	912,673	9.849	4.595	31.145	.0103
48	2,304	110,592	6.928	3.634	21.909	.0208	98	9,604	941,192	9.899	4.610	31.305	.0102
49	2,401	117,649	7.000	3.659	22.136	.0204	99	9,801	970,299	9.950	4.626	31.464	.0101
50	2,500	125,000	7.071	3.684	22.361	.0200	100	10,000	1,000,000	10.000	4.642	31.623	.0100

TABLES

Table 2 Values of Trigonometric Functions

Degrees	Radians	Sin	Cos	Tan	Cot	Sec	Csc		
0° 00'	.0000	.0000	1.0000	.0000	——	1.000	——	1.5708	**90° 00'**
10	029	029	000	029	343.8	000	343.8	679	50
20	058	058	000	058	171.9	000	171.9	650	40
30	.0087	.0087	1.0000	.0087	114.6	1.000	114.6	1.5621	30
40	116	116	.9999	116	85.94	000	85.95	592	20
50	145	145	999	145	68.75	000	68.76	563	10
1° 00'	.0175	.0175	.9998	.0175	57.29	1.000	57.30	1.5533	**89° 00'**
10	204	204	998	204	49.10	000	49.11	504	50
20	233	233	997	233	42.96	000	42.98	475	40
30	.0262	.0262	.9997	.0262	38.19	1.000	38.20	1.5446	30
40	291	291	996	291	34.37	000	34.38	417	20
50	320	320	995	320	31.24	001	31.26	388	10
2° 00'	.0349	.0349	.9994	.0349	28.64	1.001	28.65	1.5359	**88° 00'**
10	378	378	993	378	26.43	001	26.45	330	50
20	407	407	992	407	24.54	001	24.56	301	40
30	.0436	.0436	.9990	.0437	22.90	1.001	22.93	1.5272	30
40	465	465	989	466	21.47	001	21.49	243	20
50	495	494	988	495	20.21	001	20.23	213	10
3° 00'	.0524	.0523	.9986	.0524	19.08	1.001	19.11	1.5184	**87° 00'**
10	553	552	985	553	18.07	002	18.10	155	50
20	582	581	983	582	17.17	002	17.20	126	40
30	.0611	.0610	.9981	.0612	16.35	1.002	16.38	1.5097	30
40	640	640	980	641	15.60	002	15.64	068	20
50	669	669	978	670	14.92	002	14.96	039	10
4° 00'	.0698	.0698	.9976	.0699	14.30	1.002	14.34	1.5010	**86° 00'**
10	727	727	974	729	13.73	003	13.76	981	50
20	756	756	971	758	13.20	003	13.23	952	40
30	.0785	.0785	.9969	.0787	12.71	1.003	12.75	1.4923	30
40	814	814	967	816	12.25	003	12.29	893	20
50	844	843	964	846	11.83	004	11.87	864	10
5° 00'	.0873	.0872	.9962	.0875	11.43	1.004	11.47	1.4835	**85° 00'**
10	902	901	959	904	11.06	004	11.10	806	50
20	931	929	957	934	10.71	004	10.76	777	40
30	.0960	.0958	.9954	.0963	10.39	1.005	10.43	1.4748	30
40	989	987	951	992	10.08	005	10.13	719	20
50	.1018	.1016	948	.1022	9.788	005	9.839	690	10
6° 00'	.1047	.1045	.9945	.1051	9.514	1.006	9.567	1.4661	**84° 00'**
10	076	074	942	080	9.255	006	9.309	632	50
20	105	103	939	110	9.010	006	9.065	603	40
30	.1134	.1132	.9936	.1139	8.777	1.006	8.834	1.4573	30
40	164	161	932	169	8.556	007	8.614	544	20
50	193	190	929	198	8.345	007	8.405	515	10
7° 00'	.1222	.1219	.9925	.1228	8.144	1.008	8.206	1.4486	**83° 00'**
10	251	248	922	257	7.953	008	8.016	457	50
20	280	276	918	287	7.770	008	7.834	428	40
30	.1309	.1305	.9914	.1317	7.596	1.009	7.661	1.4399	30
40	338	334	911	346	7.429	009	7.496	370	20
50	367	363	907	376	7.269	009	7.337	341	10
8° 00'	.1396	.1392	.9903	.1405	7.115	1.010	7.185	1.4312	**82° 00'**
10	425	421	899	435	6.968	010	7.040	283	50
20	454	449	894	465	6.827	011	6.900	254	40
30	.1484	.1478	.9890	.1495	6.691	1.011	6.765	1.4224	30
40	513	507	886	524	6.561	012	6.636	195	20
50	542	536	881	554	6.435	012	6.512	166	10
9° 00'	.1571	.1564	.9877	.1584	6.314	1.012	6.392	1.4137	**81° 00'**
		Cos	Sin	Cot	Tan	Csc	Sec	Radians	Degrees

Table 2 (continued)

Degrees	Radians	Sin	Cos	Tan	Cot	Sec	Csc		
9° 00'	.1571	.1564	.9877	.1584	6.314	1.012	6.392	1.4137	**81° 00'**
10	600	593	872	614	197	013	277	108	50
20	629	622	868	644	084	013	166	079	40
30	.1658	.1650	.9863	.1673	5.976	1.014	6.059	1.4050	30
40	687	679	858	703	871	014	5.955	1.4021	20
50	716	708	853	733	769	015	855	992	10
10° 00'	.1745	.1736	.9848	.1763	5.671	1.015	5.759	1.3963	**80° 00'**
10	774	765	843	793	576	016	665	934	50
20	804	794	838	823	485	016	575	904	40
30	.1833	.1822	.9833	.1853	5.396	1.017	5.487	1.3875	30
40	862	851	827	883	309	018	403	846	20
50	891	880	822	914	226	018	320	817	10
11° 00'	.1920	.1908	.9816	.1944	5.145	1.019	5.241	1.3788	**79° 00'**
10	949	937	811	974	066	019	164	759	50
20	978	965	805	.2004	4.989	020	089	730	40
30	.2007	.1994	.9799	.2035	4.915	1.020	5.016	1.3701	30
40	036	.2022	793	065	843	021	4.945	672	20
50	065	051	787	095	773	022	876	643	10
12° 00'	.2094	.2079	.9781	.2126	4.705	1.022	4.810	1.3614	**78° 00'**
10	123	108	775	156	638	023	745	584	50
20	153	136	769	186	574	024	682	555	40
30	.2182	.2164	.9763	.2217	4.511	1.024	4.620	1.3526	30
40	211	193	757	247	449	025	560	497	20
50	240	221	750	278	390	026	502	468	10
13° 00'	.2269	.2250	.9744	.2309	4.331	1.026	4.445	1.3439	**77° 00'**
10	298	278	737	339	275	027	390	410	50
20	327	306	730	370	219	028	336	381	40
30	.2356	.2334	.9724	.2401	4.165	1.028	4.284	1.3352	30
40	385	363	717	432	113	029	232	323	20
50	414	391	710	462	061	030	182	294	10
14° 00'	.2443	.2419	.9703	.2493	4.011	1.031	4.134	1.3265	**76° 00'**
10	473	447	696	524	3.962	·031	086	235	50
20	502	476	689	555	914	032	039	206	40
30	.2531	.2504	.9681	.2586	3.867	1.033	3.994	1.3177	30
40	560	532	674	617	821	034	950	148	20
50	589	560	667	648	776	034	906	119	10
15° 00'	.2618	.2588	.9659	.2679	3.732	1.035	3.864	1.3090	**75° 00'**
10	647	616	652	711	689	036	822	061	50
20	676	644	644	742	647	037	782	032	40
30	.2705	.2672	.9636	.2773	3.606	1.038	3.742	1.3003	30
40	734	700	628	805	566	039	703	974	20
50	763	728	621	836	526	039	665	945	10
16° 00'	.2793	.2756	.9613	.2867	3.487	1.040	3.628	1.2915	**74° 00'**
10	822	784	605	899	450	041	592	886	50
20	851	812	596	931	412	042	556	857	40
30	.2880	.2840	.9588	.2962	3.376	1.043	3.521	1.2828	30
40	909	868	580	994	340	044	487	799	20
50	938	896	572	.3026	305	045	453	770	10
17° 00'	.2967	.2924	.9563	.3057	3.271	1.046	3.420	1.2741	**73° 00'**
10	996	952	555	089	237	047	388	712	50
20	.3025	979	546	121	204	048	356	683	40
30	.3054	.3007	.9537	.3153	3.172	1.049	3.326	1.2654	30
40	083	035	528	185	140	049	295	625	20
50	113	062	520	217	108	050	265	595	10
18° 00'	.3142	.3090	.9511	.3249	3.078	1.051	3.236	1.2566	**72° 00'**
		Cos	Sin	Cot	Tan	Csc	Sec	Radians	Degrees

(Continued)

Table 2 (continued)

Degrees	Radians	Sin	Cos	Tan	Cot	Sec	Csc		
18° 00′	.3142	.3090	.9511	.3249	3.078	1.051	3.236	1.2566	**72° 00′**
10	171	118	502	281	047	052	207	537	50
20	200	145	492	314	018	053	179	508	40
30	.3229	.3173	.9483	.3346	2.989	1.054	3.152	1.2479	30
40	258	201	474	378	960	056	124	450	20
50	287	228	465	411	932	057	098	421	10
19° 00′	.3316	.3256	.9455	.3443	2.904	1.058	3.072	1.2392	**71° 00′**
10	345	283	446	476	877	059	046	363	50
20	374	311	436	508	850	060	021	334	40
30	.3403	.3338	.9426	.3541	2.824	1.061	2.996	1.2305	30
40	432	365	417	574	798	062	971	275	20
50	462	393	407	607	773	063	947	246	10
20° 00′	.3491	.3420	.9397	.3640	2.747	1.064	2.924	1.2217	**70° 00′**
10	520	448	387	673	723	065	901	188	50
20	549	475	377	706	699	066	878	159	40
30	.3578	.3502	.9367	.3739	2.675	1.068	2.855	1.2130	30
40	607	529	356	772	651	069	833	101	20
50	636	557	346	805	628	070	812	072	10
21° 00′	.3665	.3584	.9336	.3839	2.605	1.071	2.790	1.2043	**69° 00′**
10	694	611	325	872	583	072	769	1.2014	50
20	723	638	315	906	560	074	749	985	40
30	.3752	.3665	.9304	.3939	2.539	1.075	2.729	1.1956	30
40	782	692	293	973	517	076	709	926	20
50	811	719	283	.4006	496	077	689	897	10
22° 00′	.3840	.3746	.9272	.4040	2.475	1.079	2.669	1.1868	**68° 00′**
10	869	773	261	074	455	080	650	839	50
20	898	800	250	108	434	081	632	810	40
30	.3927	.3827	.9239	.4142	2.414	1.082	2.613	1.1781	30
40	956	854	228	176	394	084	595	752	20
50	985	881	216	210	375	085	577	723	10
23° 00′	.4014	.3907	.9205	.4245	2.356	1.086	2.559	1.1694	**67° 00′**
10	043	934	194	279	337	088	542	665	50
20	072	961	182	314	318	089	525	636	40
30	.4102	.3987	.9171	.4348	2.300	1.090	2.508	1.1606	30
40	131	.4014	159	383	282	092	491	577	20
50	160	041	147	417	264	093	475	548	10
24° 00′	.4189	.4067	.9135	.4452	2.246	1.095	2.459	1.1519	**66° 00′**
10	218	094	124	487	229	096	443	490	50
20	247	120	112	522	211	097	427	461	40
30	.4276	.4147	.9100	.4557	2.194	1.099	2.411	1.1432	30
40	305	173	088	592	177	100	396	403	20
50	334	200	075	628	161	102	381	374	10
25° 00′	.4363	.4226	.9063	.4663	2.145	1.103	2.366	1.1345	**65° 00′**
10	392	253	051	699	128	105	352	316	50
20	422	279	038	734	112	106	337	286	40
30	.4451	.4305	.9026	.4770	2.097	1.108	2.323	1.1257	30
40	480	331	013	806	081	109	309	228	20
50	509	358	001	841	066	111	295	199	10
26° 00′	.4538	.4384	.8988	.4877	2.050	1.113	2.281	1.1170	**64° 00′**
10	567	410	975	913	035	114	268	141	50
20	596	436	962	950	020	116	254	112	40
30	.4625	.4462	.8949	.4986	2.006	1.117	2.241	1.1083	30
40	654	488	936	.5022	1.991	119	228	054	20
50	683	514	923	059	977	121	215	1.1025	10
27° 00′	.4712	.4540	.8910	.5095	1.963	1.122	2.203	1.0996	**63° 00′**
		Cos	Sin	Cot	Tan	Csc	Sec	Radians	Degrees

Table 2 (continued)

Degrees	Radians	Sin	Cos	Tan	Cot	Sec	Csc		
27° 00′	.4712	.4540	.8910	.5095	1.963	1.122	2.203	1.0996	63° 00′
10	741	566	897	132	949	124	190	966	50
20	771	592	884	169	935	126	178	937	40
30	.4800	.4617	.8870	.5206	1.921	1.127	2.166	1.0908	30
40	829	643	857	243	907	129	154	879	20
50	858	669	843	280	894	131	142	850	10
28° 00′	.4887	.4695	.8829	.5317	1.881	1.133	2.130	1.0821	62° 00′
10	916	720	816	354	868	134	118	792	50
20	945	746	802	392	855	136	107	763	40
30	.4974	.4772	.8788	.5430	1.842	1.138	2.096	1.0734	30
40	.5003	797	774	467	829	140	085	705	20
50	032	823	760	505	816	142	074	676	10
29° 00′	.5061	.4848	.8746	.5543	1.804	1.143	2.063	1.0647	61° 00′
10	091	874	732	581	792	145	052	617	50
20	120	899	718	619	780	147	041	588	40
30	.5149	.4924	.8704	.5658	1.767	1.149	2.031	1.0559	30
40	178	950	689	696	756	151	020	530	20
50	207	975	675	735	744	153	010	501	10
30° 00′	.5236	.5000	.8660	.5774	1.732	1.155	2.000	1.0472	60° 00′
10	265	025	646	812	720	157	1.990	443	50
20	294	050	631	851	709	159	980	414	40
30	.5323	.5075	.8616	.5890	1.698	1.161	1.970	1.0385	30
40	352	100	601	930	686	163	961	356	20
50	381	125	587	969	675	165	951	327	10
31° 00′	.5411	.5150	.8572	.6009	1.664	1.167	1.942	1.0297	59° 00′
10	440	175	557	048	653	169	932	268	50
20	469	200	542	088	643	171	923	239	40
30	.5498	.5225	.8526	.6128	1.632	1.173	1.914	1.0210	30
40	527	250	511	168	621	175	905	181	20
50	556	275	496	208	611	177	896	152	10
32° 00′	.5585	.5299	.8480	.6249	1.600	1.179	1.887	1.0123	58° 00′
10	614	324	465	289	590	181	878	094	50
20	643	348	450	330	580	184	870	065	40
30	.5672	.5373	.8434	.6371	1.570	1.186	1.861	1.0036	30
40	701	398	418	412	560	188	853	1.0007	20
50	730	422	403	453	550	190	844	977	10
33° 00′	.5760	.5446	.8387	.6494	1.540	1.192	1.836	.9948	57° 00′
10	789	471	371	536	530	195	828	919	50
20	818	495	355	577	520	197	820	890	40
30	.5847	.5519	.8339	.6619	1.511	1.199	1.812	.9861	30
40	876	544	323	661	501	202	804	832	20
50	905	568	307	703	1.492	204	796	803	10
34° 00′	.5934	.5592	.8290	.6745	1.483	1.206	1.788	.9774	56° 00′
10	963	616	274	787	473	209	781	745	50
20	992	640	258	830	464	211	773	716	40
30	.6021	.5664	.8241	.6873	1.455	1.213	1.766	.9687	30
40	050	688	225	916	446	216	758	657	20
50	080	712	208	959	437	218	751	628	10
35° 00′	.6109	.5736	.8192	.7002	1.428	1.221	1.743	.9599	55° 00′
10	138	760	175	046	419	223	736	570	50
20	167	783	158	089	411	226	729	541	40
30	.6196	.5807	.8141	.7133	1.402	1.228	1.722	.9512	30
40	225	831	124	177	393	231	715	483	20
50	254	854	107	221	385	233	708	454	10
36° 00′	.6283	.5878	.8090	.7265	1.376	1.236	1.701	.9425	54° 00′
		Cos	Sin	Cot	Tan	Csc	Sec	Radians	Degrees

(Continued)

Table 2 (continued)

Degrees	Radians	Sin	Cos	Tan	Cot	Sec	Csc		
36° 00'	.6283	.5878	.8090	.7265	1.376	1.236	1.701	.9425	**54° 00'**
10	312	901	073	310	368	239	695	396	50
20	341	925	056	355	360	241	688	367	40
30	.6370	.5948	.8039	.7400	1.351	1.244	1.681	.9338	30
40	400	972	021	445	343	247	675	308	20
50	429	995	004	490	335	249	668	279	10
37° 00'	.6458	.6018	.7986	.7536	1.327	1.252	1.662	.9250	**53° 00'**
10	487	041	969	581	319	255	655	221	50
20	516	065	951	627	311	258	649	192	40
30	.6545	.6088	.7934	.7673	1.303	1.260	1.643	.9163	30
40	574	111	916	720	295	263	636	134	20
50	603	134	898	766	288	266	630	105	10
38° 00'	.6632	.6157	.7880	.7813	1.280	1.269	1.624	.9076	**52° 00'**
10	661	180	862	860	272	272	618	047	50
20	690	202	844	907	265	275	612	.9018	40
30	.6720	.6225	.7826	.7954	1.257	1.278	1.606	.8988	30
40	749	248	808	.8002	250	281	601	959	20
50	778	271	790	050	242	284	595	930	10
39° 00'	.6807	.6293	.7771	.8098	1.235	1.287	1.589	.8901	**51° 00'**
10	836	316	753	146	228	290	583	872	50
20	865	338	735	195	220	293	578	843	40
30	.6894	.6361	.7716	.8243	1.213	1.296	1.572	.8814	30
40	923	383	698	292	206	299	567	785	20
50	952	406	679	342	199	302	561	756	10
40° 00'	.6981	.6428	.7660	.8391	1.192	1.305	1.556	.8727	**50° 00'**
10	.7010	450	642	441	185	309	550	698	50
20	039	472	623	491	178	312	545	668	40
30	.7069	.6494	.7604	.8541	1.171	1.315	1.540	.8639	30
40	098	517	585	591	164	318	535	610	20
50	127	539	566	642	157	322	529	581	10
41° 00'	.7156	.6561	.7547	.8693	1.150	1.325	1.524	.8552	**49° 00'**
10	185	583	528	744	144	328	519	523	50
20	214	604	509	796	137	332	514	494	40
30	.7243	.6626	.7490	.8847	1.130	1.335	1.509	.8465	30
40	272	648	470	899	124	339	504	436	20
50	301	670	451	952	117	342	499	407	10
42° 00'	.7330	.6691	.7431	.9004	1.111	1.346	1.494	.8378	**48° 00'**
10	359	713	412	057	104	349	490	348	50
20	389	734	392	110	098	353	485	319	40
30	.7418	.6756	.7373	.9163	1.091	1.356	1.480	.8290	30
40	447	777	353	217	085	360	476	261	20
50	476	799	333	271	079	364	471	232	10
43° 00'	.7505	.6820	.7314	.9325	1.072	1.367	1.466	.8203	**47° 00'**
10	534	841	294	380	066	371	462	174	50
20	563	862	274	435	060	375	457	145	40
30	.7592	.6884	.7254	.9490	1.054	1.379	1.453	.8116	30
40	621	905	234	545	048	382	448	087	20
50	650	926	214	601	042	386	444	058	10
44° 00'	.7679	.6947	.7193	.9657	1.036	1.390	1.440	.8029	**46° 00'**
10	709	967	173	713	030	394	435	999	50
20	738	988	153	770	024	398	431	970	40
30	.7767	.7009	.7133	.9827	1.018	1.402	1.427	.7941	30
40	796	030	112	884	012	406	423	912	20
50	825	050	092	942	006	410	418	883	10
45° 00'	.7854	.7071	.7071	1.000	1.000	1.414	1.414	.7854	**45° 00'**
		Cos	Sin	Cot	Tan	Csc	Sec	Radians	Degrees

Table 3 Common Logarithms

x	0	1	2	3	4	5	6	7	8	9
1.0	.0000	.0043	.0086	.0128	.0170	.0212	.0253	.0294	.0334	.0374
1.1	.0414	.0453	.0492	.0531	.0569	.0607	.0645	.0682	.0719	.0755
1.2	.0792	.0828	.0864	.0899	.0934	.0969	.1004	.1038	.1072	.1106
1.3	.1139	.1173	.1206	.1239	.1271	.1303	.1335	.1367	.1399	.1430
1.4	.1461	.1492	.1523	.1553	.1584	.1614	.1644	.1673	.1703	.1732
1.5	.1761	.1790	.1818	.1847	.1875	.1903	.1931	.1959	.1987	.2014
1.6	.2041	.2068	.2095	.2122	.2148	.2175	.2201	.2227	.2253	.2279
1.7	.2304	.2330	.2355	.2380	.2405	.2430	.2455	.2480	.2504	.2529
1.8	.2553	.2577	.2601	.2625	.2648	.2672	.2695	.2718	.2742	.2765
1.9	.2788	.2810	.2833	.2856	.2878	.2900	.2923	.2945	.2967	.2989
2.0	.3010	.3032	.3054	.3075	.3096	.3118	.3139	.3160	.3181	.3201
2.1	.3222	.3243	.3263	.3284	.3304	.3324	.3345	.3365	.3385	.3404
2.2	.3424	.3444	.3464	.3483	.3502	.3522	.3541	.3560	.3579	.3598
2.3	.3617	.3636	.3655	.3674	.3692	.3711	.3729	.3747	.3766	.3784
2.4	.3802	.3820	.3838	.3856	.3874	.3892	.3909	.3927	.3945	.3962
2.5	.3979	.3997	.4014	.4031	.4048	.4065	.4082	.4099	.4116	.4133
2.6	.4150	.4166	.4183	.4200	.4216	.4232	.4249	.4265	.4281	.4298
2.7	.4314	.4330	.4346	.4362	.4378	.4393	.4409	.4425	.4440	.4456
2.8	.4472	.4487	.4502	.4518	.4533	.4548	.4564	.4579	.4594	.4609
2.9	.4624	.4639	.4654	.4669	.4683	.4698	.4713	.4728	.4742	.4757
3.0	.4771	.4786	.4800	.4814	.4829	.4843	.4857	.4871	.4886	.4900
3.1	.4914	.4928	.4942	.4955	.4969	.4983	.4997	.5011	.5024	.5038
3.2	.5051	.5065	.5079	.5092	.5105	.5119	.5132	.5145	.5159	.5172
3.3	.5185	.5198	.5211	.5224	.5237	.5250	.5263	.5276	.5289	.5307
3.4	.5315	.5328	.5340	.5353	.5366	.5378	.5391	.5403	.5416	.5428
3.5	.5441	.5453	.5465	.5478	.5490	.5502	.5514	.5527	.5539	.5551
3.6	.5563	.5575	.5587	.5599	.5611	.5623	.5635	.5647	.5658	.5670
3.7	.5682	.5694	.5705	.5717	.5729	.5740	.5752	.5763	.5775	.5786
3.8	.5798	.5809	.5821	.5832	.5843	.5855	.5866	.5877	.5888	.5899
3.9	.5911	.5922	.5933	.5944	.5955	.5966	.5977	.5988	.5999	.6010
4.0	.6021	.6031	.6042	.6053	.6064	.6075	.6085	.6096	.6107	.6117
4.1	.6128	.6138	.6149	.6160	.6170	.6180	.6191	.6201	.6212	.6222
4.2	.6232	.6243	.6253	.6263	.6274	.6284	.6294	.6304	.6314	.6325
4.3	.6335	.6345	.6355	.6365	.6375	.6385	.6395	.6405	.6415	.6425
4.4	.6435	.6444	.6454	.6464	.6474	.6484	.6493	.6503	.6513	.6522
4.5	.6532	.6542	.6551	.6561	.6571	.6580	.6590	.6599	.6609	.6618
4.6	.6628	.6637	.6646	.6656	.6665	.6675	.6684	.6693	.6702	.6712
4.7	.6721	.6730	6739	.6749	.6758	.6767	.6776	.6785	.6794	.6803
4.8	.6812	.6821	.6830	.6839	.6848	.6857	.6866	.6875	.6884	.6893
4.9	.6902	.6911	.6920	.6928	.6937	.6946	.6955	.6964	.6972	.6981
5.0	.6990	.6998	.7007	.7016	.7024	.7033	.7042	.7050	.7059	.7067
5.1	.7076	.7084	.7093	.7101	.7110	.7118	.7126	.7135	.7143	.7152
5.2	.7160	.7168	.7177	.7185	.7193	.7202	.7210	.7218	.7226	.7235
5.3	.7243	.7251	.7259	.7267	.7275	.7284	.7292	.7300	.7308	.7316
5.4	.7324	.7332	.7340	.7348	.7356	.7364	.7372	.7380	.7388	.7396
x	0	1	2	3	4	5	6	7	8	9

Table 3 (continued)

x	0	1	2	3	4	5	6	7	8	9
5.5	.7404	.7412	.7419	.7427	.7435	.7443	.7451	.7459	.7466	.7474
5.6	.7482	.7490	.7497	.7505	.7513	.7520	.7528	.7536	.7543	.7551
5.7	.7559	.7566	.7574	.7582	.7589	.7597	.7604	.7612	.7619	.7627
5.8	.7634	.7642	.7649	.7657	.7664	.7672	.7679	.7686	.7694	.7701
5.9	.7709	.7716	.7723	.7731	.7738	.7745	.7752	.7760	.7767	.7774
6.0	.7782	.7789	.7796	.7803	.7810	.7818	.7825	.7832	.7839	.7846
6.1	.7853	.7860	.7868	.7875	.7882	.7889	.7896	.7903	.7910	.7917
6.2	.7924	.7931	.7938	.7945	.7952	.7959	.7966	.7973	.7980	.7987
6.3	.7993	.8000	.8007	.8014	.8021	.8028	.8035	.8041	.8048	.8055
6.4	.8062	.8069	.8075	.8082	.8089	.8096	.8102	.8109	.8116	.8122
6.5	.8129	.8136	.8142	.8149	.8156	.8162	.8169	.8176	.8182	.8189
6.6	.8195	.8202	.8209	.8215	.8222	.8228	.8235	.8241	.8248	.8254
6.7	.8261	.8267	.8274	.8280	.8287	.8293	.8299	.8306	.8312	.8319
6.8	.8325	.8331	.8338	.8344	.8351	.8357	.8363	.8370	.8376	.8382
6.9	.8388	.8395	.8401	.8407	.8414	.8420	.8426	.8432	.8439	.8445
7.0	.8451	.8457	.8463	.8470	.8476	.8482	.8488	.8494	.8500	.8506
7.1	.8513	.8519	.8525	.8531	.8537	.8543	.8549	.8555	.8561	.8567
7.2	.8573	.8579	.8585	.8591	.8597	.8603	.8609	.8615	.8621	.8627
7.3	.8633	.8639	.8645	.8651	.8657	.8663	.8669	.8675	.8681	.8686
7.4	.8692	.8698	.8704	.8710	.8716	.8722	.8727	.8733	.8739	.8745
7.5	.8751	.8756	.8762	.8768	.8774	.8779	.8785	.8791	.8797	.8802
7.6	.8808	.8814	.8820	.8825	.8831	.8837	.8842	.8848	.8854	.8859
7.7	.8865	.8871	.8876	.8882	.8887	.8893	.8899	.8904	.8910	.8915
7.8	.8921	.8927	.8932	.8938	.8943	.8949	.8954	.8960	.8965	.8971
7.9	.8976	.8982	.8987	.8993	.8998	.9004	.9009	.9015	.9020	.9025
8.0	.9031	.9036	.9042	.9047	.9053	.9058	.9063	.9069	.9074	.9079
8.1	.9085	.9090	.9096	.9101	.9106	.9112	.9117	.9122	.9128	.9133
8.2	.9138	.9143	.9149	.9154	.9159	.9165	.9170	.9175	.9180	.9186
8.3	.9191	.9196	.9201	.9206	.9212	.9217	.9222	.9227	.9232	.9238
8.4	.9243	.9248	.9253	.9258	.9263	.9269	.9274	.9279	.9284	.9289
8.5	.9294	.9299	.9304	.9309	.9315	.9320	.9325	.9330	.9335	.9340
8.6	.9345	.9350	.9555	.9360	.9365	.9370	.9375	.9380	.9385	.9390
8.7	.9395	.9400	.9405	.9410	.9415	.9420	.9425	.9430	.9435	.9440
8.8	.9445	.9450	.9455	.9460	.9465	.9469	.9474	.9479	.9484	.9489
8.9	.9494	.9499	.9504	.9509	.9513	.9518	.9523	.9528	.9533	.9538
9.0	.9542	.9547	.9552	.9557	.9562	.9566	.9571	.9576	.9581	.9586
9.1	.9590	.9595	.9600	.9605	.9609	.9614	.9619	.9624	.9628	.9633
9.2	.9638	.9643	.9647	.9652	.9657	.9661	.9666	.9671	.9675	.9680
9.3	.9685	.9689	.9694	.9699	.9703	.9708	.9713	.9717	.9722	.9727
9.4	.9731	.9736	.9741	.9745	.9750	.9754	.9759	.9763	.9768	.9773
9.5	.9777	.9782	.9786	.9791	.9795	.9800	.9805	.9809	.9814	.9818
9.6	.9823	.9827	.9832	.9836	.9841	.9845	.9850	.9854	.9859	.9863
9.7	.9868	.9872	.9877	.9881	.9886	.9890	.9894	.9899	.9903	.9908
9.8	.9912	.9917	.9921	.9926	.9930	.9934	.9939	.9943	.9948	.9952
9.9	.9956	.9961	.9965	.9969	.9974	.9978	.9983	.9987	.9991	.9996
x	0	1	2	3	4	5	6	7	8	9

Table 4 Exponential Functions

x	e^x	e^{-x}	x	e^x	e^{-x}	x	e^x	e^{-x}
0.00	1.0000	1.0000	0.55	1.7333	0.5769	3.6	36.598	0.0273
0.01	1.0101	0.9900	0.60	1.8221	0.5488	3.7	40.447	0.0247
0.02	1.0202	0.9802	0.65	1.9155	0.5220	3.8	44.701	0.0224
0.03	1.0305	0.9704	0.70	2.0138	0.4966	3.9	49.402	0.0202
0.04	1.0408	0.9608	0.75	2.1170	0.4724	4.0	54.598	0.0183
0.05	1.0513	0.9512	0.80	2.2255	0.4493	4.1	60.340	0.0166
0.06	1.0618	0.9418	0.85	2.3396	0.4274	4.2	66.686	0.0150
0.07	1.0725	0.9324	0.90	2.4596	0.4066	4.3	73.700	0.0136
0.08	1.0833	0.9231	0.95	2.5857	0.3867	4.4	81.451	0.0123
0.09	1.0942	0.9139	1.0	2.7183	0.3679	4.5	90.017	0.0111
0.10	1.1052	0.9048	1.1	3.0042	0.3329	4.6	99.484	0.0101
0.11	1.1163	0.8958	1.2	3.3201	0.3012	4.7	109.95	0.0091
0.12	1.1275	0.8869	1.3	3.6693	0.2725	4.8	121.51	0.0082
0.13	1.1388	0.8781	1.4	4.0552	0.2466	4.9	134.29	0.0074
0.14	1.1503	0.8694	1.5	4.4817	0.2231	5	148.41	0.0067
0.15	1.1618	0.8607	1.6	4.9530	0.2019	6	403.43	0.0025
0.16	1.1735	0.8521	1.7	5.4739	0.1827	7	1096.6	0.0009
0.17	1.1853	0.8437	1.8	6.0496	0.1653	8	2981.0	0.0003
0.18	1.1972	0.8353	1.9	6.6859	0.1496	9	8103.1	0.0001
0.19	1.2092	0.8270	2.0	7.3891	0.1353	10	22026	0.00005
0.20	1.2214	0.8187	2.1	8.1662	0.1225	11	59874	0.00002
0.21	1.2337	0.8106	2.2	9.0250	0.1108	12	162,754	0.000006
0.22	1.2461	0.8025	2.3	9.9742	0.1003	13	442,413	0.000002
0.23	1.2586	0.7945	2.4	11.023	0.0907	14	1,202,604	0.0000008
0.24	1.2712	0.7866	2.5	12.182	0.0821	15	3,269,017	0.0000003
0.25	1.2840	0.7788	2.6	13.464	0.0743			
0.26	1.2969	0.7711	2.7	14.880	0.0672			
0.27	1.3100	0.7634	2.8	16.445	0.0608			
0.28	1.3231	0.7558	2.9	18.174	0.0550			
0.29	1.3364	0.7483	3.0	20.086	0.0498			
0.30	1.3499	0.7408	3.1	22.198	0.0450			
0.35	1.4191	0.7047	3.2	24.533	0.0408			
0.40	1.4918	0.6703	3.3	27.113	0.0369			
0.45	1.5683	0.6376	3.4	29.964	0.0334			
0.50	1.6487	0.6065	3.5	33.115	0.0302			

Table 5 Natural Logarithms (ln x)

x	0.00	0.01	0.02	0.03	0.04	0.05	0.06	0.07	0.08	0.09
1.0	0.0000	0.0100	0.0198	0.0296	0.0392	0.0488	0.0583	0.0677	0.0770	0.0862
1.1	0.0953	0.1044	0.1133	0.1222	0.1310	0.1398	0.1484	0.1570	0.1655	0.1740
1.2	0.1823	0.1906	0.1989	0.2070	0.2151	0.2231	0.2311	0.2390	0.2469	0.2546
1.3	0.2624	0.2700	0.2776	0.2852	0.2927	0.3001	0.3075	0.3148	0.3221	0.3293
1.4	0.3365	0.3436	0.3507	0.3577	0.3646	0.3716	0.3784	0.3853	0.3920	0.3988
1.5	0.4055	0.4121	0.4187	0.4253	0.4318	0.4383	0.4447	0.4511	0.4574	0.4637
1.6	0.4700	0.4762	0.4824	0.4886	0.4947	0.5008	0.5068	0.5128	0.5188	0.5247
1.7	0.5306	0.5365	0.5423	0.5481	0.5539	0.5596	0.5653	0.5710	0.5766	0.5822
1.8	0.5878	0.5933	0.5988	0.6043	0.6098	0.6152	0.6206	0.6259	0.6313	0.6366
1.9	0.6419	0.6471	0.6523	0.6575	0.6627	0.6678	0.6729	0.6780	0.6831	0.6881
2.0	0.6931	0.6981	0.7031	0.7080	0.7130	0.7178	0.7227	0.7275	0.7324	0.7372
2.1	0.7419	0.7467	0.7514	0.7561	0.7608	0.7655	0.7701	0.7747	0.7793	0.7839
2.2	0.7885	0.7930	0.7975	0.8020	0.8065	0.8109	0.8154	0.8198	0.8242	0.8286
2.3	0.8329	0.8372	0.8416	0.8459	0.8502	0.8544	0.8587	0.8629	0.8671	0.8713
2.4	0.8755	0.8796	0.8838	0.8879	0.8920	0.8961	0.9002	0.9042	0.9083	0.9123
2.5	0.9163	0.9203	0.9243	0.9282	0.9322	0.9361	0.9400	0.9439	0.9478	0.9517
2.6	0.9555	0.9594	0.9632	0.9670	0.9708	0.9746	0.9783	0.9821	0.9858	0.9895
2.7	0.9933	0.9969	1.0006	1.0043	1.0080	1.0116	1.0152	1.0188	1.0225	1.0260
2.8	1.0296	1.0332	1.0367	1.0403	1.0438	1.0473	1.0508	1.0543	1.0578	1.0613
2.9	1.0647	1.0682	1.0716	1.0750	1.0784	1.0818	1.0852	1.0886	1.0919	1.0953
3.0	1.0986	1.1019	1.1053	1.1086	1.1119	1.1151	1.1184	1.1217	1.1249	1.1282
3.1	1.1314	1.1346	1.1378	1.1410	1.1442	1.1474	1.1506	1.1537	1.1569	1.1600
3.2	1.1632	1.1663	1.1694	1.1725	1.1756	1.1787	1.1817	1.1848	1.1878	1.1909
3.3	1.1939	1.1970	1.2000	1.2030	1.2060	1.2090	1.2119	1.2149	1.2179	1.2208
3.4	1.2238	1.2267	1.2296	1.2326	1.2355	1.2384	1.2413	1.2442	1.2470	1.2499
3.5	1.2528	1.2556	1.2585	1.2613	1.2641	1.2669	1.2698	1.2726	1.2754	1.2782
3.6	1.2809	1.2837	1.2865	1.2892	1.2920	1.2947	1.2975	1.3002	1.3029	1.3056
3.7	1.3083	1.3110	1.3137	1.3164	1.3191	1.3218	1.3244	1.3271	1.3297	1.3324
3.8	1.3350	1.3376	1.3403	1.3429	1.3455	1.3481	1.3507	1.3533	1.3558	1.3584
3.9	1.3610	1.3635	1.3661	1.3686	1.3712	1.3737	1.3762	1.3788	1.3813	1.3838
4.0	1.3863	1.3888	1.3913	1.3938	1.3962	1.3987	1.4012	1.4036	1.4061	1.4085
4.1	1.4110	1.4134	1.4159	1.4183	1.4207	1.4231	1.4255	1.4279	1.4303	1.4327
4.2	1.4351	1.4375	1.4398	1.4422	1.4446	1.4469	1.4493	1.4516	1.4540	1.4563
4.3	1.4586	1.4609	1.4633	1.4656	1.4679	1.4702	1.4725	1.4748	1.4770	1.4793
4.4	1.4816	1.4839	1.4861	1.4884	1.4907	1.4929	1.4952	1.4974	1.4996	1.5019
4.5	1.5041	1.5063	1.5085	1.5107	1.5129	1.5151	1.5173	1.5195	1.5217	1.5239
4.6	1.5261	1.5282	1.5304	1.5326	1.5347	1.5369	1.5390	1.5412	1.5433	1.5454
4.7	1.5476	1.5497	1.5518	1.5539	1.5560	1.5581	1.5602	1.5623	1.5644	1.5665
4.8	1.5686	1.5707	1.5728	1.5748	1.5769	1.5790	1.5810	1.5831	1.5851	1.5872
4.9	1.5892	1.5913	1.5933	1.5953	1.5974	1.5994	1.6014	1.6034	1.6054	1.6074
5.0	1.6094	1.6114	1.6134	1.6154	1.6174	1.6194	1.6214	1.6233	1.6253	1.6273
5.1	1.6292	1.6312	1.6332	1.6351	1.6371	1.6390	1.6409	1.6429	1.6448	1.6467
5.2	1.6487	1.6506	1.6525	1.6544	1.6563	1.6582	1.6601	1.6620	1.6639	1.6658
5.3	1.6677	1.6696	1.6715	1.6734	1.6752	1.6771	1.6790	1.6808	1.6827	1.6845
5.4	1.6864	1.6882	1.6901	1.6919	1.6938	1.6956	1.6974	1.6993	1.7011	1.7029
5.5	1.7047	1.7066	1.7084	1.7102	1.7120	1.7138	1.7156	1.7174	1.7192	1.7210
5.6	1.7228	1.7246	1.7263	1.7281	1.7299	1.7317	1.7334	1.7352	1.7370	1.7387
5.7	1.7405	1.7422	1.7440	1.7457	1.7475	1.7492	1.7509	1.7527	1.7544	1.7561
5.8	1.7579	1.7596	1.7613	1.7630	1.7647	1.7664	1.7682	1.7699	1.7716	1.7733
5.9	1.7750	1.7766	1.7783	1.7800	1.7817	1.7834	1.7851	1.7867	1.7884	1.7901

Table 5 (continued)

x	0.00	0.01	0.02	0.03	0.04	0.05	0.06	0.07	0.08	0.09
6.0	1.7918	1.7934	1.7951	1.7967	1.7984	1.8001	1.8017	1.8034	1.8050	1.8066
6.1	1.8083	1.8099	1.8116	1.8132	1.8148	1.8165	1.8181	1.8197	1.8213	1.8229
6.2	1.8245	1.8262	1.8278	1.8294	1.8310	1.8326	1.8342	1.8358	1.8374	1.8390
6.3	1.8406	1.8421	1.8437	1.8453	1.8469	1.8485	1.8500	1.8516	1.8532	1.8547
6.4	1.8563	1.8579	1.8594	1.8610	1.8625	1.8641	1.8656	1.8672	1.8687	1.8703
6.5	1.8718	1.8733	1.8749	1.8764	1.8779	1.8795	1.8810	1.8825	1.8840	1.8856
6.6	1.8871	1.8886	1.8901	1.8916	1.8931	1.8946	1.8961	1.8976	1.8991	1.9006
6.7	1.9021	1.9036	1.9051	1.9066	1.9081	1.9095	1.9110	1.9125	1.9140	1.9155
6.8	1.9169	1.9184	1.9199	1.9213	1.9228	1.9242	1.9257	1.9272	1.9286	1.9301
6.9	1.9315	1.9330	1.9344	1.9359	1.9373	1.9387	1.9402	1.9416	1.9430	1.9445
7.0	1.9459	1.9473	1.9488	1.9502	1.9516	1.9530	1.9544	1.9559	1.9573	1.9587
7.1	1.9601	1.9615	1.9629	1.9643	1.9657	1.9671	1.9685	1.9699	1.9713	1.9727
7.2	1.9741	1.9755	1.9769	1.9782	1.9796	1.9810	1.9824	1.9838	1.9851	1.9865
7.3	1.9879	1.9892	1.9906	1.9920	1.9933	1.9947	1.9961	1.9974	1.9988	2.0001
7.4	2.0015	2.0028	2.0042	2.0055	2.0069	2.0082	2.0096	2.0109	2.0122	2.0136
7.5	2.0149	2.0162	2.0176	2.0189	2.0202	2.0215	2.0229	2.0242	2.0255	2.0268
7.6	2.0282	2.0295	2.0308	2.0321	2.0334	2.0347	2.0360	2.0373	2.0386	2.0399
7.7	2.0412	2.0425	2.0438	2.0451	2.0464	2.0477	2.0490	2.0503	2.0516	2.0528
7.8	2.0541	2.0554	2.0567	2.0580	2.0592	2.0605	2.0618	2.0631	2.0643	2.0665
7.9	2.0669	2.0681	2.0694	2.0707	2.0719	2.0732	2.0744	2.0757	2.0769	2.0782
8.0	2.0794	2.0807	2.0819	2.0832	2.0844	2.0857	2.0869	2.0882	2.0894	2.0906
8.1	2.0919	2.0931	2.0943	2.0956	2.0968	2.0980	2.0992	2.1005	2.1017	2.1029
8.2	2.1041	2.1054	2.1066	2.1078	2.1090	2.1102	2.1114	2.1126	2.1133	2.1150
8.3	2.1163	2.1175	2.1187	2.1199	2.1211	2.1223	2.1235	2.1247	2.1258	2.1270
8.4	2.1282	2.1294	2.1306	2.1318	2.1330	2.1342	2.1353	2.1365	2.1377	2.1389
8.5	2.1401	2.1412	2.1424	2.1436	2.1448	2.1459	2.1471	2.1483	2.1494	2.1506
8.6	2.1518	2.1529	2.1541	2.1552	2.1564	2.1576	2.1587	2.1599	2.1610	2.1622
8.7	2.1633	2.1645	2.1656	2.1668	2.1679	2.1691	2.1702	2.1713	2.1725	2.1736
8.8	2.1748	2.1759	2.1770	2.1782	2.1793	2.1804	2.1815	2.1827	2.1838	2.1849
8.9	2.1861	2.1872	2.1883	2.1894	2.1905	2.1917	2.1928	2.1939	2.1950	2.1961
9.0	2.1972	2.1983	2.1994	2.2006	2.2017	2.2028	2.2039	2.2050	2.2061	2.2072
9.1	2.2083	2.2094	2.2105	2.2116	2.2127	2.2138	2.2148	2.2159	2.2170	2.2181
9.2	2.2192	2.2203	2.2214	2.2225	2.2235	2.2246	2.2257	2.2268	2.2279	2.2289
9.3	2.2300	2.2311	2.2322	2.2332	2.2343	2.2354	2.2364	2.2375	2.2386	2.2396
9.4	2.2407	2.2418	2.2428	2.2439	2.2450	2.2460	2.2471	2.2481	2.2492	2.2502
9.5	2.2513	2.2523	2.2534	2.2544	2.2555	2.2565	2.2576	2.2586	2.2597	2.2607
9.6	2.2618	2.2628	2.2638	2.2649	2.2659	2.2670	2.2680	2.2690	2.2701	2.2711
9.7	2.2721	2.2732	2.2742	2.2752	2.2762	2.2773	2.2783	2.2793	2.2803	2.2814
9.8	2.2824	2.2834	2.2844	2.2854	2.2865	2.2875	2.2885	2.2895	2.2905	2.2915
9.9	2.2925	2.2935	2.2946	2.2956	2.2966	2.2976	2.2986	2.2996	2.3006	2.3016

Examples.

$$\ln 96{,}700 = \ln 9.67 + 4\ln 10$$
$$= 2.2690 + 9.2103$$
$$= 11.4793.$$

$$\ln 0.00967 = \ln 9.67 - 3\ln 10$$
$$= 2.2690 - 6.9078$$
$$= -4.6388.$$

ln 10	= 2.3026	7 ln 10	= 16.1181
2 ln 10	= 4.6052	8 ln 10	= 18.4207
3 ln 10	= 6.9078	9 ln 10	= 20.7233
4 ln 10	= 9.2103	10 ln 10	= 23.0259
5 ln 10	= 11.5129	11 ln 10	= 25.3284
6 ln 10	= 13.8155	12 ln 10	= 27.6310

Note: Adapted from *Functional Approach to Precalculus*, 2nd ed., Mustafa A. Munem and James P. Yizze (New York, NY: Worth Publishers, Inc., © 1974), pp. 500–501. Reproduced by permission of the publisher.

ANSWERS

CHAPTER 1

Margin Exercises, Section 1.1

1. $\{(d, 1), (d, 2), (e, 1), (e, 2), (f, 1), (f, 2)\}$
2. $\{(1, 1), (1, 2), (1, 3), (1, 4), (2, 1), (2, 2), (2, 3), (2, 4), (3, 1), (3, 2), (3, 3), (3, 4), (4, 1), (4, 2), (4, 3), (4, 4)\}$
3. $\{(1, 1), (2, 2), (3, 3), (4, 4)\}$
4. $\{(2, 1), (3, 1), (3, 2), (4, 1), (4, 2), (4, 3)\}$
5. Domain $= \{1, 2, 3, 4\}$; range $= \{1, 2, 3, 4\}$
6. Domain $= \{2, 3, 4\}$; range $= \{1, 2, 3\}$
7. Domain $= \{1, 2\}$; range $= \{1, 2, 3\}$

Exercise Set 1.1, p. 4

1. $\{(0, a), (0, b), (0, c), (2, a), (2, b), (2, c)\}$
3. $\{(-1, 0), (-1, 1), (-1, 2), (0, 1), (0, 2), (1, 2)\}$
5. $\{(-1, -1), (-1, 0), (-1, 1), (-1, 2), (0, 0), (0, 1), (0, 2), (1, 1), (1, 2), (2, 2)\}$
7. $\{(-1, -1), (0, 0), (1, 1), (2, 2)\}$
9. Domain $\{0, 1\}$; range $\{0, 1, 2\}$

Margin Exercises, Section 1.2

1. (a), (b), (c); (d) Domain $= \{-5, -4, 3\}$; range $= \{-2, 2, 3\}$

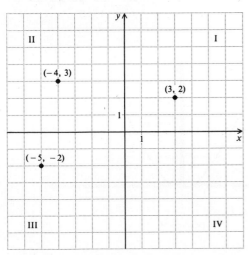

2. Yes   3. No   4. No   5. Yes

6.

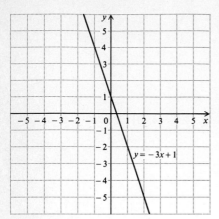

7.

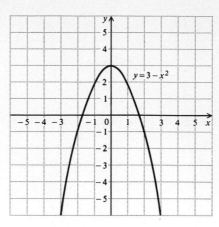

8. The shapes are the same, but this curve opens to the right instead of up.

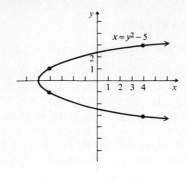

9.

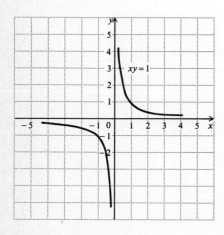

10. The shapes are the same, but this graph opens to the right instead of up.

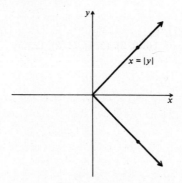

11.

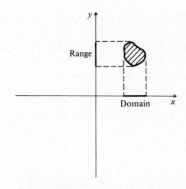

12.

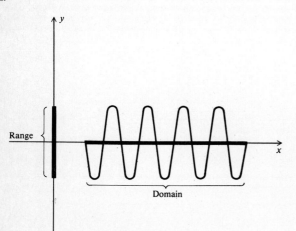

13.

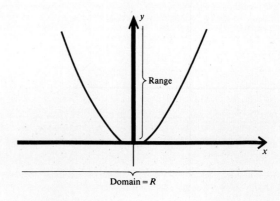

Exercise Set 1.2, pp. 10–11

1. Yes   **3.** No   **5.** No

7.

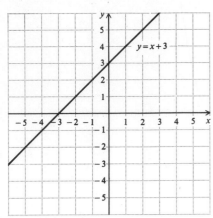

9.

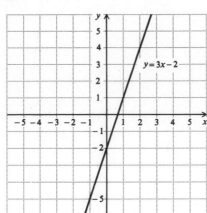

11.

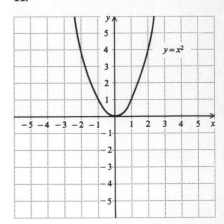

13.

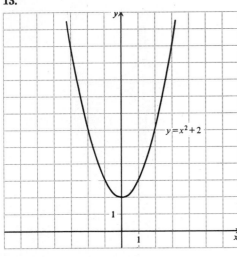

15.

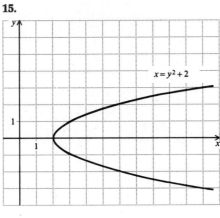

17.

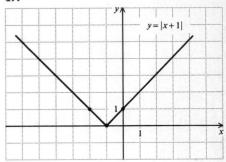

19.

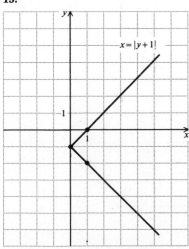

21.

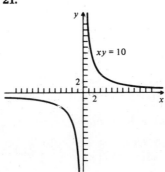

23. Same graphs.

25. Domain $\{x \mid 2 \leq x \leq 6\}$; range $\{y \mid 1 \leq y \leq 5\}$ **27.** Horizontal line through $(0, 2)$
29. Line through $(0, 1)$ and $(-1, 0)$ **31.** Line through $(0, 0)$ and $(1, 2)$ **33.** See Exercise 11.
35.

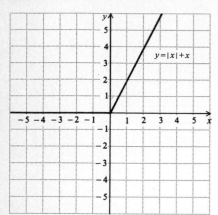

37.

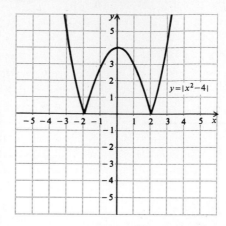

39.

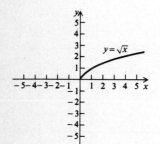

41.

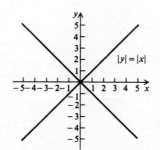

43.

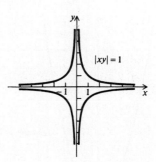

Margin Exercises, Section 1.3

1. (a), (b), (c)   **2.** 1, 4, 3, 4; domain = $\{1, 2, 3, 4\}$; range = $\{1, 3, 4\}$ **3.** 1, 3, 4.5
4. (a) 1; (b) 4; (c) 4; (d) $12a^2 + 1$; (e) $3a^2 + 6a + 4$; (f) $6a + 3h$ **5.** $\{x \mid x \neq -\frac{4}{3} \text{ and } x \neq -2\}$
6. $\{x \mid x \geq -2.5\}$ **7.** All real numbers   **8.** $f \circ g(x) = 2x^2 - 2$; $g \circ f(x) = 2x^2 - 8x + 8$
9. $u \circ v(x) = 18x^2 + 24x + 8$; $v \circ u(x) = 6x^2 + 2$

Exercise Set 1.3, pp. 17–19

1. (b), (c), (d)   **3.** (a) 0; (b) 1; (c) 57; (d) $5t^2 + 4t$; (e) $5t^2 - 6t + 1$; (f) $10a + 5h + 4$

5. (a) 5; (b) -2; (c) -4; (d) $4|y| + 6y$; (e) $2|a + h| + 3a + 3h$; (f) $\dfrac{2|a + h| + 3h - - 2|a|}{h}$

7. (a) 3.14977; (b) 55.73147; (c) 3178.20675; (d) 1116.70323   **9.** (a) $\frac{2}{3}$; (b) $\frac{10}{9}$; (c) 0; (d) not possible   **11.** All real numbers
13. $\{x \mid x \neq 0\}$ **15.** $\{x \mid x \geq -\frac{4}{7}\}$ **17.** $\{x \mid x \neq 2, -2\}$ **19.** $\{x \mid x \neq -\frac{3}{4}, 2\}$

21. $f \circ g(x) = 12x^2 - 12x + 5$; $g \circ f(x) = 6x^2 + 3$ **23.** $f \circ g(x) = \dfrac{16}{x^2} - 1$; $g \circ f(x) = \dfrac{2}{4x^2 - 1}$

25. $f \circ g(x) = x^4 - 2x^2 + 2$; $g \circ f(x) = x^4 + 2x^2$ **27.** 0, -3, 3, 2   **29.** $\dfrac{-1}{x(x + h)}$ **31.** $\dfrac{1}{\sqrt{x + h} + \sqrt{x}}$

33. $\{x \mid x \neq 2, -1 \text{ and } x \geq -3\}$ **35.** All real numbers
37. Domain of $f \circ g$ is $\{x \mid x \neq 0\}$; domain of $g \circ f$ is $\{x \mid x \neq \frac{1}{2}, -\frac{1}{2}\}$.

Margin Exercises, Section 1.4

1. (a) (−3, 2); (b) (4, −5)

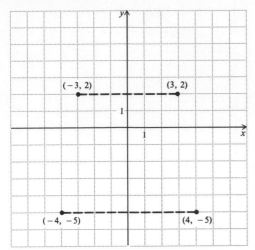

2. (a) (4, −3); (b) (3, 5)

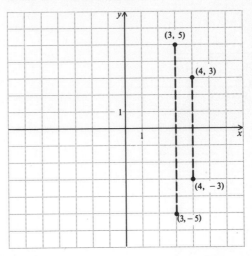

3. x-axis: no; y-axis: yes   **4.** x-axis: yes; y-axis: no   **5.** x-axis: yes; y-axis: yes
6. x-axis: yes; y-axis: yes   **7.** a-axis: no; b-axis: no   **8.** p-axis: no; q-axis; no
9. (a) (−3, −2); (b) (4, −3);
(c) (5, 7)   **10.** Yes   **11.** Yes   **12.** Yes   **13.** Yes   **14.** Yes   **15.** No

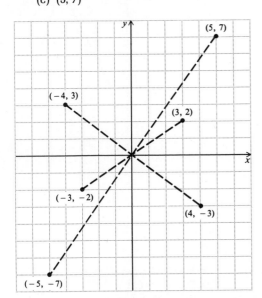

Exercise Set 1.4, p. 25

1. x-axis, no; y-axis, yes; origin, no   **3.** x-axis, no; y-axis, yes; origin, no   **5.** All yes   **7.** All yes   **9.** All no
11. All no   **13.** Yes   **15.** Yes   **17.** Yes   **19.** Yes   **21.** No   **23.** No

25.

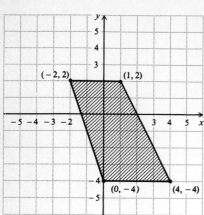

27.

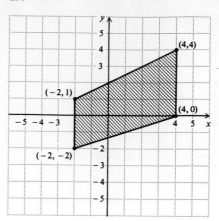

Margin Exercises, Section 1.5

1. (a) Same shape, but the second one is moved up 2 units; (b) same shape as $y = x^2$, but moved down 2 units.

2. Same shape as $y = x^2$, but moved down 3 units.

3.

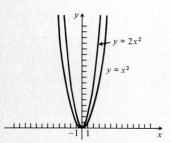

4.

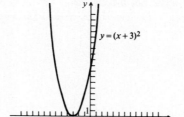

$y = (x + 3)^2$

5.

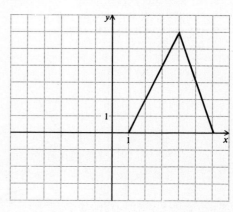

6.

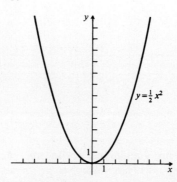

$y = 2x^2$

$y = x^2$

7.

$y = \frac{1}{2} x^2$

8.

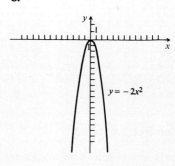

$y = -2x^2$

9.

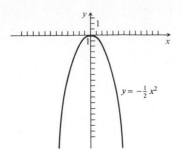

$y = -\frac{1}{2}x^2$

10.

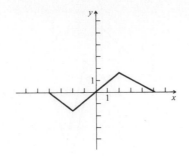

11.

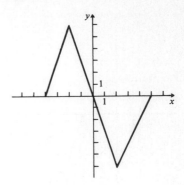

12.

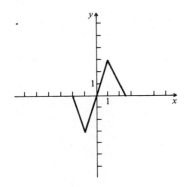

13.

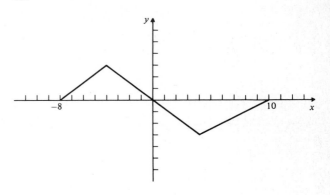

Exercise Set 1.5, pp. 32–34

1. and 3.

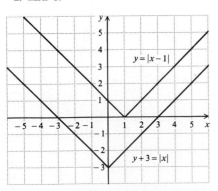

$y = |x-1|$

$y + 3 = |x|$

5. and 7.

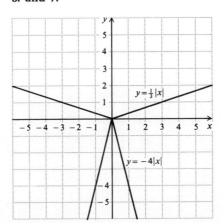

$y = \frac{1}{3}|x|$

$y = -4|x|$

9. and 11.

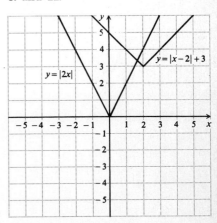

$y = |x-2| + 3$

$y = |2x|$

13.

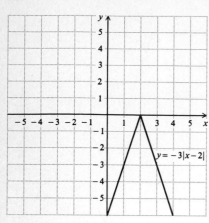

15. and 17.

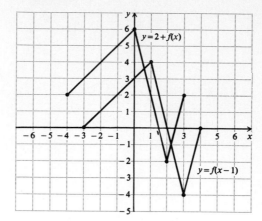

19.

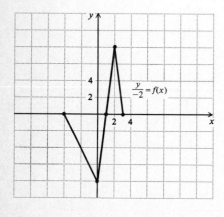

21.

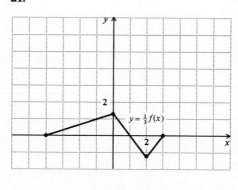

23.

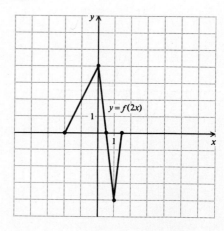

25.

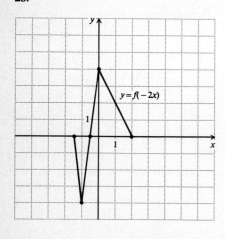

27.

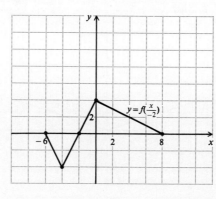

29.

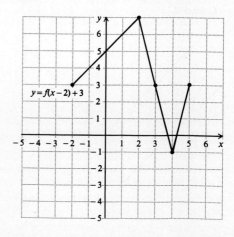

31.

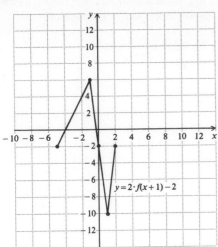

33.

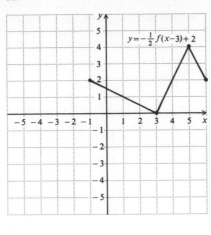

35.

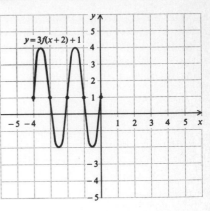

37. The graph is translated 1.8 units to the left, stretched vertically by a factor of $\sqrt{2}$, and reflected across the x-axis.

CHAPTER 2

Margin Exercises, Section 1.6

1. (a) Yes; (b) yes; (c) no; (d) yes   **2.** (a) No; (b) yes; (c) yes; (d) yes   **3.** (a) Odd; (b) odd; (c) even; (d) odd
4. (a) Neither; (b) even; (c) odd; (d) neither; (e) neither   **5.** $f(x) = f(x + 4)$, $f(x) = f(x + 6)$ **6.** (a) Yes; (b) 3
7. $t(x) = t(x + 1)$; t is periodic. It does not have a period, because there is no *smallest* p for which $t(x + p) = t(x)$.
11. (a) $[4, 5\frac{1}{2}]$; (b) $(-3, 0]$; (c) $[-\frac{1}{2}, \frac{1}{2})$; (d) $(-\pi, \pi)$ **12.** (a) No; (b) yes; (c) no; (d) yes; (e) no
13. Where $x = -3, 0, 2$ **14.** (a) Increasing; (b) increasing; (c) decreasing; (d) neither
15. Increasing: $[-3, 0]$; decreasing: $[0, 3]$; there are many answers.
16.

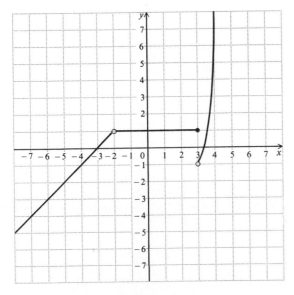

Exercise Set 1.6, pp. 40–45

1. (a) Even; (b) even; (c) odd; (d) neither **3.** Neither **5.** Even **7.** Neither **9.** Even **11.** Odd **13.** Neither
15. Odd **17.** Even and odd **19.** Even **21.** (a) No; (b) yes; (c) yes; (d) no **23.** 4
25. (a) $(-2, 2)$; (b) $(-5, -1]$; (c) $[c, d]$; (d) $[-5, 1)$ **27.** (a) $(-2, 4)$; (b) $(-\frac{1}{4}, \frac{1}{4}]$; (c) $[7, 10\pi)$; (d) $[-9, -6]$
29. (a) Yes; (b) yes; (c) no; (d) yes; (e) yes **31.** Where $x = -3$ and $x = 2$
33. (a) Increasing; (b) neither; (c) decreasing; (d) neither
35. **37.**

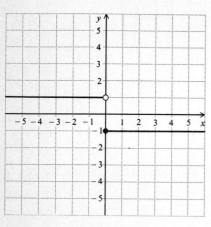 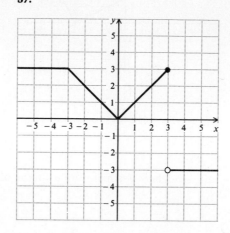

39.

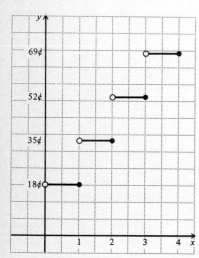

41. Increasing: $[0, 1]$; decreasing: $[-1, 0]$; many possible answers
43. Increasing: (a), (e); decreasing: (b); neither: (c), (d), (f) **45.** (a) $[2, 3]$; (b) $(0, 9]$; (c) $(-6, 1)$

Margin Exercises, Section 1.7

1. (a) (1, 4) (2, 4) (3, 4) (4, 4)   (b) {(4, 1), (4, 2), (3, 2), (2, 3), (2, 4), (1, 4)}
    (1, 3) (2, 3) (3, 3) (4, 3)   (c) (1, 4) (2, 4) (3, 4) (4, 4)
    (1, 2) (2, 2) (3, 2) (4, 2)   (1, 3) (2, 3) (3, 3) (4, 3)
    (1, 1) (2, 1) (3, 1) (4, 1)   (1, 2) (2, 2) (3, 2) (4, 2)
                                        (1, 1) (2, 1) (3, 1) (4, 1)

2. (a) $x = 3y + 2$; (b) $x = y$; (c) $y^2 + 3x^2 = 4$; (d) $x = 5y^2 + 2$; (e) $x^2 = 4y - 5$; (f) $yx = 5$
3. (d) They are reflections across the line $y = x$.

4. (a) 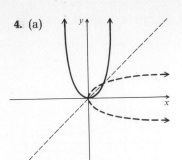   (b)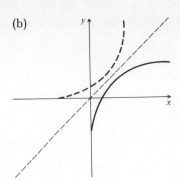

(c)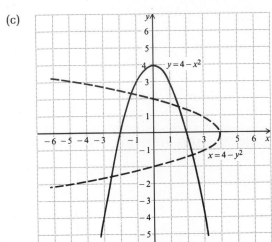

5. Yes   **6.** Yes   **7.** Yes   **8.** No   **9.** Yes   **10.** Yes   **11.** No   **12.** No   **13.** (a), (d)

14. $y = x^2 - 1$ is a function; $x = y^2 - 1$ is not a function.   **15.** $g^{-1}(x) = \dfrac{x + 2}{3}$ **16.** $f^{-1}(x) = \sqrt{x + 1}$ **17.** 5, 5

18. a; a

Exercise Set 1.7, pp. 51–52

1. $x = 4y - 5$ **3.** $y^2 - 3x^2 = 3$ **5.** $x = 3y^2 + 2$ **7.** $yx = 7$

9.   **11.**

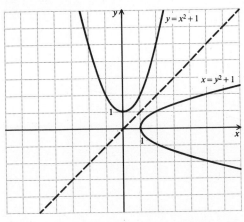   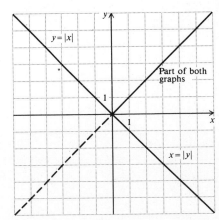

13. No   **15.** Yes   **17.** Yes   **19.** Yes   **21.** No   **23.** Yes   **25.** (a), (c)   **27.** $f^{-1}(x) = \dfrac{x - 5}{2}$ **29.** $f^{-1}(x) = x^2 - 1$

31. 3; -125 **33.** 12,053; $-17,243$ **35.** 1.8   **37.** x-axis: no; y-axis: yes; origin: (no); $y = x$: (no)

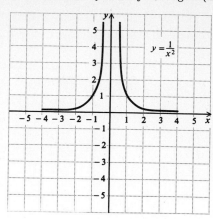

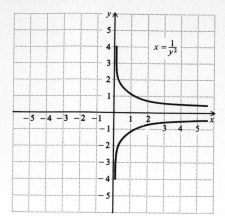

Chapter 1 Review, pp. 52–56

1. $\{(1, 1), (1, 3), (1, 5), (1, 7), (3, 1), (3, 3), (3, 5), (3, 7), (5, 1), (5, 3), (5, 5), (5, 7), (7, 1), (7, 3), (7, 5), (7, 7)\}$

2. Domain $= \{3, 5, 7\}$; range $= \{1, 3, 5, 7\}$

3.

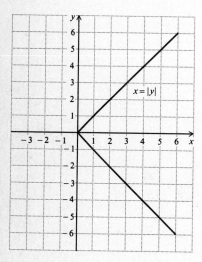

4.

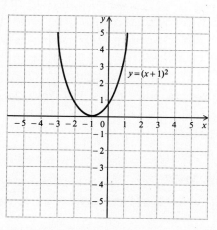

5.

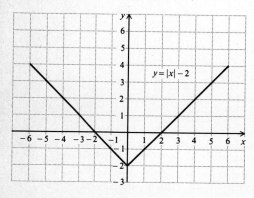

6.

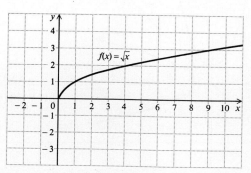

7.

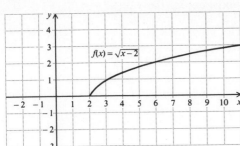

8.

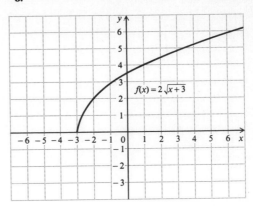

9.

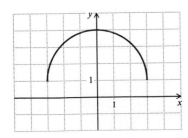

10.

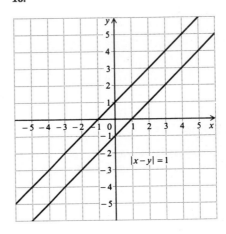

11. (b), (d), (f)   **12.** (a), (b), (d), (g)   **13.** (b), (c), (d), (h)   **14.** (b), (e)   **15.** $x = 3y^2 + 2y - 1$ **16.** $x = \sqrt{y + 2}$
17. (b), (c)   **18.** (d)   **19.** -3 **20.** 9   **21.** $a^2 + 2ah + h^2 - a - h - 3$ **22.** 0   **23.** 4   **24.** $2\sqrt{a + 1}$
25. $g^{-1}(x) = (2x - 4)^2$ **26.** $f^{-1}(x) = \sqrt{x - 2}$ **27.** $\{x \mid x \le \frac{7}{3}\}$ **28.** $\{x \mid x \ne 1, 5\}$ **29.** $\{x \mid x \ne 0, 3, -3\}$

30. $\{x \mid x < 0\}$ **31.** $f \circ g(x) = \dfrac{4}{(3 - 2x)^2}$, $g \circ f(x) = 3 - \dfrac{8}{x^2}$ **32.** $f \circ g(x) = 12x^2 - 4x - 1$, $g \circ f(x) = 6x^2 + 8x - 1$

33. a **34.** t
35. (a) $y = 1 + f(x)$ (b) $y = \frac{1}{2}f(x)$ (c) $y = f(x + 1)$

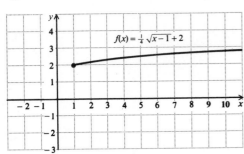
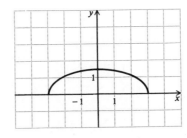
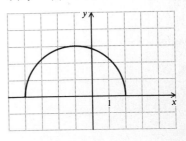

36. (a), (b), (c)   **37.** (e), (f)   **38.** (d)   **39.** (a), (c), (d)   **40.** 2   **41.** (a) Yes; (b) no   **42.** $[-\pi, 2\pi]$ **43.** $(0, 1]$
44. (c), (d)   **45.** (a)   **46.** (b)

47.

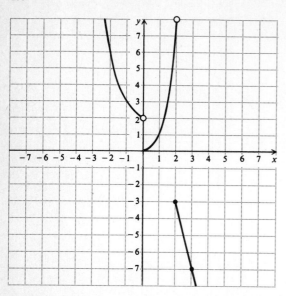

48. Graph $y = f(x)$. Then reflect that portion that lies below the x-axis, across the x-axis.

CHAPTER 2

Margin Exercises, Section 2.1

1. 43°   **2.** 7.5   **3.** 20 m   **4.** (a) $\sin\theta = \frac{12}{13}$, $\cos\theta = \frac{5}{13}$, $\tan\theta = \frac{12}{5}$, $\cot\theta = \frac{5}{12}$, $\sec\theta = \frac{13}{5}$, $\csc\theta = \frac{13}{12}$;
(b) $\sin\phi = \frac{5}{13}$, $\cos\phi = \frac{12}{13}$, $\tan\phi = \frac{5}{12}$, $\cot\phi = \frac{12}{5}$, $\sec\phi = \frac{13}{12}$, $\csc\phi = \frac{13}{5}$ **5.** 75.84 m   **6.** 25.5°

Exercise Set 2.1, pp. 62–64

1. 52°   **3.** 17.5   **5.** $\sin\theta = \frac{3}{5}$, $\cos\theta = \frac{4}{5}$, $\tan\theta = \frac{3}{4}$, $\cot\theta = \frac{4}{3}$, $\sec\theta = \frac{5}{4}$, $\csc\theta = \frac{5}{3}$
7. $\sin\theta = 0.8788$, $\cos\theta = 0.4771$, $\tan\theta = 1.8419$, $\cot\theta = 0.5429$, $\sec\theta = 2.0958$, $\csc\theta = 1.1379$
9. (a) 53.5°; (b) 16.8; (c) 22.7   **11.** (a) 52.5°; (b) 36.5 ft; (c) 46 ft
13. 3.3287   **15.** (a) $\sin 45° = \dfrac{1}{\sqrt{2}}$; (b) $\cos 45° = \dfrac{1}{\sqrt{2}}$; (c) $\tan 45° = 1$

Margin Exercises, Section 2.2

1. (a) (3, 4); (b) (−3, −4); (c) (−3, 4); (d) yes, by symmetry   **2.** (a) (−5, 2); (b) (5, −2); (c) (5, 2); (d) yes, by symmetry
3. (a) $\left(\dfrac{\sqrt{2}}{2}, -\dfrac{\sqrt{2}}{2}\right)$; (b) $\left(-\dfrac{\sqrt{2}}{2}, \dfrac{\sqrt{2}}{2}\right)$; (c) $\left(-\dfrac{\sqrt{2}}{2}, -\dfrac{\sqrt{2}}{2}\right)$; (d) yes, by symmetry   **4.** (a) $\dfrac{\pi}{2}$; (b) $\frac{3}{4}\pi$; (c) $\frac{3}{2}\pi$
5. (a) $\dfrac{\pi}{4}$; (b) $\dfrac{7\pi}{4}$; (c) $\dfrac{5\pi}{4}$; (d) $\dfrac{3\pi}{4}$ **6.** (a) $\dfrac{\pi}{6}$; (b) $\dfrac{5\pi}{6}$; (c) $\dfrac{4\pi}{3}$; (d) $\dfrac{11\pi}{6}$

7.

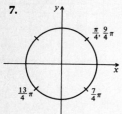

8.

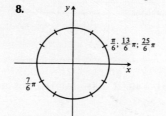

9.

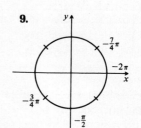

Exercise Set 2.2, pp. 68–69

1. (a) $(-5, -2)$; (b) $(5, 2)$; (c) $(5, -2)$; (d) yes, by symmetry   **3.** $M(0, 1)$; $N\left(-\dfrac{\sqrt{2}}{2}, \dfrac{\sqrt{2}}{2}\right)$; $P(0, -1)$; $Q\left(\dfrac{\sqrt{2}}{2}, -\dfrac{\sqrt{2}}{2}\right)$

5.   **7.**   **9.**

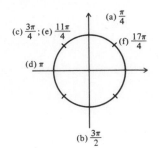   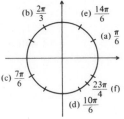   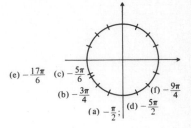

11. M: $\dfrac{2}{3}\pi$, $-\dfrac{4}{3}\pi$; N: $\dfrac{5}{6}\pi$, $-\dfrac{7}{6}\pi$; P: $\dfrac{5}{4}\pi$, $-\dfrac{3}{4}\pi$; Q: $\dfrac{11}{6}\pi$, $-\dfrac{\pi}{6}$ **13.** $\left(\dfrac{\sqrt{3}}{2}, -\dfrac{1}{2}\right)$ **15.** M: $\dfrac{8\pi}{3}$; N: $\dfrac{17\pi}{6}$; P: $\dfrac{13\pi}{4}$; Q: $\dfrac{23\pi}{6}$

17. $\pm\dfrac{\sqrt{3}}{2}$ **19.** ± 0.96649

Margin Exercises, Section 2.3

1. (a) 0; (b) -1; (c) $\dfrac{\sqrt{2}}{2}$; (d) $-\dfrac{\sqrt{2}}{2}$; (e) $\dfrac{\sqrt{3}}{2}$; (f) $-\dfrac{1}{2}$ **2.** Yes, 2π **3.** Odd   **4.** Yes   **5.** The set of all real numbers

6. The set of real numbers from -1 to 1, inclusive

7. (a) $(-a, -b)$; (b) $(-a, -b)$ **8.** (a) -1; (b) 0; (c) $\dfrac{\sqrt{2}}{2}$; (d) $\dfrac{\sqrt{2}}{2}$; (e) $-\dfrac{\sqrt{3}}{2}$; (f) $\dfrac{1}{2}$

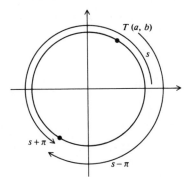

9. Yes, 2π **10.** Even   **11.** Yes   **12.** The set of all real numbers
13. The set of all real numbers from -1 to 1, inclusive

Exercise Set 2.3, pp. 75–77

1.

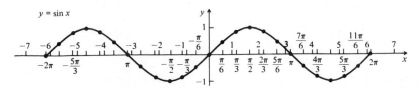

3. (a) See Exercise 1;
(b)

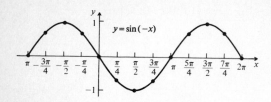

(c) same as (b); (d) same as (b)

5. (a) See Exercise 1;
(b)

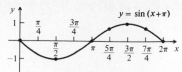

(c) same as (b); (d) the same

7. (a) See Exercise 2;
(b)

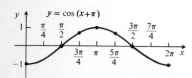

(c) same as (b); (d) same as (b)

9. $\cos x$ **11.** $-\sin x$ **13.** $-\cos x$ **15.** $\cos x$

17. $-\cos x$

19. (a) $\dfrac{\sqrt{2}}{2}$; (b) 0; (c) $\dfrac{1}{2}$; (d) $-\dfrac{\sqrt{2}}{2}$; (e) -1; (f) $\dfrac{1}{2}$; (g) $-\dfrac{\sqrt{2}}{2}$; (h) 0; (j) $\dfrac{\sqrt{2}}{2}$; (k) 1; (m) $-\dfrac{\sqrt{3}}{2}$

21. (a) $\dfrac{\sqrt{2}}{2}$; (b) -1; (c) $-\dfrac{\sqrt{2}}{2}$; (d) $\dfrac{\sqrt{3}}{2}$; (e) 0; (f) $-\dfrac{\sqrt{3}}{2}$; (g) $\dfrac{\sqrt{2}}{2}$; (h) -1; (j) $-\dfrac{\sqrt{2}}{2}$; (k) 0; (m) $\dfrac{1}{2}$

23. (a) $\dfrac{\pi}{2} + 2k\pi$, k any integer; (b) $\dfrac{3\pi}{2} + 2k\pi$, k any integer   **25.** $x = k\pi$, k any integer

27. $f \circ g(x) = \cos^2 x + 2 \cos x$, $g \circ f(x) = \cos(x^2 + 2x)$
29. (a) 0.8660; (b) 0.7071

Margin Exercises, Section 2.4

1. 0   **2.** -1 **3.** Does not exist   **4.** $\dfrac{3\pi}{2}$, $-\dfrac{\pi}{2}$, $-\dfrac{3\pi}{2}$, etc.   **5.** π

6. The set of all real numbers except $\dfrac{\pi}{2} + k\pi$, k any integer   **7.** The set of all real numbers   **8.** Odd

9. π, $-\pi$, 2π, -2π, etc.   **10.** π **11.** The set of all real numbers except $k\pi$, k any integer
12. The set of all real numbers   **13.** Positive: I, III; negative: II, IV   **14.** Odd   **15.** $\sqrt{2}$ **16.** -1
17. $\dfrac{\pi}{2}$, $-\dfrac{\pi}{2}$, $\dfrac{3\pi}{2}$, $-\dfrac{3\pi}{2}$, etc.   **18.** 2π **19.** The set of all real numbers except $k\pi$, k any integer
20. The set of real numbers 1 and greater, together with the set of real numbers -1 and less
21. Positive: I, II; negative: III, IV   **22.** Odd   **23.** Negative: sine, cosine, secant, cosecant. Others positive.
24. $\tan \dfrac{\pi}{11} = 0.29362$, $\cot \dfrac{\pi}{11} = 3.40571$, $\sec \dfrac{\pi}{11} = 1.04222$, $\csc \dfrac{\pi}{11} = 3.54950$

25. $\cot(-x) \equiv -\cot x$; $\cot(-x) \equiv \dfrac{1}{\tan(-x)} \equiv \dfrac{\cos(-x)}{\sin(-x)} \equiv \dfrac{\cos x}{-\sin x} \equiv -\cot x$; $\therefore \cot(-x) \equiv -\cot x$

26. $\tan(x - \pi) \equiv \dfrac{\sin(x - \pi)}{\cos(x - \pi)} \equiv \dfrac{-\sin x}{-\cos x} \equiv \dfrac{\sin x}{\cos x} \equiv \tan x$

27. $\csc(\pi - x) \equiv \dfrac{1}{\sin(\pi - x)} \equiv \dfrac{1}{\sin x} \equiv \csc x$

Exercise Set 2.4, pp. 83–84

1. 1 **3.** $\dfrac{\sqrt{3}}{3}$ **5.** $\sqrt{2}$ **7.** Does not exist **9.** $-\sqrt{3}$

11.

	$\dfrac{\pi}{16}$	$\dfrac{\pi}{8}$	$\dfrac{\pi}{6}$	$\dfrac{\pi}{4}$	$\dfrac{3\pi}{8}$	$\dfrac{7\pi}{16}$
sin	0.19509	0.38268	0.50000	0.70711	0.92388	0.98079
cos	0.98079	0.92388	0.86603	0.70711	0.38268	0.19509
tan	0.19891	0.41421	0.57735	1.00000	2.41424	5.02737
cot	5.02737	2.41424	1.73206	1.00000	0.41421	0.19891
sec	1.01959	1.08239	1.15469	1.41421	2.61315	5.12584
csc	5.12584	2.61315	2.00000	1.41421	1.08239	1.01959

13.

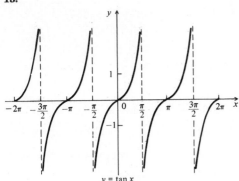

$y = \tan x$

15.

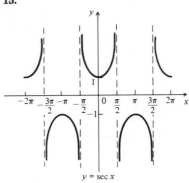

$y = \sec x$

17. cos, sec **19.** sin, cos, sec, cosec **21.** Positive: I, III; negative: II, IV **23.** Positive: I, IV; negative: II, III

25. $\sec(-x) \equiv \dfrac{1}{\cos(-x)} \equiv \dfrac{1}{\cos x} \equiv \sec x$ **27.** $\cot(x+\pi) \equiv \dfrac{\cos(x+\pi)}{\sin(x+\pi)} \equiv \dfrac{-\cos x}{-\sin x} \equiv \dfrac{\cos x}{\sin x} \equiv \cot x$

29. $\sec(x+\pi) \equiv \dfrac{1}{\cos(x+\pi)} \equiv -\dfrac{1}{\cos x} \equiv -\sec x$

31. The graph of $\sec(x-\pi)$ is like that of $\sec x$, moved π units to the right. The graph of $-\sec x$ is that of $\sec x$ reflected across the x-axis. The graphs are identical.

33. If the graph of $\tan x$ were reflected across the y-axis and then translated to the right a distance of $\dfrac{\pi}{2}$, the graph of $\cot x$ would be obtained. There are other ways to describe the relation.

35. The sine and tangent functions; the cosine and cotangent functions.

37.

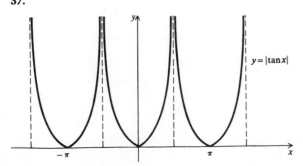

$y = |\tan x|$

Margin Exercises, Section 2.5

1. (a) $\dfrac{\sin^2 s}{\cos^2 s} + \dfrac{\cos^2 s}{\cos^2 s} \equiv \dfrac{1}{\cos^2 s}$, $\tan^2 s + 1 \equiv \sec^2 s$; (b) when $\cos^2 s = 0$ (or $\cos s = 0$); (c) yes; yes

2. $\sin^2 x + \cos^2 x \equiv 1$, $\sin^2 x \equiv 1 - \cos^2 x$ **3.** $\sin^2 x \equiv 1 - \cos^2 x$, $|\sin x| \equiv \sqrt{1 - \cos^2 x}$, or $\sin x \equiv \pm\sqrt{1 - \cos^2 x}$

4. (a) and (b)   (c)

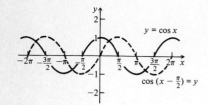

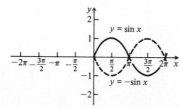

(d) the graphs of b and c are the same; (e) $\cos\left(x - \dfrac{\pi}{2}\right) \equiv \sin x$

5. (a) and (b)   (c) and (d)

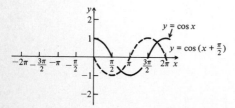

(e) the graphs are the same;

(f) $\cos\left(x + \dfrac{\pi}{2}\right) \equiv -\sin x$

6. Since the cosine function is even, $\cos\left(\dfrac{\pi}{2} - x\right) \equiv \cos\left(x - \dfrac{\pi}{2}\right)$; but $\cos\left(x - \dfrac{\pi}{2}\right) \equiv \sin x$; $\therefore \cos\left(\dfrac{\pi}{2} - x\right) \equiv \sin x$

7. By definition of the cotangent function, $\cot\left(x - \dfrac{\pi}{2}\right) \equiv \dfrac{\cos\left(x - \dfrac{\pi}{2}\right)}{\sin\left(x - \dfrac{\pi}{2}\right)} \equiv \dfrac{\sin x}{-\cos x} \equiv \dfrac{-\sin x}{\cos x} \equiv -\tan x;$

$$\therefore \cot\left(x - \dfrac{\pi}{2}\right) \equiv -\tan x$$

Exercise Set 2.5, pp. 88–89

1. $\cot^2 x \equiv \csc^2 x - 1$, $\csc^2 x - \cot^2 x \equiv 1$ **3.** (a) $\csc x \equiv \pm\sqrt{1 + \cot^2 x}$; (b) $\cot x \equiv \pm\sqrt{\csc^2 x - 1}$

5. $\tan\left(x - \dfrac{\pi}{2}\right) \equiv -\cot x$ **7.** $\sec\left(\dfrac{\pi}{2} - x\right) \equiv \csc x$

9. $\sin\left(x \pm \dfrac{\pi}{2}\right) \equiv \pm\cos x$, $\cos\left(x \pm \dfrac{\pi}{2}\right) \equiv \mp\sin x$, $\tan\left(x \pm \dfrac{\pi}{2}\right) \equiv -\cot x$, $\cot\left(x \pm \dfrac{\pi}{2}\right) \equiv -\tan x$, $\sec\left(x \pm \dfrac{\pi}{2}\right) \equiv \mp\csc x$,

$$\csc\left(x \pm \dfrac{\pi}{2}\right) \equiv \pm\sec x$$

11. $\cos x = -0.9898$, $\tan x = -0.1440$, $\cot x = -6.9460$, $\sec x = -1.0103$, $\csc x = 7.0175$

13. $\sin x = 0.7987$, $\cos x = 0.6018$, $\tan x = 1.3273$, $\sec x = 1.6617$, $\csc x = 1.2520$ **15.** $\begin{vmatrix} \sin x & \cos x \\ -\cos x & \sin x \end{vmatrix} = 1$

Margin Exercises, Section 2.6

1.

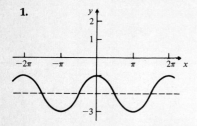

2.

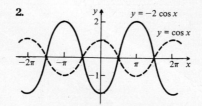

3.

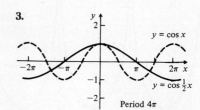

4.

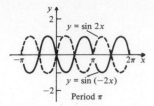

5.

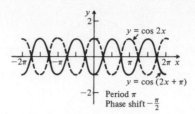

6.

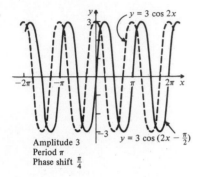

7.

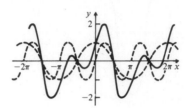

Exercise Set 2.6, pp. 93–94

1.

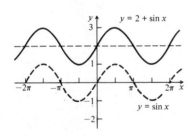

3.

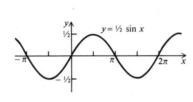

5.

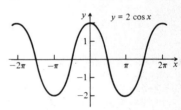

7.

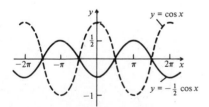

9.

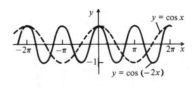

11.

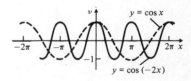

13.

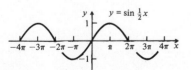

15.

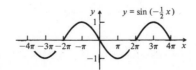

17.

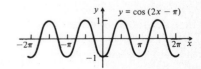

19.

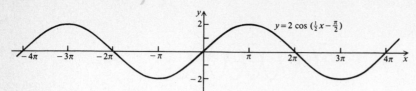

$y = 2 \cos\left(\tfrac{1}{2}x - \tfrac{\pi}{2}\right)$

21.

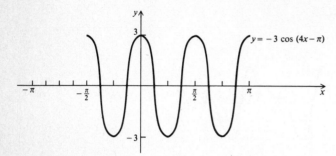

$y = -3 \cos(4x - \pi)$

23.   **25.**   **27.**

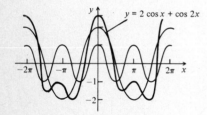   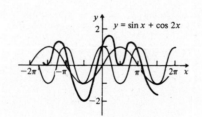   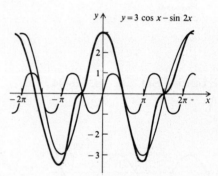

$y = 2 \cos x + \cos 2x$ $y = \sin x + \cos 2x$ $y = 3 \cos x - \sin 2x$

29. Amplitude, 3; period, $\dfrac{2\pi}{3}$; phase shift, $\dfrac{\pi}{6}$ **31.** Amplitude, 5; period, $\dfrac{\pi}{2}$; phase shift, $-\dfrac{\pi}{12}$

33. Amplitude, $\tfrac{1}{2}$; period, 1; phase shift, $-\tfrac{1}{2}$

35.   **37.**

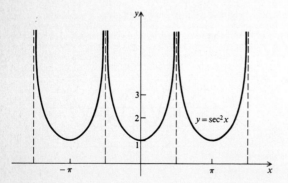   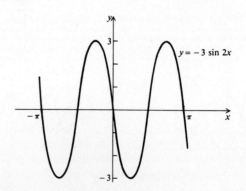

$y = \sec^2 x$ $y = -3 \sin 2x$

Margin Exercises, Section 2.7

1. $\cos x + 1$ **2.** $\sin x$ **3.** $\cos x\left(\dfrac{1}{\sin x} - 1\right)$ or $\cos x(\csc x - 1)$ **4.** $\dfrac{1 + \sin x}{1 - \cot x}$ **5.** $\dfrac{5 \sin x - \cos x}{\sin^2 x - \cos^2 x}$

6. $\dfrac{2}{\cos^2 x(\cot x - 2)}$ **7.** $\tan x \sin x \sqrt{\sin x}$ **8.** $-\dfrac{\cos x}{\sqrt{3 \cos x}}$ **9.** $3, -4$ **10.** $\dfrac{1 \pm \sqrt{97}}{8}$ **11.** $\cos x = \dfrac{-2}{3}, \cos x = \dfrac{1}{2}$

Exercise Set 2.7A, pp. 97–98

1. $\sin^2 x - \cos^2 x$ **3.** $\sin x - \sec x$ **5.** $\sin y + \cos y$ **7.** $\cot x - \tan x$ **9.** $1 - 2 \sin y \cos y$ **11.** $2 \tan x + \sec^2 x$
13. $\sin^2 y + \csc^2 y - 2$ **15.** $\cos^3 x - \sec^3 x$ **17.** $\cot^3 x - \tan^3 x$ **19.** $\sin^2 x$ **21.** $\cos x(\sin x + \cos x)$
23. $(\sin y - \cos y)(\sin y + \cos y)$ **25.** $\sin x(\sec x + 1)$ **27.** $(\sin x + \cos x)(\sin x - \cos x)$ **29.** $3(\cot y + 1)^2$
31. $(\csc^2 x + 5)(\csc x + 1)(\csc x - 1)$ **33.** $(\sin y + 3)(\sin^2 y - 3 \sin y + 9)$ **35.** $(\sin y - \csc y)(\sin^2 y + 1 + \csc^2 y)$
37. $\sin x(\cos y + \tan y)$ **39.** $-\cos x(1 + \cot x)$

Exercise Set 2.7B, pp. 98–99

1. $\tan x$ **3.** $\dfrac{2 \cos^2 x}{9 \sin x}$ **5.** $\cos x - 1$ **7.** $\cos x + 1$ **9.** $\dfrac{2 \tan x + 1}{3 \sin x + 1}$ **11.** $\cos x + 1$ **13.** $\dfrac{\cos x - 1}{1 + \tan x}$ **15.** 1

17. $\cos x - 1$ **19.** $-\tan^2 y$ **21.** $\dfrac{1}{2 \cos x}$ **23.** $\dfrac{3 \tan x}{\cos x - \sin x}$ **25.** $\frac{1}{3} \cot y$ **27.** $\dfrac{1 - 2 \sin y + 2 \cos y}{\sin^2 y - \cos^2 y}$ **29.** -1

31. $\dfrac{5(\sin x - 3)}{3}$

Exercise Set 2.7C, p. 99

1. $\sin x \cos x$ **3.** $\sqrt{\sin y}(\sin y + \cos y)$ **5.** $\sin x + \cos x$ **7.** $1 - \sin y$ **9.** $\sin x(\sqrt{2} + \sqrt{\cos x})$ **11.** $\dfrac{\sqrt{\sin x \cos x}}{\cos x}$

13. $\dfrac{\sqrt{\cos x}}{\cot x}$ **15.** $\dfrac{\sqrt{2} \cot x}{2}$ **17.** $\dfrac{\cos x}{1 - \sin x}$ **19.** $3, -7$ **21.** $\frac{3}{4}, -\frac{1}{2}$ **23.** $-10, 1$ **25.** $2, -3$ **27.** $3 \pm \sqrt{13}$

29. $\dfrac{5 \pm \sqrt{73}}{12}$

Chapter 2 Review, pp. 100–101

1. $(-3, -2)$ **2.** $(-3, 2)$ **3.** $(3, 2)$
4.–7.

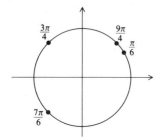

9. Set of all real numbers

8., 12.

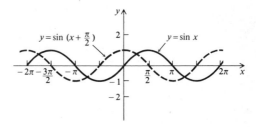

10.

11. 2π **12.** See graph for Exercise 8.

13.

	$\dfrac{\pi}{6}$	$\dfrac{\pi}{4}$	$\dfrac{\pi}{3}$	$\dfrac{\pi}{2}$	$\dfrac{3\pi}{4}$	$\dfrac{5\pi}{4}$
sin x	$\dfrac{1}{2}$	$\dfrac{\sqrt{2}}{2}$	$\dfrac{\sqrt{3}}{2}$	1	$\dfrac{\sqrt{2}}{2}$	$-\dfrac{\sqrt{2}}{2}$
cos x	$\dfrac{\sqrt{3}}{2}$	$\dfrac{\sqrt{2}}{2}$	$\dfrac{1}{2}$	0	$-\dfrac{\sqrt{2}}{2}$	$-\dfrac{\sqrt{2}}{2}$

14.

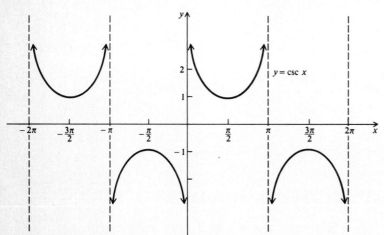

15. 2π **16.** $x \geq 1$ and $x \leq -1$ **17.** I, IV

18. The points for s and $s - \pi$ are symmetric with respect to the origin. $\cot s = \dfrac{x}{y}$ and

$\cot(s - \pi) = \dfrac{-x}{-y} = \dfrac{x}{y}$, so $\cot s = \cot(s - \pi)$.

19. 1   **20.** $\csc^2 x$ **21.** $-\sin x$ **22.** $\sin x$ **23.** $-\cos x$

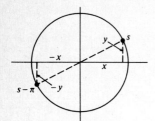

24. $\pm\sqrt{\sec^2 x - 1}$

25.

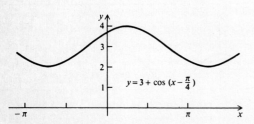

26. $\dfrac{\pi}{4}$ **27.** 2π

28.

29. $\dfrac{1}{\sin x}$ **30.** 1

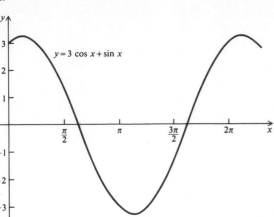

$y = 3 \cos x + \sin x$

31. $\dfrac{\sqrt{\sec x \tan x}}{\sec x}$ **32.** $\dfrac{1 \pm \sqrt{7}}{3}$ **33.** No, $\sin x = \frac{7}{5}$, but sines are never greater than 1.   **34.** All values

CHAPTER 3

Margin Exercises, Section 3.1

1. (a) I; (b) III; (c) IV; (d) III; (e) I; (f) IV; (g) I   **2.** (a) 360°; (b) 180°; (c) 90°; (d) 45°; (e) 60°; (f) 30°
3. (a) $\frac{5}{4}\pi$; (b) $\frac{7}{4}\pi$; (c) -4π **4.** (a) 1.26; (b) 5.23; (c) -5.50 **5.** (a) 240°; (b) 450°; (c) $-144°$
6. (a) III; (b) I; (c) IV; (d) I   **7.** 57.6 cm   **8.** 6 radians   **9.** $\frac{3}{2}$ radians

Exercise Set 3.1, pp. 107–108

1. I   **3.** III   **5.** I   **7.** II   **9.** $\dfrac{\pi}{6}$ **11.** $\dfrac{\pi}{3}$ **13.** $\dfrac{5\pi}{12}$ **15.** 0.2095π **17.** 1.1922π **19.** 2.093   **21.** 5.582   **23.** 3.489
25. 2.0550   **27.** 0.0236   **29.** 57.32°   **31.** 1440°   **33.** 135°   **35.** 74.69°   **37.** 135.6°   **39.** 1.1 radians, 63°
41. 5.233 radians   **43.** 16 m   **45.** $30° = \dfrac{\pi}{6}$, $60° = \dfrac{\pi}{3}$, $135° = \dfrac{3\pi}{4}$, $180° = \pi$, $225° = \dfrac{5\pi}{4}$, $270° = \dfrac{3\pi}{2}$, $315° = \dfrac{7\pi}{4}$
47. (a) 53.33; (b) 170; (c) 25; (d) 142.86   **49.** 111.7 km, 69.81 mi   **51.** $\frac{1}{30}$ radian

Margin Exercises, Section 3.2

1. 377 cm/sec   **2.** r in cm, ω in radians/sec   **3.** km/yr   **4.** 3.6 radians/sec   **5.** 70.4 radians, or 11.21 revolutions

Exercise Set 3.2, pp. 111–112

1. 3150 cm/min   **3.** 52.3 cm/sec   **5.** 1047 mph   **7.** 68.8 radians/sec   **9.** 10 mph   **11.** 75.4 ft   **13.** 377 radians
15. 2.093 radians/sec   **17.** (a) 395.35 rpm/sec, angular acc $= \dfrac{\Delta v}{t}$; (b) 41.4 radians/sec^2

Margin Exercises, Section 3.3

1. $\sin \theta = -\frac{3}{5}$, $\cos \theta = -\frac{4}{5}$, $\tan \theta = \frac{3}{4}$, $\cot \theta = \frac{4}{3}$, $\sec \theta = -\frac{5}{4}$, $\csc \theta = -\frac{5}{3}$

2. $\sin 120° = \dfrac{\sqrt{3}}{2}$, $\cos 120° = -\dfrac{1}{2}$, $\tan 120° = -\sqrt{3}$, $\cot 120° = -\dfrac{1}{\sqrt{3}}$ or $-\dfrac{\sqrt{3}}{3}$, $\sec 120° = -2$, $\csc 120° = +\dfrac{2}{\sqrt{3}}$ or $+\dfrac{2\sqrt{3}}{3}$ **3.** $\sin(-135°) = -\dfrac{\sqrt{2}}{2}$, $\cos(-135°) = -\dfrac{\sqrt{2}}{2}$, $\tan(-135°) = 1$, $\cot(-135°) = 1$, $\sec(-135°) = -\sqrt{2}$, $\csc(-135°) = -\sqrt{2}$ **4.** $\sin 270° = -1$, $\cos 270° = 0$, $\tan 270°$: undefined, $\cot 270° = 0$, $\sec 270°$: undefined, $\csc 270° = -1$ **5.** 15.3 ft **6.** $\sin 324° = -0.6283$, $\cos 324° = 0.8090$, $\tan 324° = -0.7265$, $\cot 324° = -1.3765$, $\sec 324° = 1.2361$, $\csc 324° = -1.5916$

Exercise Set 3.3, pp. 121–122

1. $\sin \theta = \frac{5}{13}$, $\cos \theta = -\frac{12}{13}$, $\tan \theta = -\frac{5}{12}$, $\cot \theta = -\frac{12}{5}$, $\sec \theta = -\frac{13}{12}$, $\csc \theta = \frac{13}{5}$
3. $\sin \theta = -\dfrac{3}{4}$, $\cos \theta = -\dfrac{\sqrt{7}}{4}$, $\tan \theta = \dfrac{3}{\sqrt{7}}$, $\cot \theta = \dfrac{\sqrt{7}}{3}$, $\sec \theta = -\dfrac{4}{\sqrt{7}}$, $\csc \theta = -\dfrac{4}{3}$
5. $\sin 30° = 0.5$, $\cos 30° = 0.866$, $\tan \theta = 0.577$, $\cot \theta = 1.732$, $\sec 30° = 1.155$, $\csc 30° = 2$ **7.** -1 **9.** Does not exist.
11. 1 **13.** -0.707 **15.** 0.5 **17.** 1.732 **19.** 1.414 **21.** 1 **23.** 0.5 **25.** 0.577 **27.** 0.707 **29.** -0.577
31.

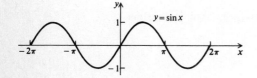

33.

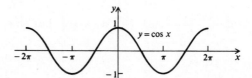

Margin Exercises, Section 3.4

1. $\sin \theta = -\dfrac{\sqrt{7}}{4}$, $\tan \theta = -\dfrac{\sqrt{7}}{3}$, $\cot \theta = -\dfrac{3}{\sqrt{7}}$, $\sec \theta = \dfrac{4}{3}$, $\csc \theta = -\dfrac{4}{\sqrt{7}}$
2. $\sin \theta = \dfrac{1}{\sqrt{10}}$, $\cos \theta = -\dfrac{3}{\sqrt{10}}$, $\tan \theta = -\dfrac{1}{3}$, $\sec \theta = -\dfrac{\sqrt{10}}{3}$, $\csc \theta = \sqrt{10}$
3. $\sin 15° = 0.2588$, $\cos 15° = 0.9659$, $\tan 15° = 0.2679$, $\cot 15° = 3.732$, $\sec 15° = 1.035$, $\csc 15° = 3.864$

Exercise Set 3.4, p. 124

1. $\cos \theta = -\dfrac{2\sqrt{2}}{3}$, $\tan \theta = \dfrac{\sqrt{2}}{4}$, $\cot \theta = 2\sqrt{2}$, $\sec \theta = -\dfrac{3\sqrt{2}}{4}$, $\csc \theta = -3$
3. $\sin \theta = -\frac{4}{5}$, $\tan \theta = -\frac{4}{3}$, $\cot \theta = -\frac{3}{4}$, $\sec \theta = \frac{5}{3}$, $\csc \theta = -\frac{5}{4}$
5. $\sin \theta = -\dfrac{\sqrt{5}}{5}$, $\cos \theta = \dfrac{2\sqrt{5}}{5}$, $\tan \theta = -\dfrac{1}{2}$, $\sec \theta = \dfrac{\sqrt{5}}{2}$, $\csc \theta = -\sqrt{5}$
7. $\sin 25° = 0.4226$, $\cos 25° = 0.9063$, $\tan 25° = 0.4663$, $\cot 25° = 2.145$, $\sec 25° = 1.103$, $\csc 25° = 2.366$
9. (a) $\cos \theta = -\dfrac{\sqrt{8}}{3}$, $\tan \theta = -\dfrac{1}{\sqrt{8}}$, $\cot \theta = -\sqrt{8}$, $\sec \theta = -\dfrac{3}{\sqrt{8}}$, $\csc \theta = 3$;
(b) $\sin(\pi + \theta) = -\dfrac{1}{3}$, $\cos(\pi + \theta) = \dfrac{\sqrt{8}}{3}$, $\tan(\pi + \theta) = -\dfrac{1}{\sqrt{8}}$, $\cot(\pi + \theta) = -\sqrt{8}$, $\sec(\pi + \theta) = \dfrac{3}{\sqrt{8}}$, $\csc(\pi + \theta) = -3$;
(c) $\sin(\pi - \theta) = \dfrac{1}{3}$, $\cos(\pi - \theta) = \dfrac{\sqrt{8}}{3}$, $\tan(\pi - \theta) = \dfrac{1}{\sqrt{8}}$, $\cot(\pi - \theta) = \sqrt{8}$, $\sec(\pi - \theta) = \dfrac{3}{\sqrt{8}}$, $\csc(\pi - \theta) = 3$;
(d) $\sin(2\pi - \theta) = -\dfrac{1}{3}$, $\cos(2\pi - \theta) = -\dfrac{\sqrt{8}}{3}$, $\tan(2\pi - \theta) = \dfrac{1}{\sqrt{8}}$, $\cot(2\pi - \theta) = \sqrt{8}$, $\sec(2\pi - \theta) = -\dfrac{3}{\sqrt{8}}$, $\csc(2\pi - \theta) = -3$ **11.** (a) $\cos 27° = 0.89101$, $\tan 27° = 0.50952$, $\cot 27° = 1.96262$, $\sec 27° = 1.12232$, $\csc 27° = 2.20269$;
(b) $\sin 63° = 0.89101$, $\cos 63° = 0.45399$, $\tan 63° = 1.96262$, $\cot 63° = 0.50952$, $\sec 63° = 2.20269$, $\csc 63° = 1.2232$
13. $\sin 128° = 0.78801$, $\cos 128° = -0.61566$, $\tan 128° = -1.27994$, $\cot 128° = -0.78129$, $\sec 128° = -1.62427$, $\csc 128° = 1.26902$

Margin Exercises, Section 3.5

1. 0.2644   **2.** 0.4699   **3.** 0.9026   **4.** 1.098   **5.** 0.0495 radian, or 2°50′; 0.0931 radian, or 5°20′   **6.** 37°27′   **7.** 43.917°
8. −0.6264   **9.** 1.901   **10.** 67°20′ or 1.1752 radians   **11.** 28°48′, or 0.5027 radian   **12.** −0.9182   **13.** 0.9853
14. 226°55′

Exercise Set 3.5, pp. 131–132

1. 0.2306   **3.** 0.5519   **5.** 0.4176   **7.** 0.7720   **9.** 46°23′   **11.** 67°50′   **13.** 45.75°   **15.** 76.88°   **17.** 0.4728   **19.** 0.5894
21. 0.2563   **23.** 0.2995   **25.** 0.7824   **27.** 13°40′   **29.** 68°10′   **31.** 69°17′   **33.** 0.4392   **35.** 0.8145   **37.** −0.9228
39. −1.590   **41.** 166°20′   **43.** 248°10′   **45.** 1.9324   **47.** 5.5245   **49.** 2°30′   **51.** 25°42′51″, 61.63944°   **53.** 0.48083
55. 865,000 miles

Margin Exercises, Section 3.6

1. $\cos \theta = 0.88$, $\tan \theta = -0.53$, $\cot \theta = -1.87$, $\sec \theta = 1.14$, $\csc \theta = -2.13$
2. $\sec \theta = -1.75$, $\cos \theta = -0.57$, $\sin \theta = -0.82$, $\csc \theta = -1.22$, $\cot \theta = 0.69$

3.   **4.**   **5.** $\cot \left(\dfrac{\pi}{2} - x \right) \equiv \tan x$

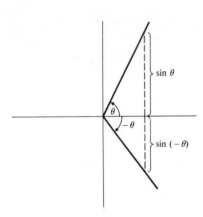   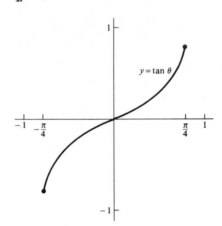

Exercise Set 3.6, pp. 137–138

1. $\cos x = -0.9898$, $\tan x = -0.1440$, $\cot x = -6.9444$, $\sec x = -1.0103$, $\csc x = 7.0175$
3. $\sin x = 0.7987$, $\cos x = 0.6018$, $\tan x = 1.3273$, $\csc x = 1.2520$, $\sec x = 1.6618$ **5.** Odd

7. $\sin \theta = \dfrac{-y}{1}$, $\sin(-\theta) = \dfrac{y}{1}$

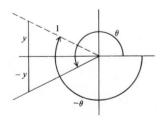

9. Same as Exercise 31 in Exercise Set 8.3   **11.** Domain: set of all real numbers; range: $[-1, 1]$; period: 2π
13. Domain: all real numbers except odd multiples of $\dfrac{\pi}{2}$; range: set of all real numbers; period: π

15. $\cos \left(\dfrac{\pi}{2} - x \right) \equiv \sin x$ **17.** $\tan \left(x - \dfrac{\pi}{2} \right) \equiv -\cot x$ **19.** $\sec \left(x - \dfrac{\pi}{2} \right) \equiv \csc x$

21.

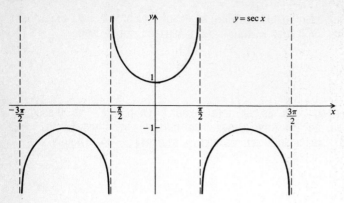

23.

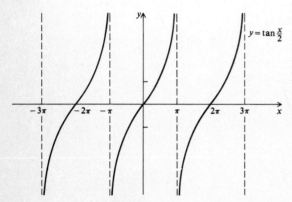

25. By similar triangles, $\dfrac{x}{\cos s} = \dfrac{y}{\sin s} = \dfrac{r}{1}$. Thus $\begin{cases} x = r\cos s \\ y = r\sin s \end{cases}$.

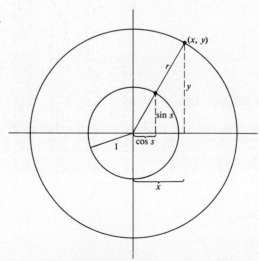

Chapter 3 Review, pp. 138–139

1. I, 0.483π, 1.52 **2.** II, 0.806π, 2.53 **3.** I, 0.167π, 0.525 **4.** IV, -0.167π, -0.525 **5.** 270° **6.** 720° **7.** $\dfrac{7\pi}{4}$ or 5.5 cm

8. 2.25, 129° **9.** 1130 cm/min **10.** 146,215 radians/hr

11. $\sin\theta = \dfrac{3}{\sqrt{13}}$, $\cos\theta = -\dfrac{2}{\sqrt{13}}$, $\tan\theta = -\dfrac{3}{2}$, $\cot\theta = -\dfrac{2}{3}$, $\sec\theta = -\dfrac{\sqrt{13}}{2}$, $\csc\theta = \dfrac{\sqrt{13}}{3}$

12.

13. $\dfrac{\sqrt{2}}{2}$ **14.** 1 **15.** $\sqrt{3}$

	0°	30°	45°	60°	90°	270°
$\sin\theta$	0	$\dfrac{1}{2}$	$\dfrac{\sqrt{2}}{2}$	$\dfrac{\sqrt{3}}{2}$	1	-1
$\cos\theta$	1	$\dfrac{\sqrt{3}}{2}$	$\dfrac{\sqrt{2}}{2}$	$\dfrac{1}{2}$	0	0
$\tan\theta$	0	$\dfrac{\sqrt{3}}{3}$	1	$\sqrt{3}$	—	—
$\cot\theta$	—	$\sqrt{3}$	1	$\dfrac{\sqrt{3}}{3}$	0	0
$\sec\theta$	1	$\dfrac{2}{\sqrt{3}}$	$\sqrt{2}$	2	—	—
$\csc\theta$	—	2	$\sqrt{2}$	$\dfrac{2}{\sqrt{3}}$	1	-1

16.

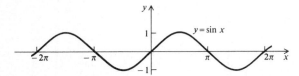

 Domain: set of all real numbers; range: $[-1, 1]$; period: 2π

17. $-\dfrac{2}{3}$ **18.** $-\dfrac{\sqrt{5}}{3}$ **19.** $\dfrac{\sqrt{5}}{2}$ **20.** $-\dfrac{3}{\sqrt{5}}$ or $-\dfrac{3\sqrt{5}}{5}$ **21.** $-\dfrac{3}{2}$ **22.** 0.7314 **23.** 0.6820 **24.** 1.0724 **25.** 0.9325

26. 1.4663 **27.** 1.3673 **28.** 22°12′ **29.** 47.55° **30.** 0.9894 **31.** 0.5147 **32.** 1.402 **33.** 16°10′ **34.** 39°

35. $\sin x = 0.9898$, $\tan x = -0.1440$, $\cot x = -6.9460$, $\sec x = 7.0175$, $\csc x = -1.0103$

36. (a) cosine, secant; (b) sine, cosecant, tangent, cotangent

37.

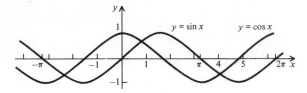

38. $\csc\left(x - \dfrac{\pi}{2}\right) \equiv \sec x$

39. Domain: Set of all reals; range: $[-3, 3]$; period: 4π

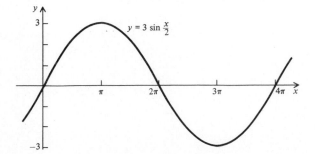

CHAPTER 4

Margin Exercises, Section 4.1

1. $\dfrac{1}{2}$ **2.** $\dfrac{-\sqrt{6}+\sqrt{2}}{4}$, or $\dfrac{\sqrt{2}}{4}(1-\sqrt{3})$ **3.** $\cos\dfrac{7\pi}{12}$, or $\cos\left(-\dfrac{7\pi}{2}\right)$ **4.** $\cos 25°$ **5.** $\cos(\alpha+\beta)$

6. $\dfrac{\sqrt{2}+\sqrt{6}}{4}$, or $\dfrac{\sqrt{2}}{4}(1+\sqrt{3})$ **7.** $\sin(\alpha-\beta)$

8. $\tan(\alpha-\beta) \equiv \dfrac{\sin(\alpha-\beta)}{\cos(\alpha-\beta)} \equiv \dfrac{\sin\alpha\cos\beta - \cos\alpha\sin\beta}{\cos\alpha\cos\beta + \sin\alpha\sin\beta} \cdot \dfrac{\dfrac{1}{\cos\alpha\cos\beta}}{\dfrac{1}{\cos\alpha\cos\beta}}$

$$\equiv \dfrac{\dfrac{\sin\alpha\cos\beta}{\cos\alpha\cos\beta} - \dfrac{\cos\alpha\sin\beta}{\cos\alpha\cos\beta}}{\dfrac{\cos\alpha\cos\beta}{\cos\alpha\cos\beta} + \dfrac{\sin\alpha\sin\beta}{\cos\alpha\cos\beta}} \equiv \dfrac{\tan\alpha-\tan\beta}{1+\tan\alpha\tan\beta}$$

9. $\dfrac{3+\sqrt{3}}{3-\sqrt{3}}$ **10.** $\sin\dfrac{\pi}{6}$, or $\dfrac{1}{2}$

11. $30°$

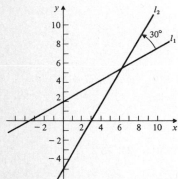

12. $150°$

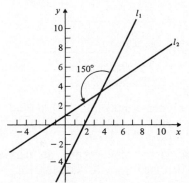

13. $46.4°$

Exercise Set 4.1, pp. 148–149

1. 0.9659 **3.** 0.2588 **5.** 0.9659 **7.** 3.7322 **9.** 1 **11.** 0 **13.** Undefined   **15.** 0.8448 **17.** $\sin 59°$, or 0.8572

19. $\tan 52°$ or 1.280 **21.** $2\sin\alpha\cos\beta$ **23.** $2\cos\alpha\cos\beta$ **25.** $\cos u$ **27.** $\dfrac{2\pi}{3}$ **29.** $0°$ **31.** $86°40'$ **33.** $135°$ **35.** $2\sin\theta\cos\theta$

37. $\cot(\alpha+\beta) = \dfrac{\cos(\alpha+\beta)}{\sin(\alpha+\beta)} = \dfrac{\cos\alpha\cos\beta - \sin\alpha\sin\beta}{\sin\alpha\cos\beta + \cos\alpha\sin\beta} \cdot \dfrac{\dfrac{1}{\sin\alpha\sin\beta}}{\dfrac{1}{\sin\alpha\sin\beta}}$

$$= \dfrac{\dfrac{\cos\alpha\cos\beta}{\sin\alpha\sin\beta} - 1}{\dfrac{\sin\alpha\cos\beta}{\sin\alpha\sin\beta} + \dfrac{\cos\alpha\sin\beta}{\sin\alpha\sin\beta}} = \dfrac{\cot\alpha\cot\beta - 1}{\cot\alpha + \cot\beta}$$

39. $\cos x$ **41.** $\dfrac{1}{7}$ **43.** $\dfrac{1}{6}$ **45.** $173°$, approximately   **47.** $22.8°$

Margin Exercises, Section 4.2

1. $24/25$ **2.** II, $\cos 2\theta = -119/169$, $\tan 2\theta = -120/119$, $\sin 2\theta = 120/169$ **3.** $\cos\theta - 4\sin^2\theta\cos\theta$

4. $\sin^3 x \equiv \sin x \dfrac{1 - \cos 2x}{2}$ **5.** $\dfrac{\sqrt{2 + \sqrt{3}}}{2}$

6. $\dfrac{1}{2 + \sqrt{3}}$, or $2 - \sqrt{3}$

Exercise Set 4.2, pp. 155–156

1. $\sin 2\theta = \frac{24}{25}$, $\cos 2\theta = -\frac{7}{25}$, $\tan 2\theta = -\frac{24}{7}$, II   **3.** $\cos 2\theta = \frac{7}{25}$, $\sin 2\theta = \frac{24}{25}$, $\tan 2\theta = \frac{24}{7}$, I

5. $\sin 2\theta = \frac{24}{25}$, $\cos 2\theta = -\frac{7}{25}$, $\tan 2\theta = -\frac{24}{7}$, II

7. $8\sin\theta\cos^3\theta - 4\sin\theta\cos\theta$, or $4\sin\theta\cos^3\theta - 4\sin^3\theta\cos\theta$, or $4\sin\theta\cos\theta - 8\sin^3\theta\cos\theta$

9. $\dfrac{3 - 4\cos 2\theta + \cos 4\theta}{8}$ **11.** $\dfrac{\sqrt{2 + \sqrt{3}}}{2}$ **13.** $2 + \sqrt{3}$ **15.** $\dfrac{\sqrt{2 + \sqrt{2}}}{2}$ **17.** 0.6421

19. 0.9844 **21.** 0.1734 **23.** $\cos x$ **25.** $\cos x$ **27.** $\sin x$

29. $\cos x$ **31.** 1 **33.** 1 **35.** 8 **37.** $\sin 2x$

39. (i) $\frac{1}{2}[\sin(u + v) + \sin(u - v)] = \frac{1}{2}[\sin u\cos v + \cos u\sin v + \sin u\cos v - \cos u\sin v]$
$= \sin u \cdot \cos v$. The other formulas follow similarly.

Margin Exercises, Section 4.3

1.

$\dfrac{\cos^2 x}{\sin^2 x} - \cos^2 x$	$\cos^2 x \cdot \dfrac{\cos^2 x}{\sin^2 x}$
$\dfrac{\cos^2 x - \sin^2 x\cos^2 x}{\sin^2 x}$	$\dfrac{\cos^4 x}{\sin^2 x}$
$\dfrac{\cos^2 x(1 - \sin^2 x)}{\sin^2 x}$	
$\dfrac{\cos^4 x}{\sin^2 x}$	

2.

$\dfrac{2\sin\theta\cos\theta + \sin\theta}{(2\cos^2\theta - 1) + \cos\theta + 1}$	$\dfrac{\sin\theta}{\cos\theta}$
$\dfrac{\sin\theta(2\cos\theta + 1)}{2\cos^2\theta + \cos\theta}$	
$\dfrac{\sin\theta(2\cos\theta + 1)}{\cos\theta(2\cos\theta + 1)}$	
$\dfrac{\sin\theta}{\cos\theta}$	

3. $\sqrt{2}\sin\left(2x + \dfrac{\pi}{4}\right)$ **4.** $13\sin(x + b)$, where $\cos b = \frac{12}{13}$ and $\sin b = -\frac{5}{13}$ **5.** $\sqrt{3}\sin x + \cos x \equiv 2\sin\left(x + \dfrac{\pi}{6}\right)$

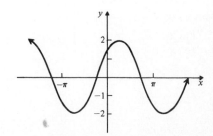

Exercise Set 4.3, pp. 160–161

1.

$$
\begin{array}{c|c}
\csc x - \cos x \cot x & \sin x \\
\hline
\dfrac{1}{\sin x} - \cos x \,\dfrac{\cos x}{\sin x} & \sin x \\[2mm]
\dfrac{1 - \cos^2 x}{\sin x} & \\[2mm]
\dfrac{\sin^2 x}{\sin x} & \\[2mm]
\sin x &
\end{array}
$$

3.

$$
\begin{array}{c|c}
\dfrac{1 + \cos \theta}{\sin \theta} + \dfrac{\sin \theta}{\cos \theta} & \dfrac{\cos \theta + 1}{\sin \theta \cos \theta} \\[3mm]
\hline
\dfrac{1 + \cos \theta}{\sin \theta} \cdot \dfrac{\cos \theta}{\cos \theta} + \dfrac{\sin \theta \cdot \sin \theta}{\cos \theta \sin \theta} & \dfrac{1 + \cos \theta}{\sin \theta \cos \theta} \\[3mm]
\dfrac{\cos^2 \theta + \cos \theta + \sin^2 \theta}{\sin \theta \cos \theta} & \\[3mm]
\dfrac{1 + \cos \theta}{\sin \theta \cos \theta} &
\end{array}
$$

5.

$$
\begin{array}{c|c}
\dfrac{1 - \sin x}{\cos x} & \dfrac{\cos x}{1 + \sin x} \\[3mm]
\hline
\dfrac{1 - \sin x}{\cos x} \cdot \dfrac{\cos x}{\cos x} & \dfrac{\cos x}{1 + \sin x} \cdot \dfrac{1 - \sin x}{1 - \sin x} \\[3mm]
\dfrac{\cos x - \sin x \cos x}{\cos^2 x} & \dfrac{\cos x - \sin x \cos x}{1 - \sin^2 x} \\[3mm]
& \dfrac{\cos x - \sin x \cos x}{\cos^2 x}
\end{array}
$$

7.

$$
\begin{array}{c|c}
\dfrac{1 + \tan \theta}{1 + \cot \theta} & \dfrac{\sec \theta}{\csc \theta} \\[3mm]
\hline
\dfrac{1 + \dfrac{\sin \theta}{\cos \theta}}{1 + \dfrac{\cos \theta}{\sin \theta}} & \dfrac{\sin \theta}{\cos \theta} \\[4mm]
\dfrac{\dfrac{\cos \theta + \sin \theta}{\cos \theta}}{\dfrac{\sin \theta + \cos \theta}{\sin \theta}} & \tan \theta \\[4mm]
\dfrac{\cos \theta + \sin \theta}{\cos \theta} \cdot \dfrac{\sin \theta}{\sin \theta + \cos \theta} & \\[3mm]
\dfrac{\sin \theta}{\cos \theta} & \\[3mm]
\tan \theta &
\end{array}
$$

9.

$$
\begin{array}{c|c}
\dfrac{\sin x + \cos x}{\sec x + \csc x} & \dfrac{\sin x}{\sec x} \\[3mm]
\hline
\dfrac{\sin x + \cos x}{\dfrac{1}{\cos x} + \dfrac{1}{\sin x}} & \dfrac{\sin x}{\sec x} \\[4mm]
\dfrac{\sin x + \cos x}{\dfrac{\sin x + \cos x}{\sin x \cos x}} & \\[4mm]
(\sin x + \cos x) \cdot \dfrac{\sin x \cos x}{\sin x + \cos x} & \\[3mm]
\sin x \cos x & \\[3mm]
\dfrac{\sin x}{\sec x} &
\end{array}
$$

11.

$$
\begin{array}{c|c}
\dfrac{1 + \tan \theta}{1 - \tan \theta} + \dfrac{1 + \cot \theta}{1 - \cot \theta} & 0 \\[3mm]
\hline
\dfrac{1 + \dfrac{\sin \theta}{\cos \theta}}{1 - \dfrac{\sin \theta}{\cos \theta}} + \dfrac{1 + \dfrac{\cos \theta}{\sin \theta}}{1 - \dfrac{\cos \theta}{\sin \theta}} & 0 \\[4mm]
\dfrac{\dfrac{\cos \theta + \sin \theta}{\cos \theta}}{\dfrac{\cos \theta - \sin \theta}{\cos \theta}} + \dfrac{\dfrac{\sin \theta + \cos \theta}{\sin \theta}}{\dfrac{\sin \theta - \cos \theta}{\sin \theta}} & \\[4mm]
\dfrac{\cos \theta + \sin \theta}{\cos \theta} \cdot \dfrac{\cos \theta}{\cos \theta - \sin \theta} + \dfrac{\sin \theta + \cos \theta}{\sin \theta} \cdot \dfrac{\sin \theta}{\sin \theta - \cos \theta} & \\[3mm]
\dfrac{\cos \theta + \sin \theta}{\cos \theta - \sin \theta} + \dfrac{\sin \theta + \cos \theta}{\sin \theta - \cos \theta} & \\[3mm]
\dfrac{\cos \theta + \sin \theta}{\cos \theta - \sin \theta} - \dfrac{\cos \theta + \sin \theta}{\cos \theta - \sin \theta} & \\[3mm]
0 &
\end{array}
$$

13.

$$\frac{1 + \cos 2\theta}{\sin 2\theta} \qquad\qquad \cot\theta$$

$$\frac{1 + 2\cos^2\theta - 1}{2\sin\theta\cos\theta} \qquad\qquad \frac{\cos\theta}{\sin\theta}$$

$$\frac{\cos\theta}{\sin\theta}$$

15.

$$\sec 2\theta \qquad\qquad \frac{\sec^2\theta}{2 - \sec^2\theta}$$

$$\frac{1}{\cos 2\theta} \qquad\qquad \frac{1 + \tan^2\theta}{2 - 1 - \tan^2\theta}$$

$$\frac{1}{\cos^2\theta - \sin^2\theta} \qquad\qquad \frac{1 + \tan^2\theta}{1 - \tan^2\theta}$$

$$\frac{1 + \dfrac{\sin^2\theta}{\cos^2\theta}}{1 - \dfrac{\sin^2\theta}{\cos^2\theta}}$$

$$\frac{\cos^2\theta + \sin^2\theta}{\cos^2\theta - \sin^2\theta}$$

$$\frac{1}{\cos^2\theta - \sin^2\theta}$$

17.

$$\frac{\sin\alpha\cos\beta + \cos\alpha\sin\beta}{\cos\alpha\cos\beta} \qquad\qquad \frac{\sin\alpha}{\cos\alpha} + \frac{\sin\beta}{\cos\beta}$$

$$\frac{\sin\alpha\cos\beta + \cos\alpha\sin\beta}{\cos\alpha\cos\beta}$$

19.

$$1 - [\cos 5\theta \cos 3\theta + \sin 5\theta \sin 3\theta] \qquad\qquad 1 - \cos 2\theta$$

$$1 - \cos(5\theta - 3\theta)$$

$$1 - \cos 2\theta$$

21.

$$\frac{1}{2}\left[\frac{\dfrac{\sin\theta + \sin\theta\cos\theta}{\cos\theta}}{\dfrac{\sin\theta}{\cos\theta}}\right] \qquad\qquad \frac{1 + \cos\theta}{2}$$

$$\frac{1}{2}\left[\frac{\sin\theta(1 + \cos\theta)}{\cos\theta}\cdot\frac{\cos\theta}{\sin\theta}\right]$$

$$\frac{1 + \cos\theta}{2}$$

23.

$$(\cos^2 x - \sin^2 x)(\cos^2 x + \sin^2 x) \qquad\qquad \cos^2 x - \sin^2 x$$

$$\cos^2 x - \sin^2 x$$

25.

$$\tan(3\theta - \theta) \qquad\qquad \tan 2\theta$$

$$\tan 2\theta$$

27.

$$\frac{(\cos x - \sin x)(\cos^2 x + \cos x \sin x + \sin^2 x)}{\cos x - \sin x} \qquad\qquad \frac{2 + 2\sin x \cos x}{2}$$

$$1 + \cos x \sin x \qquad\qquad 1 + \sin x \cos x$$

29.

$$(\sin\alpha\cos\beta + \cos\alpha\sin\beta)(\sin\alpha\cos\beta - \cos\alpha\sin\beta) \qquad\qquad 1 - \cos^2\alpha - (1 - \cos^2\beta)$$

$$\sin^2\alpha\cos^2\beta - \cos^2\alpha\sin^2\beta \qquad\qquad \cos^2\beta - \cos^2\alpha$$

$$\cos^2\beta(1 - \cos^2\alpha) - \cos^2\alpha(1 - \cos^2\beta)$$

$$\cos^2\beta - \cos^2\beta\cos^2\alpha - \cos^2\alpha + \cos^2\alpha\cos^2\beta$$

$$\cos^2\beta - \cos^2\alpha$$

31.

$$\cos\alpha\cos\beta - \sin\alpha\sin\beta + \cos\alpha\cos\beta + \sin\alpha\sin\beta \qquad\qquad 2\cos\alpha\cos\beta$$

$$2\cos\alpha\cos\beta$$

33. $2 \sin\left(2x + \dfrac{\pi}{3}\right)$ **35.** $5 \sin(x + b)$, where $\cos b = \frac{4}{5}$, $\sin b = \frac{3}{5}$

37. $7.89 \sin(0.374x + 31.2°)$

39. $y = \sqrt{2} \sin\left(2x - \dfrac{\pi}{4}\right)$

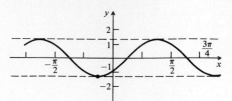

42. $\dfrac{1}{\omega C(\tan\theta + \tan\phi)} = \dfrac{1}{\omega C\left(\dfrac{\sin\theta}{\cos\theta} + \dfrac{\sin\phi}{\cos\phi}\right)} = \dfrac{1}{\omega C\left(\dfrac{\sin\theta\cos\phi + \sin\phi\cos\theta}{\cos\theta\cos\phi}\right)} = \dfrac{\cos\theta\cos\phi}{\omega C \sin(\theta + \phi)}$

Margin Exercises, Section 4.4

1. Not a function

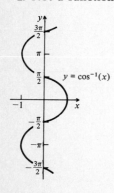

2. Not a function

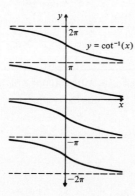

3. $\dfrac{\pi}{4} + 2\pi k$, $-\dfrac{\pi}{4} + 2\pi k$, where k is an integer

4. $\dfrac{\pi}{3} + 2\pi k$, $\dfrac{2\pi}{3} + 2k\pi$, where k is an integer   **5.** $25° + 360°k$, $155° + 360°k$ **6.** $\dfrac{3\pi}{4} + k\pi$ **7.** $\dfrac{\pi}{3}$

8. $\dfrac{3\pi}{4}$ **9.** $\dfrac{3\pi}{4}$ **10.** $-\dfrac{\pi}{4}$

Exercise Set 4.4, p. 166

1. $\dfrac{\pi}{4} + 2k\pi$, $\dfrac{3\pi}{4} + 2k\pi$ **3.** $\dfrac{\pi}{4} + 2k\pi$, $-\dfrac{\pi}{4} + 2k\pi$ **5.** $\dfrac{5\pi}{4} + 2k\pi$, $-\dfrac{\pi}{4} + 2k\pi$ **7.** $\dfrac{3\pi}{4} + 2k\pi$, $\dfrac{5\pi}{4} + 2k\pi$ **9.** $\dfrac{\pi}{3} + k\pi$

11. $\dfrac{\pi}{4} + k\pi$ **13.** $\dfrac{5\pi}{6} + k\pi$ **15.** $\dfrac{3\pi}{4} + k\pi$ **17.** $0 + 2k\pi$ **19.** $\dfrac{\pi}{2} + 2k\pi$ **21.** $23° + k \cdot 360°$, $157° + k \cdot 360°$

23. $39° + k \cdot 360°$, $141° + k \cdot 360°$ **25.** $36°58' + k \cdot 360°$, $323°02' + k \cdot 360°$ **27.** $21°25' + k \cdot 360°$, $338°35' + k \cdot 360°$

29. $20°10' + k \cdot 180°$ **31.** $38°20' + k \cdot 180°$ **33.** $31° + k \cdot 360°$, $329° + k \cdot 360°$

35. $9°10' + k \cdot 360°$, $170°50' + k \cdot 360°$ **37.** $\dfrac{\pi}{4}$ **39.** $\dfrac{\pi}{3}$ **41.** $-\dfrac{\pi}{3}$ **43.** $\dfrac{3\pi}{4}$ **45.** $-\dfrac{\pi}{6}$ **47.** $\dfrac{2\pi}{3}$ **49.** $13°30'$

51. $-39°30'$ **53.** $152°50'$ **55.** $-22°10'$ **57.** $169°30'$

Margin Exercises, Section 4.5

1. $-\dfrac{\pi}{2}$ **2.** $\dfrac{\pi}{6}$ **3.** $-\dfrac{\pi}{4}$ **4.** $\dfrac{3\pi}{4}$ **5.** $\dfrac{5\pi}{6}$ **6.** $\dfrac{\sqrt{2}}{2}$ **7.** $\dfrac{\sqrt{3}}{2}$ **8.** $\dfrac{\pi}{3}$ **9.** $\dfrac{\pi}{2}$ **10.** $\dfrac{3}{\sqrt{b^2+9}}$ **11.** $\dfrac{2}{\sqrt{5}}$ **12.** $\dfrac{t}{\sqrt{1-t^2}}$
13. 1   **14.** $\frac{1}{3}$

Exercise Set 4.5, pp. 170–171

1. 0.3   **3.** -4.2 **5.** $\dfrac{\pi}{3}$ **7.** $-\dfrac{\pi}{4}$ **9.** $\dfrac{\pi}{5}$ **11.** $-\dfrac{\pi}{3}$ **13.** $\dfrac{\sqrt{3}}{2}$ **15.** $\dfrac{1}{2}$ **17.** 1   **19.** $\dfrac{\pi}{6}$ **21.** $\dfrac{\pi}{3}$ **23.** $\dfrac{\pi}{2}$ **25.** $\dfrac{x}{\sqrt{x^2+4}}$
27. $\dfrac{\sqrt{x^2-9}}{3}$ **29.** $\dfrac{\sqrt{b^2-a^2}}{a}$ **31.** $\dfrac{3}{\sqrt{11}}$ **33.** $\dfrac{1}{3\sqrt{11}}$ **35.** $-\dfrac{1}{2\sqrt{6}}$ **37.** $\dfrac{1}{\sqrt{1+y^2}}$ **39.** $\dfrac{1}{\sqrt{1+t^2}}$ **41.** $\dfrac{\sqrt{1-y^2}}{y}$
43. $\sqrt{1-x^2}$ **45.** $\dfrac{1}{2}$ **47.** $\dfrac{\sqrt{2+\sqrt{3}}}{2}$ **49.** $\dfrac{24}{25}$ **51.** $\dfrac{119}{169}$ **53.** $\dfrac{3+4\sqrt{3}}{10}$ **55.** $-\dfrac{\sqrt{2}}{10}$ **57.** $xy+\sqrt{(1-x^2)(1-y^2)}$
59. $y\sqrt{1-x^2}-x\sqrt{1-y^2}$ **61.** 0.9861   **63.** $\theta = \text{Arctan}\dfrac{y+h}{x}-\text{Arctan}\dfrac{y}{x}$

Margin Exercises, Section 4.6

1. $60°+360°k,\ 300°+360°k,\ \dfrac{\pi}{3}+2k\pi,\ \dfrac{5\pi}{3}+2k\pi$ **2.** $\dfrac{\pi}{6},\dfrac{5\pi}{6},\dfrac{7\pi}{6},\dfrac{11\pi}{6};\ 30°,150°,210°,330°$ **3.** $\dfrac{\pi}{6},\dfrac{5\pi}{6},\dfrac{7\pi}{6},\dfrac{11\pi}{6}$
4. $40°,320°$ **5.** $140°,220°$ **6.** $120°,240°,75°30',284°30'$ **7.** $120°,240°,90°,270°$ **8.** $241°40',118°20'$

Exercise Set 4.6, pp. 175–176

1. $\dfrac{\pi}{3}+2k\pi,\dfrac{2\pi}{3}+2k\pi$ **3.** $\dfrac{\pi}{4}+2k\pi,\dfrac{-\pi}{4}+2k\pi$ **5.** $20°10'+k\cdot360°,159°50'+k\cdot360°$ **7.** $236°40',123°20'$ **9.** $\dfrac{4\pi}{3},\dfrac{5\pi}{3}$
11. $123°40',303°40'$ **13.** $\dfrac{\pi}{6},\dfrac{5\pi}{6},\dfrac{7\pi}{6},\dfrac{11\pi}{6}$ **15.** $\dfrac{\pi}{6},\dfrac{5\pi}{6},\dfrac{7\pi}{6},\dfrac{11\pi}{6}$ **17.** $\dfrac{\pi}{6},\dfrac{5\pi}{6},\dfrac{3\pi}{2}$ **19.** 0   **21.** $0,\dfrac{\pi}{6},\dfrac{5\pi}{6},\pi,\dfrac{7\pi}{6},\dfrac{11\pi}{6}$
23. $\dfrac{\pi}{6},\dfrac{5\pi}{6}$ **25.** $109°30',250°30',120°,240°$ **27.** $\dfrac{\pi}{6},\dfrac{5\pi}{6},\pi$ **29.** $0,\pi,\dfrac{\pi}{2},\dfrac{3\pi}{2}$ **31.** $60°,120°,240°,300°$ **33.** $60°,240°$
35. $30°,60°,120°,150°,210°,240°,300°,330°$ **37.** $0,\pi$ **39.** $0,\pi$ **41.** $0,\pm4.8,\pm14.3,\pm17.1,$ etc.

Margin Exercises, Section 4.7

1. $\dfrac{\pi}{2},\pi$ **2.** $\dfrac{\pi}{2},\pi$ **3.** $\dfrac{\pi}{2},\pi$ **4.** $\dfrac{\pi}{2},\dfrac{3\pi}{2},\dfrac{\pi}{4},\dfrac{3\pi}{4},\dfrac{5\pi}{4},\dfrac{7\pi}{4}$ **5.** $\dfrac{\pi}{2},\dfrac{3\pi}{2},\dfrac{7\pi}{6},\dfrac{11\pi}{6}$ **6.** $\dfrac{\pi}{12},\dfrac{5\pi}{12},\dfrac{13\pi}{12},\dfrac{17\pi}{12}$ **7.** $\dfrac{\pi}{3},\pi$

Exercise Set 4.7, pp. 179–180

1. $0,\pi$ **3.** $\dfrac{3\pi}{4},\dfrac{7\pi}{4}$ **5.** $\dfrac{\pi}{2},\dfrac{3\pi}{2},\dfrac{\pi}{6},\dfrac{5\pi}{6}$ **7.** $\dfrac{\pi}{2},\dfrac{3\pi}{2},\dfrac{\pi}{4},\dfrac{3\pi}{4},\dfrac{5\pi}{4},\dfrac{7\pi}{4}$ **9.** $0,\dfrac{\pi}{2},\pi,\dfrac{3\pi}{2}$ **11.** 0   **13.** $0,\dfrac{\pi}{2},\pi,\dfrac{3\pi}{2}$
15. $\dfrac{\pi}{6},\dfrac{5\pi}{6},\pi$ **17.** $\dfrac{\pi}{6},\dfrac{5\pi}{6},\dfrac{7\pi}{6},\dfrac{11\pi}{6}$ **19.** $63°30',243°30',101°20',281°20'$ **21.** $\dfrac{\pi}{6},\dfrac{\pi}{2},\dfrac{5\pi}{6},\dfrac{7\pi}{6},\dfrac{3\pi}{2},\dfrac{11\pi}{6}$ **23.** $\dfrac{2\pi}{3},\dfrac{4\pi}{3}$
25. $\dfrac{\pi}{4},\dfrac{7\pi}{4}$ **27.** $\dfrac{\pi}{12},\dfrac{5\pi}{12},\dfrac{13\pi}{12},\dfrac{17\pi}{12}$ **29.** $\dfrac{\pi}{6},\dfrac{3\pi}{2}$ **31.** 1   **33.** 1.15, 5.65, $-0.65,$ etc.

Chapter 4 Review, p. 180

1. $\sin x$ **2.** $\dfrac{\tan 45° - \tan 30°}{1 + \tan 45° \tan 30°}$ **3.** $\cos(27° - 16°)$ or $\cos 11°$ **4.** $2 - \sqrt{3}$ **5.** $\frac{1}{2}\sqrt{2 - \sqrt{2}}$ **6.** $2 \cot \theta$

7.

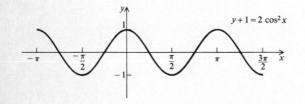

$\tan 2\theta$	$\dfrac{2\tan\theta}{1 - \tan^2\theta}$
$\dfrac{\sin 2\theta}{\cos 2\theta}$	$\dfrac{2\dfrac{\sin\theta}{\cos\theta}}{\dfrac{\cos^2\theta}{\cos^2\theta} - \dfrac{\sin^2\theta}{\cos^2\theta}}$
$\dfrac{2\sin\theta\cos\theta}{\cos^2\theta - \sin^2\theta}$	$\dfrac{2\sin\theta}{\cos\theta} \cdot \dfrac{\cos^2\theta}{\cos^2\theta - \sin^2\theta}$
	$\dfrac{2\sin\theta\cos\theta}{\cos^2\theta - \sin^2\theta}$

8. $\dfrac{\pi}{6} + 2k\pi,\ \dfrac{5\pi}{6} + 2k\pi$ **9.** $81° + k \cdot 180°$ **10.** $-45°$ or $-\dfrac{\pi}{4}$ **11.** $\dfrac{7}{8}$ **12.** $18°30'$ **13.** $0, \pi$

14. $\sqrt{40}\sin\left(3x + \text{Arcsin }\dfrac{\sqrt{10}}{10}\right)$

15.

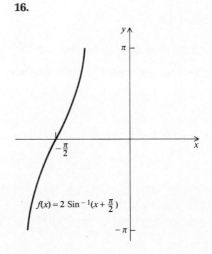

16.

CHAPTER 5

Margin Exercises, Section 5.1

1. $a = 38.43,\ b = 54.89$ **2.** $A = 56°19',\ B = 33°41',\ c = 7.211$ **3.** 35.7 ft   **4.** 240.4 ft   **5.** 136 km east, 63.4 km south
6. 22.1 km   **7.** 16 mi

Exercise Set 5.1, pp. 186–187

1. $B = 53°50',\ b = 37.2,\ c = 46.1$ **3.** $A = 77°20',\ a = 436,\ c = 447$ **5.** $B = 72°40',\ a = 4.24,\ c = 14.3$
7. $A = 66°50',\ b = 0.0148,\ c = 0.0375$ **9.** $B = 42°30',\ a = 35.6,\ b = 32.6$ **11.** $A = 7°40',\ a = 0.131,\ b = 0.973$
13. $c = 21.6,\ A = 33°40',\ B = 56°20'$ **15.** $b = 12.0,\ A = 53°10',\ B = 36°50'$ **17.** $a = 3.57,\ A = 63°20',\ B = 26°40'$
19. 47.9 ft   **21.** $1°40'$ **23.** $30°10'$ **25.** 3.52 mi   **27.** 109 km   **29.** 23.9 km   **31.** 7.92 km   **33.** 3.45 km

35.

$$A = \frac{1}{2}ab = \frac{1}{2}\frac{c^2}{c^2}ab = \frac{1}{2}c^2\frac{a}{c}\frac{b}{c}$$

$$= \frac{1}{2}c^2 \sin A \cos A = \frac{1}{4}c^2\, 2 \sin A \cos A$$

$$= \frac{1}{4}c^2 \sin 2A$$

Margin Exercises, Section 5.2

1. $C = 29°$, $c = 5.93$, $b = 11.1$ **2.** $\sin A = 2.8$; impossible   **3.** $B = 90°$, $C = 41°25'$, $c = 5$
4. (a) $A = 42°50'$, $C = 104°10'$, $c = 35.6$; (b) $A = 137°10'$, $C = 9°50'$, $c = 6.27$ **5.** $C = 18°$, $A = 124°$, $a = 26.9$

Exercise Set 5.2, p. 192

1. $C = 17°$, $a = 26.3$, $c = 10.5$ **3.** $A = 121°$, $a = 33.4$, $c = 14.0$ **5.** $B = 68°50'$, $a = 32.3$, $b = 32.3$
7. $B = 56°20'$, $C = 87°40'$, $c = 40.8$, and $B = 123°40'$, $C = 20°20'$, $c = 14.2$ **9.** $B = 19°$, $C = 44°40'$, $b = 6.25$
11. $A = 74°30'$, $B = 44°20'$, $a = 33.3$ **13.** 76.3 m   **15.** 50.8 ft   **17.** 1470 km   **19.** 10.6 km
21. $A = bh$, $h = a \sin \theta$, so $A = ab \sin \theta$

Margin Exercises, Section 5.3

1. $a = 40.5$, $B = 22°10'$, $C = 35°50'$ **2.** $A = 108°10'$, $B = 22°20'$, $C = 49°30'$

Exercise Set 5.3, pp. 195–196

1. $a = 14.9$, $B = 23°40'$, $C = 126°20'$ **3.** $a = 24.8$, $B = 20°40'$, $C = 26°20'$ **5.** $b = 74.8$, $A = 95°30'$, $C = 11°50'$
7. $A = 36°10'$, $B = 43°30'$, $C = 100°20'$ **9.** $A = 73°40'$, $B = 51°50'$, $C = 54°30'$ **11.** $A = 25°40'$, $B = 126°$, $C = 28°20'$
13. 28.2 nautical mi, S 55°20'E   **15.** 37 nautical mi   **17.** 59.4 ft   **19.** 68°, 68°, 44°   **21.** 42.6 ft
23. (a) 15.73 ft; (b) 120.8 ft²   **25.** 3424 yd²   **27.** 116.6 ft   **29.** $A = \frac{1}{2}a^2 \sin \theta$; when $\theta = 90°$

Margin Exercises, Section 5.4

1. 14.9 kg, 19°40'   **2.** 75°, 171 km/h

Exercise Set 5.4, p. 199

1. 57, 38°   **3.** 18.4, 37°   **5.** 20.9, 59°   **7.** 68.3, 18°   **9.** 13 kg, 67°   **11.** 655 kg, 21°   **13.** 21.6 ft/sec, 34°
15. 726 lb, 47°   **17.** 174 nautical mi, S 15°E   **19.** An angle of 12° upstream

Margin Exercises, Section 5.5

1. E: $50\sqrt{2}$, S: $50\sqrt{2}$ **2.** S: 36.9 lb; W: 25.8 lb   **3.** 18.9 km/h from S 32°E
4. (a) 4 up, 21 left; (b) 21.4, 10°50' with horizontal   **5.** $(12, -8)$ **6.** $(12, 4)$ **7.** $(8.54, 159°30')$ **8.** $(14.2, -4.88)$
9. (a) $(1, 7)$; (b) $(7, -3)$; (c) $(-23, 21)$; (d) 19.03

Exercise Set 5.5, p. 205

1. (5, 16) **3.** (4.8, 13.7) **5.** (−662, −426) **7.** (5, 53°10′) **9.** (18, 303°40′) **11.** (5, 233°10′) **13.** (18, 123°40′)
15. (3.46, 2) **17.** (−5.74, −8.19) **19.** (17.3, −10) **21.** (70.7, −70.7) **23.** Vertical 118, horizontal 92.4
25. S. 192 km/h, W. 161 km/h **27.** 43 kg, S 35°30′W **29.** (a) N. 28, W. 7; (b) 28.8, N 14°W **31.** (19, 36) **33.** (14, 8)
35. 18 **37.** 17.89 **39.** 35.78 **41.** 0

Margin Exercises, Section 5.6

1.

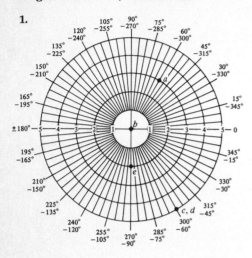

2. Many answers possible. A: (4, 30°), (4, 390°), (−4, 210°), etc.; B: (5, −60°), (5, 300°); C: (2, 150°), (2, −210°); D: (3, 225°), (3, −135°); E: (5, 60°), (−5, −120°)

3. (a) $(3\sqrt{2}, 45°)$; (b) (4, 270°); (c) (6, 120°); (d) (4, 330°)

4. (a) $\left(\dfrac{5}{2}\sqrt{3}, \dfrac{5}{2}\right)$; (b) $(5, 5\sqrt{3})$; (c) $\left(-\dfrac{5}{\sqrt{2}}, -\dfrac{5}{\sqrt{2}}\right)$; (d) $(4\sqrt{3}, 4)$

5. $2r\cos\theta + 5r\sin\theta = 9$

6. $r^2 + 8r\cos\theta = 0$ **7.** $x^2 + y^2 = 49$ **8.** $y = 5$ **9.** $x^2 + y^2 - 3x = 5y$ **10.**

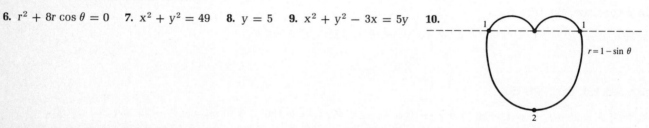

$r = 1 - \sin\theta$

Exercise Set 5.6, p. 209

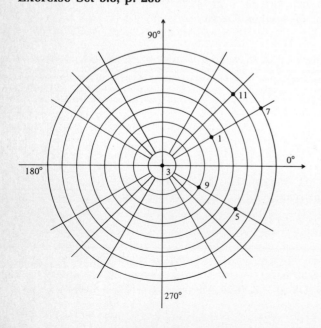

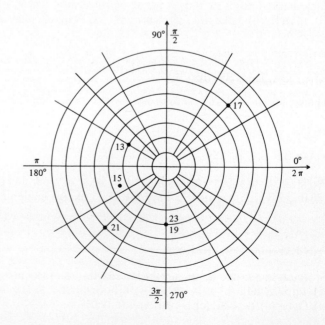

25. $(4\sqrt{2}, 45°)$ **27.** $(5, 90°)$ **29.** $(4, 0°)$ **31.** $(6, 60°)$ **33.** $(2, 30°)$ **35.** $(6, 30°)$

37. $\left(\dfrac{4}{\sqrt{2}}, \dfrac{4}{\sqrt{2}}\right)$ or $(2.83, 2.83)$ **39.** $(0, 0)$ **41.** $\left(\dfrac{-3}{\sqrt{2}}, \dfrac{-3}{\sqrt{2}}\right)$ **43.** $(3, -3\sqrt{3})$ **45.** $(5\sqrt{3}, 5)$

47. $(4.33, -2.5)$ **49.** $3r\cos\theta + 4r\sin\theta = 5$ **51.** $r\cos\theta = 5$ **53.** $r^2 = 36$ **55.** $r^2(\cos^2\theta - 4\sin^2\theta) = 4$

57. $x^2 + y^2 = 25$ **59.** $y = x$ **61.** $y = 2$ **63.** $x^2 + y^2 = 4x$ **65.** $x^2 - 4y = 4$ **67.** $x^2 - 2x + y^2 - 3y = 0$

69.

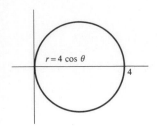

71.

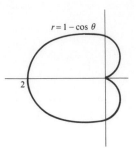

73.

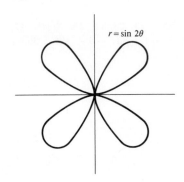

Margin Exercises, Section 5.7

1. Horizontal rope $300\sqrt{3}$ lb; other rope 600 lb
2. Parallel to incline 50 kg; perpendicular to incline $50\sqrt{3}$ or 86.6 kg
3. 500 kg on left, 866 kg on right

Exercise Set 5.7, p. 212–214

1. Cable 224-lb tension, boom 166-lb compression
3. Horizontal rod 168-kg tension, other rod 261-kg compression
5. Lift 2472 lb, drag 1315 lb
7. 60 kg
9. 80°20′
11. Horizontal rope 400 kg, other rope 566 kg
13. 2000 kg in each
15. 2242 kg on left, 1830 kg on right

Chapter 5 Review, p. 214

1. $A = 58°10'$, $B = 31°50'$, $c = 4.55$ **2.** $A = 38°50'$, $b = 37.9$, $c = 48.6$ **3.** 1748 cm
4. $A = 34°10'$, $a = 0.621$, $c = 0.511$ **5.** 420 cm **6.** 8.0 **7.** 13.95 ft **8.** 13.72 cm^2 **9.** 75.8 mph; N 20°10′E
10. Parallel: 106 lb; perpendicular: 106 lb **11.** $(-18.79, 6.84)$ **12.** $(\sqrt{13}, 123°40')$ **13.** $r + 2\cos\theta - 3\sin\theta = 0$
14. 23 lb down, 17 lb left; 28.6 lb, 53.5° downward from horizontal (to left) **15.** (a) $(25, 0)$; (b) $\sqrt{50}$ **16.** 577 kg
17. 50.52°, 129.48°

CHAPTER 6

Margin Exercises, Section 6.1

1. $i\sqrt{6}$ **2.** $-i\sqrt{10}$ **3.** $2i$ **4.** $-5i$ **5.** $-\sqrt{10}$ **6.** $\sqrt{11}$ **7.** $i\sqrt{7}$ **8.** $7i$ **9.** $3i$ **10.** $(\sqrt{17} + 3)i$ **11.** i **12.** -1
13. $-i$ **14.** $12 + i$ **15.** $5 - i$ **16.** $2 + 14i$ **17.** 8 **18.** $-6 + 8i$ **19.** $3i$ **20.** $(x + 2i)(x - 2i)$ **21.** $(3 + yi)(3 - yi)$
22. Yes **23.** $x = -1$, $y = 2$

Exercise Set 6.1, pp. 219–220

1. $i\sqrt{15}$ **3.** $4i$ **5.** $-2i\sqrt{3}$ **7.** $9i$ **9.** $i(\sqrt{7} - \sqrt{10})$ **11.** $-\sqrt{55}$ **13.** $2\sqrt{5}$ **15.** $\sqrt{\frac{5}{2}}i$ **17.** $-\frac{3}{2}i$ **19.** -2
21. $6 + 5i$ **23.** 8 **25.** $2 + 4i$ **27.** $-4 - i$ **29.** $-5 + 5i$ **31.** $7 - i$ **33.** $-6 + 12i$ **35.** $-5 + 12i$ **37.** i
39. $(2x + 5yi)(2x - 5yi)$ **41.** Yes **43.** $x = -\frac{3}{2}$, $y = 7$ **45.** $-4 + 3i$
47. For example, $\sqrt{-1}\sqrt{-1} = i^2 = -1$, but $\sqrt{(-1)(-1)} = \sqrt{1} = 1$.

Margin Exercises, Section 6.2

1. $7 - 2i$ **2.** $6 + 4i$ **3.** $5i$ **4.** $-3i$ **5.** -3 **6.** 8 **7.** $\frac{9}{13} + \frac{7}{13}i$ **8.** $\frac{4}{13} + \frac{7}{13}i$ **9.** $\frac{1}{3 + 4i}, \frac{3}{25} - \frac{4}{25}i$
10. Both are $7 + 3i$ **11.** Both are $-13 - 11i$ **12.** $\overline{z^3} = \overline{z \cdot z \cdot z} = \bar{z} \cdot \bar{z} \cdot \bar{z} = \bar{z}^3$ **13.** $5\bar{z}^3 + 4\bar{z}^2 - 2\bar{z} + 1$
14. $7\bar{z}^5 - 3\bar{z}^3 + 8\bar{z}^2 + \bar{z}$

Exercise Set 6.2, p. 224

1. $\frac{1}{2} + \frac{7}{2}i$ **3.** $\frac{1}{3} + \frac{2}{3}i\sqrt{2}$ **5.** $2 - 3i$ **7.** $\frac{1}{5} + \frac{2}{5}i$ **9.** $-\frac{1}{2} - \frac{i}{2}$ **11.** $\frac{28}{65} - \frac{29}{65}i$ **13.** $-\frac{1}{2} + \frac{3}{2}i$ **15.** $\frac{5}{2} + \frac{13}{2}i$ **17.** $\frac{4}{25} - \frac{3}{25}i$
19. $\frac{5}{29} + \frac{2}{29}i$ **21.** $-i$ **23.** $\frac{i}{4}$ **25.** $3\bar{z}^5 - 4\bar{z}^2 + 3\bar{z} - 5$ **27.** $4\bar{z}^7 - 3\bar{z}^5 + 4\bar{z}$ **29.** $z = 1$ **31.** a
33. $\frac{3 - i}{2 + i}$, or $1 - i$

Margin Exercises, Section 6.3

1. $2 + 5i$ **2.** $\dfrac{-1 + i \pm \sqrt{-18i}}{4}$ **3.** $\dfrac{-3 \pm 4i}{5}$ **4.** $x^2 - (1 + 2i)x + i - 1 = 0$ **5.** $x^3 - 2x^2 + x - 2 = 0$
6. $2 - i$, $-2 + i$

Exercise Set 6.3, pp. 227–228

1. $\frac{2}{5} + \frac{6}{5}i$ **3.** $\frac{8}{5} - \frac{9}{5}i$ **5.** $2 - i$ **7.** $\frac{11}{25} + \frac{2}{25}i$ **9.** $\dfrac{-1 + i \pm \sqrt{-6i}}{2}$ **11.** $-i$, $\frac{i}{2}$ **13.** $\dfrac{-1 - 2i \pm \sqrt{-15 + 16i}}{6}$

15. $1 \pm 2i$ **17.** $2 \pm 3i$ **19.** $-\dfrac{3}{2} \pm \dfrac{\sqrt{7}}{2}i$ **21.** $x^2 + 4 = 0$ **23.** $x^2 - 2x + 2 = 0$ **25.** $x^2 - 4x + 13 = 0$

27. $x^2 - 3x - ix + 3i = 0$ **29.** $x^3 - x^2 + 9x - 9 = 0$ **31.** $x^3 - 2x^2i - 3x^2 + 5ix + x - 2i + 2 = 0$

33. $x^3 + x - 2x^2i - 2i = 0$ **35.** $\sqrt{2} + \sqrt{2}i, \; -\sqrt{2} - \sqrt{2}i$ **37.** $2 + i, \; -2 - i$ **39.** $2, \; -1 \pm \sqrt{3}i$

41. $x = 2 + i, \; y = 1 - 3i$

Margin Exercises, Section 6.4

1.

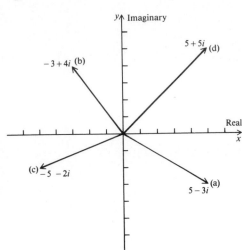

2.

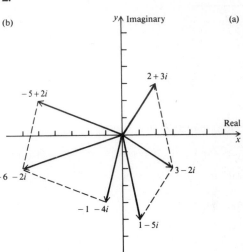

3. (a) 5; (b) 13   **4.** $1 - i$ **5.** $\sqrt{3} - i$ **6.** $\sqrt{2} \text{ cis } 315°$ **7.** $6 \text{ cis } 225°$ **8.** $20 \text{ cis } 55°$ **9.** $4 \text{ cis } \dfrac{5\pi}{4}$ **10.** $4 \text{ cis } 90°$

11. $2 \text{ cis } \dfrac{\pi}{4}$ **12.** $\sqrt{2} \text{ cis } 285°$

Exercise Set 6.4, pp. 233–234

1.

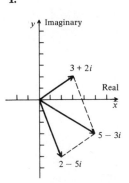

3.

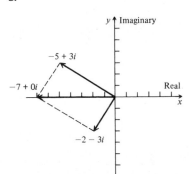

5.

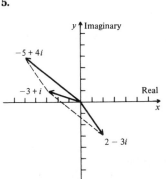

7.

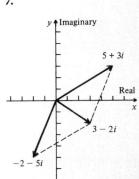

9. $\dfrac{3\sqrt{3}}{2} + \dfrac{3}{2}i$ **11.** $-10i$ **13.** $2 + 2i$ **15.** $-2 - 2i$ **17.** $\sqrt{2} \text{ cis } 315°$ **19.** $20 \text{ cis } 330°$ **21.** $5 \text{ cis } 180°$

23. $4 \text{ cis } 0°$, or 4   **25.** $8 \text{ cis } 120°$ **27.** $\text{cis } 270°$, or $-i$ **29.** $2 \text{ cis } 270°$, or $-2i$

31. $z = a + bi, \; |z| = \sqrt{a^2 + b^2}; \; -z = -a - bi, \; |-z| = \sqrt{(-a)^2 + (-b)^2} = \sqrt{a^2 + b^2}, \; \therefore \; |z| = |-z|$

33. $|(a + bi)(a - bi)| = |a^2 + b^2| = a^2 + b^2$; $|(a + bi)^2| = |a^2 + 2abi - b^2| = |a^2 - b^2 + 2abi| = \sqrt{(a^2 - b^2)^2 + (2ab)^2} = \sqrt{a^4 + 2a^2b^2 + b^4} = a^2 + b^2$ **35.** $z \cdot w = (r_1 \text{ cis } \theta_1)(r_2 \text{ cis } \theta_2) = r_1r_2 \text{ cis } (\theta_1 + \theta_2)$, $|z \cdot w| = \sqrt{[r_1r_2 \cos (\theta_1 + \theta_2)]^2 + [r_1r_2 \sin (\theta_1 + \theta_2)]^2} = \sqrt{(r_1r_2)^2} = |r_1r_2|$, $|z| = \sqrt{(r_1 \cos \theta_1)^2 + (r_1 \sin \theta_1)^2} = \sqrt{r_1^2} = |r_1|$, $|w| = \sqrt{(r_2 \cos \theta_2)^2 + (r_2 \sin \theta_2)^2} = \sqrt{r_2^2} = |r_2|$. Then $|z| \cdot |w| = |r_1| \cdot |r_2| = |r_1r_2| = |z \cdot w|$

37.

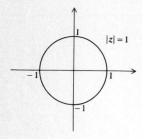

Margin Exercises, Section 6.5

1. 32 cis 270°, or $-32i$ **2.** 16 cis 120°, or $-8 + 8i\sqrt{3}$ **3.** (a) $1 + i$, $-1 - i$; (b) $\sqrt{5} + i\sqrt{5}$, $-\sqrt{5} - i\sqrt{5}$

4. 1 cis 60°, 1 cis 180°, 1 cis 300°; or $\dfrac{1}{2} + \dfrac{\sqrt{3}}{2}i$, -1, $\dfrac{1}{2} - \dfrac{\sqrt{3}}{2}i$

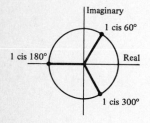

Exercise Set 6.5, pp. 236–237

1. 8 cis π **3.** 64 cis π **5.** 8 cis 270°   **7.** $-8 - 8\sqrt{3}i$ **9.** $-8 - 8\sqrt{3}i$ **11.** i **13.** 1

15. $\sqrt{2}$ cis 60°, $\sqrt{2}$ cis 240°; or $\dfrac{\sqrt{2}}{2} + \dfrac{\sqrt{6}}{2}i$, $\dfrac{-\sqrt{2}}{2} - \dfrac{\sqrt{6}}{2}i$ **17.** cis 30°, cis 150°, cis 270°; or $\dfrac{\sqrt{3}}{2} + \dfrac{1}{2}i$, $\dfrac{-\sqrt{3}}{2} + \dfrac{1}{2}i$, $-i$

19. 2 cis 0°, 2 cis 90°, 2 cis 180°, 2 cis 270°; or 2, 2i, -2, $-2i$

21. $-1.366 + 1.366i$, $0.366 - 0.366i$

Chapter 6 Review, p. 237

1. $14 + 2i$ **2.** $1 - 4i$ **3.** $2 - i$ **4.** $\dfrac{11}{10} + \dfrac{3}{10}i$ **5.** $x = 2$, $y = -4$ **6.** $3\bar{z}^3 + \bar{z} - 7$ **7.** $x^2 - 2x + 5 = 0$ **8.** $\dfrac{2}{5} \pm \dfrac{1}{5}i$
9. $\dfrac{(-3 \pm \sqrt{5})i}{2}$ **10.** $\sqrt{2} + \sqrt{2}i$, $-\sqrt{2} - \sqrt{2}i$

11.

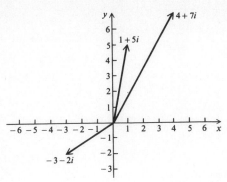

12. $-\sqrt{2} + \sqrt{2}i$ **13.** $\sqrt{2}$ cis $45°$ **14.** 70 cis $50°$
15. $\sqrt[6]{2}$ cis $15°$, $\sqrt[6]{2}$ cis $135°$, $\sqrt[6]{2}$ cis $255°$ **16.** $5 - 9i$
17. $1 + 2i$ **18.** $x = 2 - i$, $y = -1 - 3i$
19. $3, \frac{3}{2}(-1 \pm \sqrt{3}i)$
20.

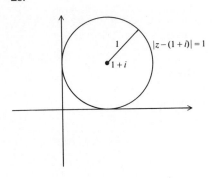

CHAPTER 7

Margin Exercises, Section 7.1

1. (a) increasing; (b) set of all real numbers; (c) set of all positive numbers; (d) 1; (e) 3.32
2. (a)–(d) all the same as Exercise 1; (e) 7.10; (f) 4^x

3.

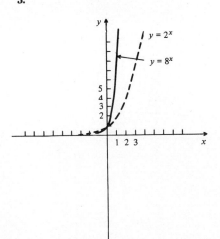

4.

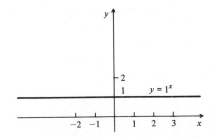

5.

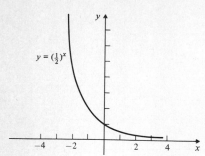

6.

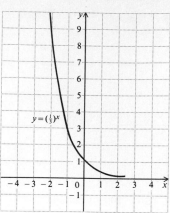

7. Domain: positive real numbers; range: all real numbers

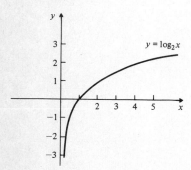

8. Domain: positive real numbers; range: all real numbers

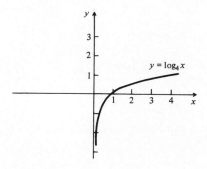

9. $\log_6 1 = 0$ **10.** $\log_{10} 0.001 = -3$ **11.** $\log_{16} 2 = \frac{1}{4}$ **12.** $\log_{6/5} \frac{25}{36} = -2$ **13.** $2^5 = 32$ **14.** $10^3 = 1000$
15. $10^{-2} = 0.01$ **16.** $(\sqrt{5})^2 = 5$ **17.** 10,000   **18.** 3   **19.** 4   **20.** 1   **21.** -2 **22.** 3   **23.** π **24.** 42   **25.** 37   **26.** M
27. 3.2

Exercise Set 7.1, p. 246

1. (a) and (b)

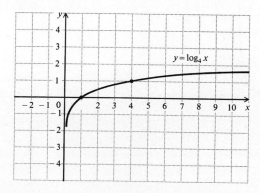

(c)

3.

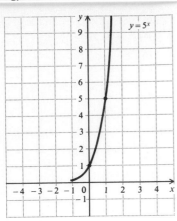

5.

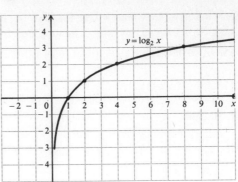

7. $2^5 = 32$ **9.** $10^{-2} = 0.01$ **11.** $6^1 = 6$ **13.** $\log_6 1 = 0$ **15.** $\log_{6/5} \frac{25}{36} = -2$ **17.** $\log_5 \frac{1}{25} = -2$ **19.** $\log_e 1.0833 = 0.08$
21. 10,000   **23.** $\frac{1}{2}$ **25.** 4   **27.** $\frac{1}{2}$ **29.** 2   **31.** 1   **33.** -3 **35.** $4x$ **37.** $\sqrt{5}$

39.

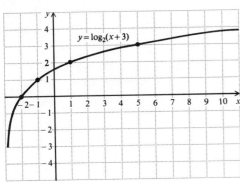

41.

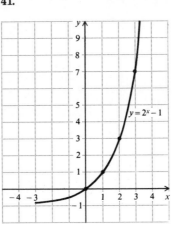

43.

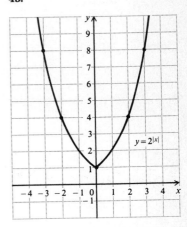

45.

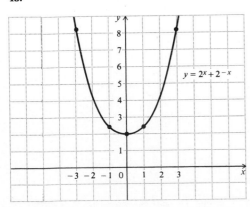

47. All real numbers   **49.** $\{x \mid x \neq 0\}$
51. $\{x \mid x > \frac{4}{3}\}$ **53.** $\{x \mid x < -3 \text{ or } x > 3\}$
55. $\{x \mid x \leq 0\}$ **57.** $\{x \mid x \geq 16\}$
59. $\{x \mid x > -3\}$ **61.** π^5

63.

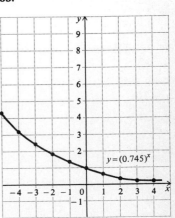

Margin Exercises, Section 7.2

1. $\log_a M + \log_a N$ **2.** $\log_5 25 + \log_5 5$ **3.** $\log_3 35$ **4.** $\log_a CABIN$ **5.** $5 \log_7 4$ **6.** $\frac{1}{2} \log_a 5$
7. (a) $\log_a M - \log_a N$; (b) $\log_c 1 - \log_c 4$ **8.** $\log_{10} 4 + \log_{10} \pi - \frac{1}{2} \log_{10} 23$ **9.** $\frac{1}{2}[3 \log_a z - \log_a x - \log_a y]$
10. $\log_a \frac{x^5 \sqrt[4]{z}}{y}$ **11.** (a) 0.954; (b) 0.1505; (c) 0.1003; (d) 0.176; (e) 1.585 **12.** 1 **13.** 0 **14.** 0 **15.** 1

Exercise Set 7.2, p. 250

1. $2 \log_a x + 3 \log_a y + \log_a z$ **3.** $\log_b x + 2 \log_b y - 3 \log_b z$ **5.** $\log_a 4$ **7.** $\log_a \frac{2x^4}{y^3}$ **9.** $\log_a \frac{\sqrt{a}}{x}$ or $\frac{1}{2} - \log_a x$
11. $\log_a (x^2 - xy + y^2)$ **13.** $\frac{1}{2}[\log_a (1 - x) + \log_a (1 + x)]$ **15.** 0.602 **17.** 1.699 **19.** 1.778 **21.** -0.088 **23.** 1.954
25. -0.046 **27.** False **29.** True **31.** False **33.** False **35.** $\frac{1}{2}$ **37.** $\sqrt{7}$ **39.** $-2, 0$ **41.** $\{x \mid x > 0\}$

Margin Exercises, Section 7.3

1. (a) 8; (b) 16 **2.** 8 **3.** 5 **4.** 2 **5.** 9 **6.** 0.4969 **7.** 0.9996 **8.** 0.6021 **9.** 5.74 **10.** 1.00 **11.** 3.62
12. $0.4609 + 2$ **13.** $0.4609 + (-4)$ **14.** 4.8312 **15.** 5.9504 **16.** 1.6618 **17.** 8.7846 **18.** $8.8932 - 10$ **19.** $6.0453 - 10$
20. $7.8976 - 10$ **21.** 64,100 **22.** 64,100 **23.** 8,560 **24.** 0.000425 **25.** 0.0105 **26.** 0.0601 **27.** 797,000

Exercise Set 7.3, pp. 256–257

1. 0.3909 **3.** 2.5403 **5.** 1.7202 **7.** 5.7952 **9.** $8.8463 - 10$ **11.** $6.3345 - 10$ **13.** 233 **15.** 0.018 **17.** 0.00000105
19. 25.2 **21.** 0.0973 **23.** 4.49 **25.** 0.0133 **27.** 190 **29.** 272 **31.** 8.77 **33.** 3.64 **35.** 25.7 **37.** 4.754264
39. -0.321371 **41.** 78,397,100 **43.** 0.000583

Margin Exercises, Section 7.4

1. 3.6592 **2.** $8.3779 - 10$ **3.** 2856 **4.** 0.0005956

Exercise Set 7.4, p. 259

1. 1.6194 **3.** 0.4689 **5.** 2.8130 **7.** $9.1538 - 10$ **9.** $7.6291 - 10$ **11.** $9.2494 - 10$ **13.** 2.7786 **15.** $9.8445 - 10$
17. 224.5 **19.** 14.53 **21.** 70,030 **23.** 0.09245 **25.** 0.5343 **27.** 0.007295 **29.** 0.8268

Margin Exercises, Section 7.5

1. 2.8076 **2.** 2.9999 **3.** $x = \frac{\log (t + \sqrt{t^2 + 1})}{\log e}$ or $\log_e (t + \sqrt{t^2 + 1})$ **4.** 125 **5.** 8.75 **6.** 2 **7.** 10 yr **8.** 34 db
9. 60 db **10.** 7.8 **11.** (a) 68; (b) 54; (c) 40

Exercise Set 7.5, pp. 264–266

1. 5 **3.** $\frac{12}{5}$ **5.** $\frac{1}{2}, -3$ **7.** 2.7093 **9.** 10 **11.** 1 **13.** 5 **15.** 1; 100 **17.** 4 **19.** $x = \log_e (t + \sqrt{t^2 + 1})$
21. $x = \frac{1}{2} \log_5 \frac{t + 1}{1 - t}$ **23.** 11.9 years **25.** 65 db **27.** 140 db **29.** 6.7 **31.** $10^5 \cdot I_0$ **33.** (a) 82; (b) 68 **35.** 9 months

37. 4.2   **39.** 1; 10,000   **41.** $\emptyset$ **43.** $-9; 9$ **45.** $\frac{7}{4}$ **47.** $\log_x y - \log_x a$ **49.** $t = \dfrac{100(\log_e P - \log_e P_0)}{r}$

51. $t = -\dfrac{1}{k}\log\left[\dfrac{T - T_0}{T_1 - T_0}\right]$ **53.** $Q = a^b \cdot \sqrt[3]{y}$ **55.** 10; 100

Margin Exercises, Section 7.6

1. 2.48832   **2.** 2.59374246

3.

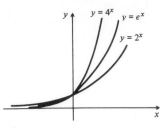

4.

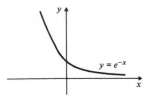

5. 0.693147   **6.** 4.605170   **7.** -2.599375 **8.** -0.000100 **9.** 2.0956   **10.** 11.3059   **11.** -2.5096 **12.** 7.6009
13. -9.2103 **14.** $t \approx 4$ **15.** $t \approx 44$ **16.** (a) 0.08; (b) 170,524   **17.** 3.5 g   **18.** 689 millibars   **19.** 3   **20.** 4.7385
21. 0.4343   **22.** 6.937   **23.** -0.783 **24.** 2.8076

Exercise Set 7.6, pp. 273–275

1. and 3.

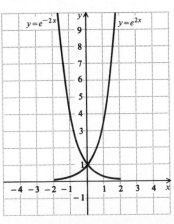

5.

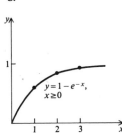

7. 0.6313   **9.** -3.9739 **11.** 0.7561   **13.** -1.5465 **15.** 8.4119   **17.** -7.4875 **19.** -2.5258 **21.** 13.7953   **23.** 4.6052
25. 4.0944   **27.** 2.3026   **29.** 140.67   **31.** $k = 0.0201$; $P = 1,243,000$ **33.** 0.4 gram   **35.** 1.2 days
37. (a) 3000 yr; (b) 5100 yr   **39.** 587 mb   **41.** (a) $k = 0.061$, $P = \$100e^{0.061t}$; (b) \$338.72; (c) 1978   **43.** 2.1610
45. -0.1544 **47.** 2.4849   **49.** $t = \dfrac{\ln P - \ln P_0}{k}$ **51.** By Theorem 7, $\ln x = \dfrac{\log x}{\log e} \approx \dfrac{\log x}{0.4343}$; by Table 2, $\approx 2.3026 \log x$
53. e^{π} **55.** 2; 2.25; 2.48832; 2.593742; 2.704814; 2.716924

Chapter 7 Review, pp. 275–276

1.

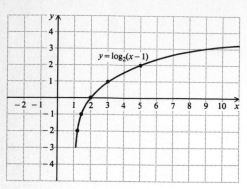

$y = \log_2(x-1)$

2.

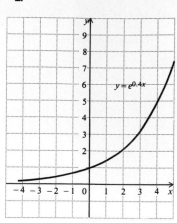

$y = e^{0.4x}$

3.

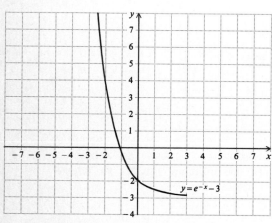

$y = e^{-x} - 3$

4. $8^{-2/3} = \frac{1}{4}$ **5.** $\log_7 x = 2.3$ **6.** 4 **7.** $\frac{1}{2}$ **8.** 3 **9.** $\log_b \dfrac{a^{1/2}c^{3/2}}{d^4}$ **10.** 1.255 **11.** 0.544 **12.** -0.602 **13.** 0.2385

14. $\frac{2}{3}\log M - \frac{1}{3}\log N$ **15.** 1.4200 **16.** $7.9063 - 10$ **17.** 73.9 **18.** 0.0276 **19.** 3.5 **20.** 0.0006934 **21.** 2.1861

22. 6.328 **23.** -3.0748 **24.** $x^2 + 1$ **25.** $\frac{1}{5}$ **26.** 9 **27.** $T = \dfrac{\log 2}{\log 1.13} \approx 5.7$ **28.** 30 **29.** 6.2 g **30.** 1 **31.** 3 **32.** 3

33. 1 **34.** 64, $\frac{1}{64}$

35.

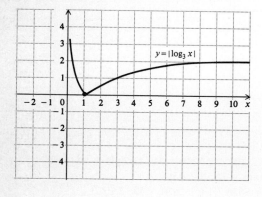

$y = |\log_3 x|$

36.

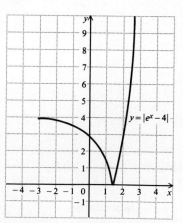

$y = |e^x - 4|$

37. $\{x \mid x > e^{6/5}\}$ **38.** $\{x \mid x \neq \frac{1}{4}\ln 10\}$

APPENDIX 2. ALGEBRA REVIEW EXERCISES

Exercise Set R-1, p. 282

1. 2^{-1} **2.** 1   **3.** 4^3 **4.** $6x^5$ **5.** $15a^{-1}b^5$ **6.** $72x^5$ **7.** $-18x^7yz$ **8.** b^3 **9.** x^3y^{-3} **10.** 20   **11.** $3ab^2$ **12.** $\frac{4}{7}xyz^{-5}$

13. $8a^3b^6$ **14.** $16x^{12}$ **15.** $-16x^{12}$ **16.** $36a^4b^6c^2$ **17.** $\frac{1}{25}c^2d^4$ **18.** 1   **19.** 32   **20.** $\frac{27}{4}a^8b^{-10}c^{18}$ **21.** $\frac{3}{4}xy$

22. 5.8×10^7 **23.** 3.65×10^5 **24.** 2.7×10^{-6} **25.** 2.7×10^{-2} **26.** 5.0×10^{-2} **27.** .0005   **28.** 7,800,000   **29.** .000000854

Exercise Set R-2, p. 282

1. $3x^2y - 5xy^2 + 7xy + 2$ **2.** $-10pq^2 - 5p^2q + 7pq - 4p + 2q + 3$ **3.** $3x + 2y - 2z - 3$ **4.** $5x\sqrt{y} - 4y\sqrt{x} - \frac{2}{5}$

5. $-.8p^2\sqrt{q} + .1\sqrt[3]{pq} + .88\sqrt{p}$ **6.** $-2xy^{-2} - 2xy - 3\sqrt{xy} - 1$ **7.** $-2x^2 + 6x - 2$ **8.** $6a - 5b - 2c + 4d$

9. $x^4 - 3x^3 - 4x^2 + 9x - 3$ **10.** $9x\sqrt{y} - 3y\sqrt{x} + 9.1$ **11.** $8xy^{-2} - 3\sqrt{xy} - 2xy - 1$

Exercise Set R-3, p. 282

1. $6x^3 + 4x^2 + 32x - 64$ **2.** $4a^3b^2 - 10a^2b^2 + 3ab^3 + 4ab^2 - 6b^3 + 4a^2b - 2ab + 3b^2$ **3.** $6y^2 + 5y - 6$

4. $4x^2 + 4xy + 3y^2$ **5.** $12x^3 + x^2y - \frac{3}{2}xy - \frac{1}{8}y^2$ **6.** $2x^3 - 2\sqrt{2}x^2y - \sqrt{2}xy^2 + 2y^3$ **7.** $2x^2 + 12xy + 9y^2$

8. $4x^4 - 12x^2y + 9y^2$ **9.** $4x^6 + 12x^3y^2 + 9y^4$ **10.** $\frac{1}{4}x^4 - \frac{3}{5}x^2y + \frac{9}{25}y^2$ **11.** $.25x^2 + .70xy^2 + .49y^4$ **12.** $9x^2 - 4y^2$

13. $x^4 - y^2z^2$ **14.** $9x^4 - 2$ **15.** $4x^2 + 12xy + 9y^2 - 16$ **16.** $x^4 + 6x^2y + 9y^2 - y^4$ **17.** $x^4 - 1$ **18.** $16x^4 - y^4$

Exercise Set R-4, p. 283

1. $3ab(6a - 5b)$ **2.** $(a + c)(b - 2)$ **3.** $(x + 6)(x + 3)$ **4.** $(3x - 5)(3x + 5)$ **5.** $4x(y^2 - z)(y^2 + z)$ **6.** $(y + 3)^2$
7. $(1 - 4x)^2$ **8.** $(2x - \sqrt{5})(2x + \sqrt{5})$ **9.** $(xy - 7)^2$ **10.** $4a(x + 7)(x - 2)$ **11.** $(a + b + c)(a + b - c)$
12. $(x + y - a - b) \cdot (x + y + a + b)$ **13.** $5(y^2 + 4x^2)(y - 2x)(y + 2x)$ **14.** $(x + 2)(x^2 - 2x + 4)$
15. $3\left(x - \frac{1}{2}\right)\left(x^2 + \frac{1}{2}x + \frac{1}{4}\right)$ **16.** $(x + 1)(x^2 - .1x + .01)$ **17.** $3(z - 2)(z^2 + 2z + 4)$

Exercise Set R-5, p. 283

1. 12   **2.** -6 **3.** 8   **4.** $\frac{4}{5}$ **5.** 2   **6.** $-\frac{3}{2}$ **7.** -2

8. $\frac{3}{2}, \frac{2}{3}$ **9.** 0, 1, -2 **10.** $\frac{2}{3}, -1$ **11.** 4, 1   **12.** $3 < x$ **13.** $xz - \frac{5}{12}$

Exercise Set R-6, p. 283

1. $\sqrt{7}, -\sqrt{7}$ **2.** $\frac{\sqrt{5}}{3}, -\frac{\sqrt{5}}{3}$ **3.** $\sqrt{\frac{b}{a}}, -\sqrt{\frac{b}{a}}$ **4.** $7 + \sqrt{5}, 7 - \sqrt{5}$ **5.** $h + \sqrt{a}, h - \sqrt{a}$ **6.** $-3 + \sqrt{5}, -3 - \sqrt{5}$

7. 3, -10 **8.** $\frac{2 \pm \sqrt{14}}{5}$ **9.** $-5, \frac{3}{2}$ **10.** 1, -5 **11.** 2, $-\frac{1}{2}$ **12.** $-1, -\frac{5}{3}$ **13.** $6 \pm \sqrt{33}$ **14.** 2, $-\frac{3}{2}$ **15.** $\frac{-3 \pm \sqrt{41}}{2}$

16. $\frac{3}{2}, \frac{2}{3}$ **17.** $-0.1 \pm \sqrt{.31}$

Exercise Set R-7, p. 284

1. $\dfrac{5x}{2}$ **2.** $\dfrac{x-2}{x+3}$ **3.** $\dfrac{x-2}{x+2}$ **4.** $\dfrac{1}{x+y}$ **5.** $\dfrac{(x+5)(2x+3)}{7x}$ **6.** $\dfrac{a+2}{a-5}$ **7.** $m+n$ **8.** $\dfrac{3(x-4)}{2(x+4)}$ **9.** $\dfrac{1}{x+y}$
10. $\dfrac{x-y-z}{x+y+z}$

Exercise Set R-8, p. 284

1. 1 **2.** $\dfrac{y-2}{y-1}$ **3.** $\dfrac{x+y}{2x-3y}$ **4.** $\dfrac{3x-4}{x^2-4}$ **5.** $\dfrac{3y-10}{(y-5)(y+4)}$ **6.** $\dfrac{4x-8y}{x^2-y^2}$ **7.** $\dfrac{3x-4}{(x-2)(x+1)}$
8. $\dfrac{5a^2+10ab-4b^2}{(a-b)(a+b)}$ **9.** $\dfrac{8-18x+11x^2}{(2+x)(2-x)^2}$ **10.** 0

Exercise Set R-9, p. 284

1. $\dfrac{x+y}{x}$ **2.** $\dfrac{x^2-1}{x^2+1}$ **3.** $\dfrac{c^2-2c+4}{c}$ **4.** $\dfrac{xy}{x-y}$ **5.** $x-y$ **6.** $\dfrac{x^2-y^2}{xy}$ **7.** $\dfrac{1+a}{1+a}$ **8.** $\dfrac{b+a}{b-a}$

Exercise Set R-10, p. 285

1. 11 **2.** $|4x|$ **3.** $|b+1|$ **4.** $-3x$ **5.** $|x-2|$ **6.** 2 **7.** -2 **8.** $4\sqrt{3}$ **9.** $3\sqrt[3]{5}$ **10.** $\dfrac{|8c|}{d^2}$ **11.** $3\sqrt{2}$
12. $2x^2|y|$ **13.** $-3x\sqrt[3]{4y}$ **14.** $2(x+4)\sqrt[3]{(x+4)^2}$ **15.** $2(x+1)\sqrt[3]{9(x+1)}$ **16.** $\sqrt{7b}$ **17.** 2 **18.** $\dfrac{1}{2x}$ **19.** $\sqrt{a+b}$
20. $\left|\dfrac{3a}{4b}\right|\sqrt{ab}$

Exercise Set R-11, p. 285

1. $-12\sqrt{5}-2\sqrt{2}$ **2.** $3x+11\sqrt[3]{x^2}$ **3.** $|14y|\sqrt{3}-2y\sqrt{6}$ **4.** 1 **5.** $3m^2y-5m\sqrt{xy}-2x$ **6.** $|x+3|-3$
7. $\dfrac{18-6\sqrt{5}}{-2}$ **8.** $\dfrac{2\sqrt{6}}{3}$ **9.** $\dfrac{8|x|-20\sqrt{xy}-6|x|\sqrt{y}+15|y|}{2|x|-5|y|}$ **10.** $\dfrac{2-5|a|}{6\sqrt{2}-6\sqrt{5a}}$ **11.** $\dfrac{|x+1|-1}{|x+1|-2\sqrt{x+1}+1}$
12. $\dfrac{|a+3|-3}{3\sqrt{a+3}+3\sqrt{3}}$

Exercise Set R-12, p. 285

1. $\sqrt[4]{x^3}$ **2.** $\sqrt[4]{16^3}$ **3.** $\sqrt[4]{x^5y^{-3}}$ **4.** $\sqrt{a^3b^{-1}}$ or $\dfrac{a}{b}\sqrt{ab}$ **5.** $20^{2/3}$ **6.** $13^{5/4}$ **7.** $11^{1/6}$ **8.** $5^{5/6}$ **9.** 4 **10.** $2y^2$
11. $(a^2+b^2)^{1/3}$ **12.** $3ab^3$ **13.** $\dfrac{m^2n^4}{2}$ **14.** $8a^2$ **15.** $\dfrac{3}{x^3b^2}$ **16.** $xy^{1/3}$ **17.** $\sqrt[6]{288}$ **18.** $\sqrt[12]{x^{11}y^7}$ **19.** $|a|\sqrt[6]{a^5}$
20. $|a+x|\sqrt[12]{(a+x)^{11}}$

INDEX

Flexibility, readability, and understandability — the Keedy/Bittinger keynotes — are made evident in this text with step-by-step explanations, clearly marked examples, guidelines for solutions, and signposts for common stumbling blocks. Suggestions for topic sequencing are found in the preface.

A work-text format, featuring margin exercises, provides opportunity for immediate reinforcement of concepts by offering developmental exercises right next to relevant textual explanation. Many new exercises and a new chapter on exponential and logarithmic functions are included in this edition.

In this text the concept of transformations is introduced early and provides an important basis for later development. Transformations are used in developing identities and establishing properties of the trigonometric functions. An introduction to polar coordinates is also included.

In keeping with the Keedy/Bittinger approach of learning by doing, verbiage is kept to a minimum. Exercises are keyed to the objectives by lowercase Roman numerals, and there are exercises, some of them challenging, that require the student to go beyond the objectives. The latter are indicated by appropriate symbols, as are calculator exercises. A complete supplementary package including Answer Booklet, Student Solutions Book, and Test Booklet is available.

The Authors

Mervin L. Keedy is Professor of Mathematics at Purdue University. Dr. Keedy earned his B.S. degree from the University of Chicago, and his M.A. and Ph.D. degrees from the University of Nebraska. He is a member of the Mathematics Association of America, the National Council of Teachers of Mathematics, and the Mathematics Association of Two Year Colleges. He is the author or coauthor of many textbooks in mathematics, for schools and colleges.

Marvin L. Bittinger is Professor of Mathematical Sciences at Indiana University–Purdue University at Indianapolis. Dr. Bittinger received his B.A. degree from Manchester College, his M.S. degree from Ohio State University, and his Ph.D. degree from Purdue University. He was Distinguished Visiting Professor at the United States Air Force Academy in 1978. Professor Bittinger is the author of *Calculus: A Modeling Approach, Second Edition* (Addison-Wesley, 1980), *Logic and Proof, Second Edition* (Addison-Wesley, 1982), and coauthor of *Finite Mathematics: A Modeling Approach, Second Edition* (Addison-Wesley, 1981).

Together Keedy and Bittinger have written and Addison-Wesley has published many mathematics books including *Arithmetic, Third Edition* (1979), *Introductory Algebra, Third Edition* (1979), *Essential Mathematics, Third Edition* (1980), and *Intermediate Algebra, Third Edition* (1979).

ADDISON-WESLEY PUBLISHING COMPANY

ISBN 0-201-13408-x